Sedimentology: A Modern Synthesis

Sedimentology:
A Modern Synthesis

Edited by Aiden Williams

SYRAWOOD
PUBLISHING HOUSE

New York

Published by Syrawood Publishing House,
750 Third Avenue, 9ᵗʰ Floor,
New York, NY 10017, USA
www.syrawoodpublishinghouse.com

Sedimentology: A Modern Synthesis
Edited by Aiden Williams

International Standard Book Number: 978-1-68286-788-4 (Hardback)

Cataloging-in-Publication Data

Sedimentology : a modern synthesis / edited by Aiden Williams.
    p. cm.
Includes bibliographical references and index.
ISBN 978-1-68286-788-4
1. Sedimentology. 2. Geology. I. Williams, Aiden.
QE471 .S43 2019
551.3--dc23

TABLE OF CONTENTS

PREFACE

The field of sedimentology is primarily concerned with the study of sediments and the processes that cause their formation. Sedimentary rocks can be divided into four primary classes, namely carbonates, chemical, clastics and evaporites. Human society has always benefitted from the use of sedimentary rocks. They play a crucial role in art and architecture, deriving various building materials, procurement of various precious metals and minerals, and for the generation of energy. Studies in sedimentology involve investigations in sequence stratigraphy, isotope geochemistry, petrology, etc. This book is a valuable compilation of topics, ranging from the basic to the most complex advancements in the field of sedimentology. Different approaches, evaluations, methodologies and advanced studies in this field have been included in this book. With state-of-the-art inputs by acclaimed experts of this field, this book targets students and professionals.

This book is a comprehensive compilation of works of different researchers from varied parts of the world. It includes valuable experiences of the researchers with the sole objective of providing the readers (learners) with a proper knowledge of the concerned field. This book will be beneficial in evoking inspiration and enhancing the knowledge of the interested readers.

In the end, I would like to extend my heartiest thanks to the authors who worked with great determination on their chapters. I also appreciate the publisher's support in the course of the book. I would also like to deeply acknowledge my family who stood by me as a source of inspiration during the project.

Editor

Lowstand wedges in carbonate platform slopes (Quaternary, Maldives, Indian Ocean)

CHRISTIAN BETZLER*, CHRISTIAN HÜBSCHER†, SEBASTIAN LINDHORST*, THOMAS LÜDMANN*, JOHN J. G. REIJMER‡ and JUAN-CARLOS BRAGA§

*Institut für Geologie, CEN, Universität Hamburg, Bundesstrasse 55, 20146 Hamburg, Germany (E-mail: christian.betzler@uni-hamburg.de)
†Institut für Geophysik, CEN, Universität Hamburg, Bundesstrasse 55, 20146 Hamburg, Germany
‡King Fadh University of Petroleum & Minerals, KFUPM Box 2263, Dhahran 31261, Saudi Arabia
§Departamento de Estratigrafía y Paleontología, Universidad de Granada,Campus de Fuentenueva s.n., 18002 Granada, Spain

Keywords
Drift deposits, large benthic foraminifers, rhodoliths, sea-level, sequence stratigraphy.

ABSTRACT

Seismic, hydroacoustic and sedimentological data were used to analyse the response of atoll-slope sedimentation in the Maldives to the late Quaternary sea-level change. The slope deposits, as imaged in multichannel seismic profiles, are arranged into stacked aggrading to backstepping basinward thinning wedges. In a piston core recovered at the lower slope of one of the atolls, the sediment texture ranges from packstone to rudstone. Major components are blackened bioclasts, the large benthic foraminifers *Operculina* and *Amphistegina*, together with *Halimeda* debris and red algae. Radiocarbon dating at a core depth of 66 cm indicates that the wedge sedimentation stopped or was largely reduced after 16 ka BP. Therefore, the atoll-slope deposits largely consist of sediment formed *in situ* and deposited during the last glacial lowstand in sea-level. This is in apparent contradiction to the concept of highstand shedding of tropical carbonate platforms, which requires slope sedimentation during sea-level highstands, when the platform is flooded. Rather than intrinsic factors, such as sediment bypass along the steep slope, the extrinsic process of current winnowing of the slope appears to be a major controlling factor in the production of this feature. This process may be relevant for other case studies of carbonate platforms, as currents may be accelerated around such edifices, leading to slope winnowing and sediment deposition in more current-protected zones. The study results also have consequences for the interpretation of outcrop and seismic subsurface data of carbonate platform slope series, because such slope sediment wedges are not necessarily formed during sea-level highstands, but can consist of lowstand wedges only.

INTRODUCTION

When tropical carbonate platforms are flooded and the platform interior is occupied by a neritic carbonate factory, sediment is exported into the adjacent basins, especially towards the leeward flanks of the platform where this sediment is redistributed through wind-driven currents and waves (Eberli & Ginsburg, 1987). This process has been defined as highstand shedding (Schlager *et al.*, 1994), which is the main mechanism to form the slope sediments that accumulate into wedge-shaped bodies. Because the amount of sediment deposited during sea-level highstands is larger than the amount formed during sea-level lowstands (Grammer & Ginsburg, 1992), carbonate platform slope wedges mainly consist of highstand deposits. Intrinsic factors, such as an oversteepening of the slope or differences in the grain size of the exported particles, were evoked as controlling factors to allow slope sediment bypass or even erosion (Schlager & Camber, 1986; Kenter, 1990; Rendle-Bühring & Reijmer, 2005). This concept of a highstand origin for slope deposits is well established and applied when interpreting subsurface and outcrop data from carbonate platform slopes.

There is, however, growing evidence that gravitationally controlled off-bank transport mechanisms are only one factor controlling carbonate platform slope deposits and that alongslope contour currents are another major driver of slope deposition. This has been shown at the different slopes of the Bahamas carbonate platform, where the highstand sediments accumulate in periplatform drifts, i.e. drift bodies lining the platform flanks (Betzler *et al.*, 2015; Tournadour *et al.*, 2015; Chabaud *et al.*, 2016; Principaud *et al.*, 2016; Wunsch *et al.*, 2016), irrespective of the windward or leeward exposure of the slope. Slope segments with elevated contour current velocities show no sediment cover or reduced sedimentary thickness (Neumann & Ball, 1970; Mulder *et al.*, 2012).

The Maldives archipelago consists of atolls bathed by vigorously flowing and seasonally reversing currents (Betzler *et al.*, 2009, 2013a, 2016; Lüdmann *et al.*, 2013). Therefore, the atoll slopes appear to be a good location to study the interaction of currents and sea-level controlled slope sedimentation and to expand the insights into the interaction of carbonate slope sedimentation and contour currents. It will be shown that the established lowstand–highstand partition is not applicable because virtually no highstand material is deposited along the slope.

GEOLOGICAL SETTING

The Maldives archipelago south-west of India, in the central equatorial Indian Ocean, is an isolated tropical carbonate platform (Fig. 1). The north–south-oriented double row of atolls encloses the Inner Sea of the Maldives (Fig. 1A). Atolls are separated from each other by inter-atoll channels, which deepen towards the Indian Ocean (Purdy & Bertram, 1993). The Inner Sea is a bank-internal basin with water depths of up to 550 m (Fig. 1B). The Maldives carbonate sedimentary succession is almost 3 km thick and has accumulated since the Eocene, away from any terrigenous input (Aubert & Droxler, 1992; Purdy & Bertram, 1993; Belopolsky & Droxler, 2004; Betzler *et al.*, 2009, 2013b).

The archipelago comprises about 1200 smaller atolls. Discontinuous marginal rims formed by smaller atolls (faros) surround lagoons with water depths of up to 50 to 60 m (Betzler *et al.*, 2015). The oceanward margins of the Maldives archipelago are generally steeply inclined, with dips of 20 to 30° down to 2000 m of water depth. On the Inner Sea side, stepped atoll slopes have the same dip angles, but reach down to water depths of a few hundred metres, where the gradient rapidly declines (Fürstenau *et al.*, 2010). The Inner Sea is characterized by periplatform ooze deposition (Droxler *et al.*, 1990), locally accumulated into sediment drift bodies (Betzler *et al.*, 2009, 2013a,b; Lüdmann *et al.*, 2013).

The climate and oceanographic setting of the Maldives is dictated by the seasonally reversing Indian monsoon system (Tomczak & Godfrey, 2003). South-western winds prevail during the northern hemisphere summer (April to November), whereas northeastern winds prevail during winter (December to March). Winds generate ocean currents, which are directed westwards in the winter and eastwards in the summer. Interseasonally, a band of Indian Ocean Equatorial Westerlies establishes strong, eastward-flowing surface currents with velocities of up to $1 \cdot 3$ m s^{-1}. Currents reach down to the sea floor (Lüdmann *et al.*, 2013), especially in the inter-atoll passages, where submarine dunes and moats occur (Betzler *et al.*, 2009, 2013b).

METHODS

Seismic signals were generated by means of two clustered GI-Guns. The volume of each GI-Gun was 45 cin for the generator with a 105 cin injector volume. The GI-Guns were operated in 'true GI mode' and synchronized by a SureShot trigger system, which displays the source signal of each airgun. The digital streamer used for the survey was a Hydroscience Technologies SeaMUX 144-channel array with an active length of 600 m and an asymmetric group interval. The selected shooting distance during the entire cruise was 25 m.

Sub-bottom data were recorded with the RV METEOR parametric sediment echo sounder (PARASOUND P70; Atlas Elektronik, Bremen, Germany). The system was operated with two frequencies (18 kHz and 22 kHz). The software PS32segy (Hanno Keil, University of Bremen, Germany) was used to cut and convert the data. Data processing was performed with the software package ReflexW (Sandmeier Software, Karlsruhe, Germany), comprising automatic gain control (AGC) and along-profile amplitude normalization.

Sediment samples were acquired with a piston corer during Cruise M74/4 with RV METEOR and a videograb sampler during Cruise SO236 with RV SONNE. Components in the sediments were analysed quantitatively: Samples were wet-sieved and 200 components from the 500 to 1000 µm and the >1000 µm fraction were counted. The following components were differentiated: *Amphistegina*, *Homotrema*, *Operculina*, *Heterostegina*, *Gyrodinoides*, miliolids, other benthic foraminifers, planktonic foraminifers, bryozoans (encrusting, robust branching, vagrant), bivalves, gastropods, echinoderms, serpulids, crustaceans, bioclasts and lithoclasts. Especially bioclasts in some samples have a dark grey to black stain. Such components were counted separately. Tables with the counts are deposited at www.pangaea.de.

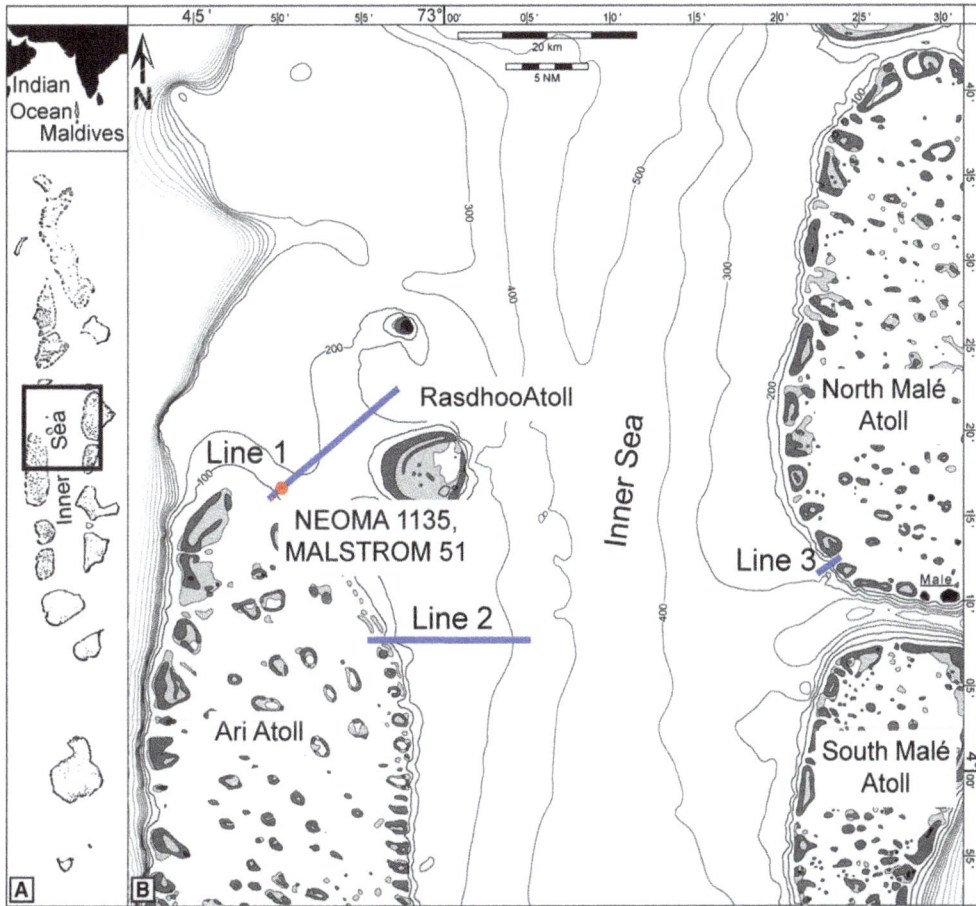

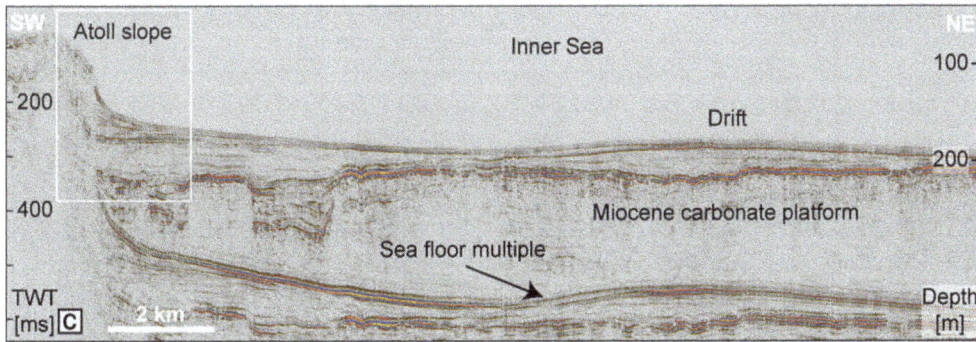

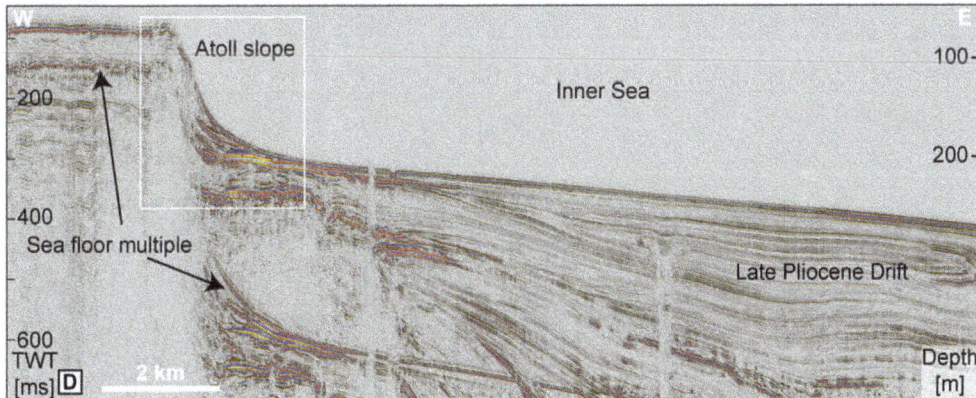

Fig. 1. (A) Double row of Maldives atolls enclosing the Inner Sea. The rectangle indicates location of B with seismic lines shown and discussed herein. (C) Overview seismic line from Ari Atoll into the Inner Sea. Note the basinward thinning of the stacked slope wedges. Rectangle indicates the position of the detailed view in Fig. 2A. (D) Overview seismic line from Ari Atoll into the Inner Sea. Note the basinward thinning of the stacked slope wedges and that the succession in the Inner Sea consists of drifts. Rectangle indicates the position of the detailed view in Fig. 2B. [Colour figure can be viewed at wileyonlinelibrary.com]

Radiocarbon dating was performed by Beta Analytics Inc. (Miami, FL, USA) on selected calcareous microfossils and macrofossils (Table 1). Samples were ultrasonically cleaned in deionized water and visually inspected for cements, overgrowths and fills. Conventional radiocarbon ages were calibrated using Calib (v.7.0.4, Stuiver & Reimer, 1993) and the calibration curve Marine13 (Reimer et al., 2013) with no local reservoir correction applied. Calibrated ages were rounded to the next decade, and in the text, the median of the probability distribution is used in conjunction with the two-sigma range (95·4% probability).

RESULTS

Seismic data

During the M74/4 cruise, the rims of the Ari and Male atolls were crossed three times at different positions in passages separating the faros lining the atoll's borders (Fig. 1B); overview lines are presented in Fig. 1C and D. Both atolls are around 50 m deep, and the passages between the faros have a slightly shallower sill before the stepped slopes dip at 10 to 50° into the Inner Sea. The Inner Sea is between 200 and 430 m deep, with the deepest area in its central part. Whereas seismic imaging of the lagoonal stratigraphy is reduced because of the sea-floor multiple, data from the slopes and the Inner Sea give a good insight into the stratigraphic succession (Figs 1C, D, and 2). The Inner Sea succession consists of a drowned carbonate bank lined and overlain by drift deposits (Fig. 1C), similar to the succession of the

Kardiva Channel located further north in the Maldives archipelago (Betzler et al., 2009, 2013a, 2016; Lüdmann et al., 2013).

Line 1, located at the northeastern margin of Ari Atoll is oriented NE-SW and crosses the position of Core M74/4-1135 and of videograb sample SO236-51 (Figs 1B, C and 2A). Down to a water depth of 100 m, the line images an irregular sea-floor relief on top of a succession with discontinuous to chaotic reflections. From ca 97 to 128 m, there is a submarine cliff. Basinwards, the slope progressively flattens out to water depths of 180 m. Here, reflections are laterally more or less continuous with moderate amplitudes. Strong, slightly inclined reflections occur in the upper part of the succession, which appear truncated downslope. Around 8 km NE of the platform edge, the deposits are arranged into drift bodies (Fig. 1C).

Line 2 (Figs 1B, D and 2B) is located at the eastern margin of Ari Atoll. Similar to the northeastern margin of the atoll (Line 1), there is a sea-floor step at ca 97 m. In front of this step, there is a sediment wedge with strong basinward reflections, which are also truncated downslope. This wedge overlies a horizon with a strong acoustic impedance, which corresponds to the limit between the Inner Sea drift units 8 and 9 (Lüdmann et al., 2013), which was formed at ca 2·3 Ma (Betzler et al., 2016). Drifts sediments form the succession further away from the atoll margin (Fig. 1D).

A sediment wedge is also imaged in Line 3 (Figs 1B and 2C), which crosses the south-western margin of North Malé Atoll. Towards the Inner Sea, the smooth pattern of reflection changes into a discontinuous to wavy

Table 1. Results of radiocarbon dating. Calibration was performed using Calib (v7.0.4, Stuiver & Reimer, 1993) and the calibration curve Marine13 (Reimer et al., 2013). No local reservoir correction was applied

| | Depth | | | ^{14}C age | $^{13}C/^{12}C$ ratio | Calibrated age ($\Delta R = 0$) | |
| | | | | | | cal BP (2σ ranges, 95·4% prob.) | |
	mbsf	Lab ID	Material	yrs BP	‰	Range (years)	Median (ka)
Core 1135	0·66	Beta-271838	Bivalve	13 800 ± 60	+1·1	15 900 to 16 310	16·12 ± 0·21
Core 1135	3·66	Beta-265298	Coral	14 400 ± 70	−2·0	16 670 to 17 260	16·99 ± 0·30
Core 1135	5·66	Beta-265299	Coral	15 230 ± 70	−2·5	17 840 to 18 250	18·03 ± 0·21
Core 1135	6·99	Beta-265297	Bivalve	15 490 ± 70	+1·5	18 100 to 18 540	18·33 ± 0·22
Core 1135	9·16	Beta-271839	Bivalve	18 740 ± 70	−1·0	21 950 to 22 410	22·22 ± 0·23
SO236-51	0	Beta-432465	*Amphistegina*	11 850 ± 40	−0·6	13 210 to 13 430	13·32 ± 0·11

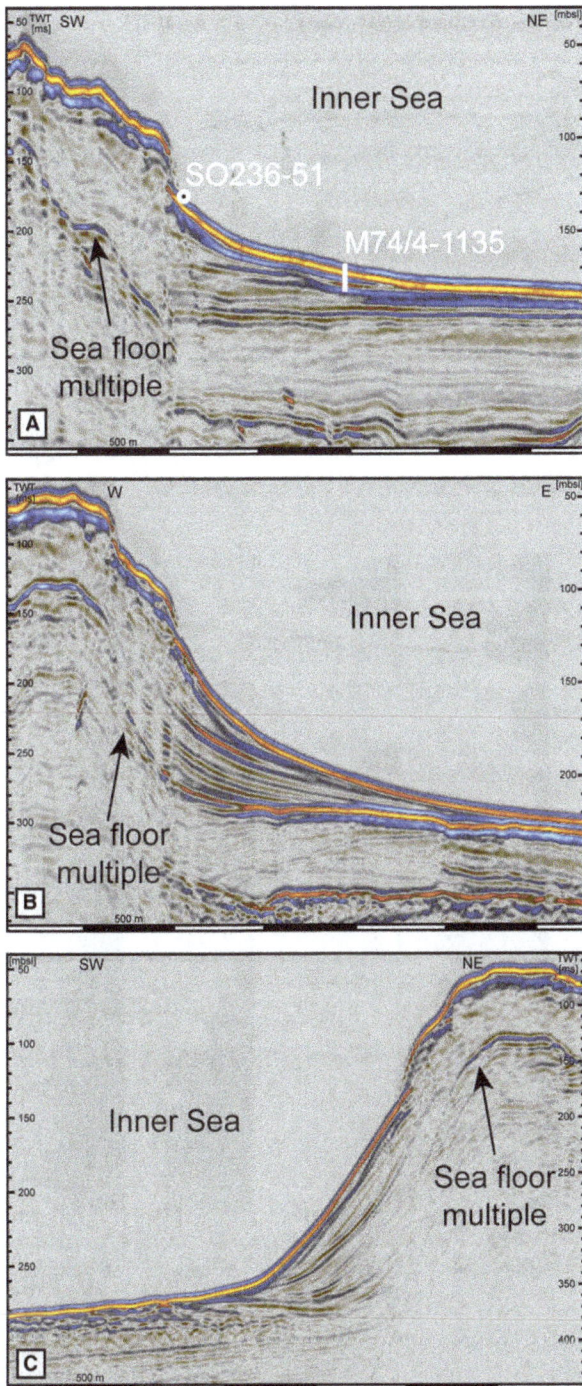

Fig. 2. (A) Seismic line 1 from the Inner Sea across the flank of Ari Atoll with location of piston core M74/4-1135 and videograb sample SO236-51 (B). (C) Seismic line 2 from the Inner Sea across the flank of Ari Atoll (D). (E) Seismic line 1 from the Inner Sea across the flank of North Malé Atoll (B). Vertical exaggeration: A – D = 10 X, F, F = 8 X. [Colour figure can be viewed at wileyonlinelibrary.com]

stratification which corresponds to an area where the sea floor is covered by submarine dunes moved by bottom currents (Betzler *et al.*, 2013a; Lüdmann *et al.*, 2013).

Sedimentology

Piston Core M74/4-1135 was retrieved at 4°16'59·592"N, 72°50'4·992"E, i.e. around 2 km NE of the margin of Ari Atoll at a water depth of 172 m (Figs 2A and 3A, B). It recovered 12·22 m of unlithified carbonates with a rudstone to grainstone texture (Fig. 4) deposited in the sediment wedge located in front of the 30 m high submarine cliff which forms the seaward limit of the terrace at a water depth of *ca* 97 m.

Figure 3 depicts the linkage between sedimentological and high-resolution parasound data. The parasound profile images a surface with a high impedance contrast at a depth of 7 mbsf (metre below sea floor). In the core, this surface correlates with a colour change from dark greenish grey to grey. The darker core colours are due to the high abundance of dark components giving the sediment a 'salt-and-pepper' texture (Fig. 5A). The dark colour of the grains is a consequence of an elevated organic and clay content, as resolved in smear slides of acid residues.

The lower 30 cm of the succession consists of a rhodolith-rich rudstone with a fining-upward trend of the components. Between 11·92 mbsf and 8·80 mbsf, the core is a grey rudstone with some dispersed rhodoliths up to 5 cm in size. Above 8·8 mbsf, a rudstone with some rhodoliths is mottled with a greenish grey to grey colour. A platy coral was recovered at 6·7 mbsf. Large components disappear upcore, with the last large rhodoliths registered at 5·5 mbsf. The top of the succession is a light grey packstone to grainstone with *Halimeda* flakes, planktic foraminifers, pteropods, benthic foraminifers, serpulids and echinoid debris (Fig. 5B).

Out of the differentiated and counted components in the fraction >1 mm, some show distinct trends with depth. The variation in abundance of these components is presented in Fig. 4. From the bottom of the core, *Amphistegina* and *Operculina* decrease in abundance upcore from around 15% and 25% of the components, respectively, to almost disappearing in the upper part of the sequence. Red algae and the green alga *Halimeda* show an inverse trend increasing upcore to reach 10% to 15% and 40% of the particles, respectively. Bioclasts decrease in abundance upcore, as is the case for the stained dark bioclasts, which make up 40% to 45% of the samples below 9·6 mbsf. In contrast to the trend described for the other components which extend over

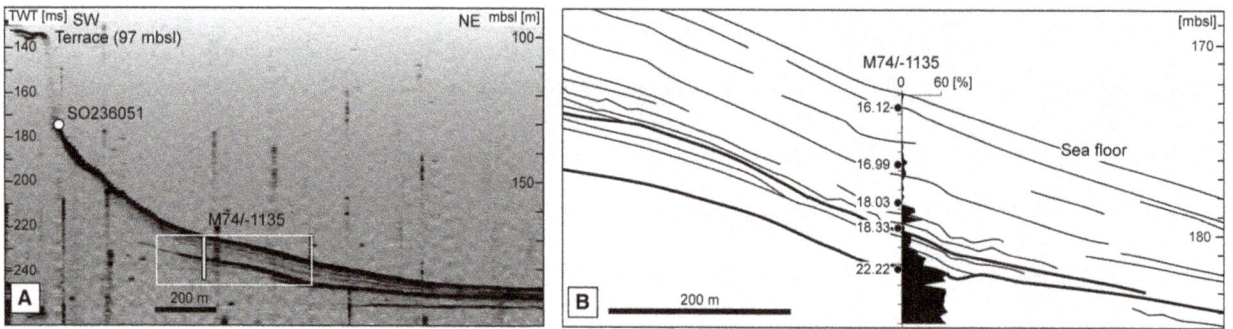

Fig. 3. (A) Parasound line covering the submarine terrace, the submarine cliff and the sediment wedge on the northern flank of Ari Atoll. (B) Detail view of the same parasound line showing the succession around Site M74/4-1135. The abundance of the dark grains (see also Fig. 4) is shown on the curve as an orientation to allow comparison of variations in abundance of components (Fig. 4) with the physical stratigraphy. [Colour figure can be viewed at wileyonlinelibrary.com]

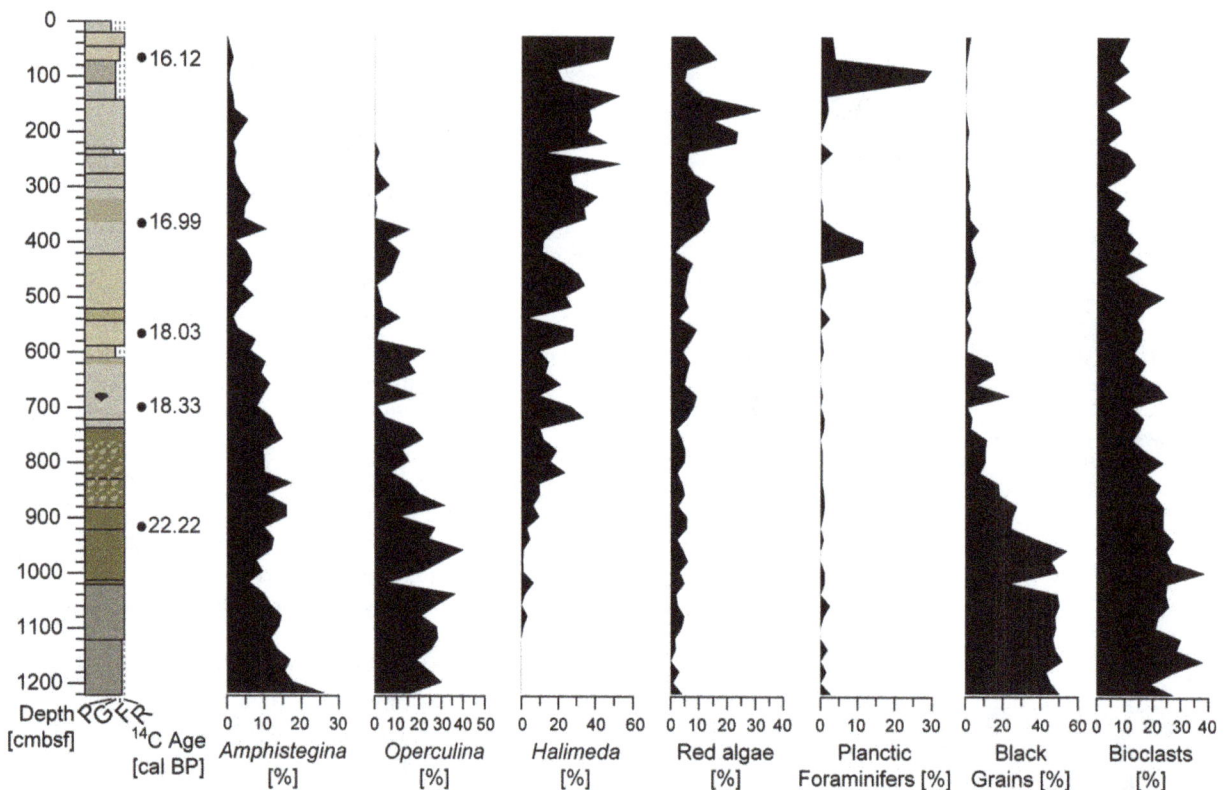

Fig. 4. Lithological column of core M74/4-1135 with variations in the relative abundance of selected components and position of samples used for radiocarbon dating. [Colour figure can be viewed at wileyonlinelibrary.com]

the entire core, the abundance of the dark bioclasts rapidly decreases between 9·6 mbsf and 7·4 mbsf to values of 0% to 5% above, except for the interval between 6·8 mbsf and 6·2 mbsf, where 10% to 20% were counted.

Red algae from selected rhodolith samples (Table 2) were analysed in thin section for identification at the most precise taxonomic level possible. Among coralline

red algae, the dominating genus is *Lithothamnion*, followed by *Lithophyllum* and minor *Lithoporella*. The aragonitic red alga *Peyssonnelia* occurs in several samples as well.

Five samples were used for age dating the succession (Fig. 4, Table 1) suggesting that the interval between 0·66 mbsf and 9·16 mbsf was deposited between

Fig. 5. Microphotograph of components >500 μm at Site M74/4-1135. (A) Sample from a core depth of 11·95 mbsf with abundant darkened bioclasts, *Amphistegina*, *Operculina*, red algal debris and minor planktic foraminifers. (B) Sample from a core depth of 0·65 mbsf with planktic foraminifers, pteropods, serpulids, echinoid debris and benthic foraminifers. [Colour figure can be viewed at wileyonlinelibrary.com]

Table 2. Red algae in selected samples of Core M74/4-1135. The samples in grey include components common in the 'intermediate water assemblage' of Webster *et al.* (2009), namely *Lithophyllum acrocamptum* and *Lithothamnion prolifer*

Sample	Depth (mbsf)	Algae
1135-2-5H-90-100	5·16	*Lithothamnion prolifer, Lithophyllum acrocamptum, Hydrolithon* sp., *Lithophyllum* gr. *pustulatum, Mesophyllum* sp.
1135-2-6G-30-35.1	5·56	Laminar *Lithothamnion, Lithoporella* sp., *Lithophyllum acrocamptum*?
1135-2-6G-30-35.2	5·56	Laminar *Lithothamnion, Lithoporella* sp., *Peyssonnelia* sp.
1135-2-7D-60-65	6·86	Laminar *Lithothamnion, Lithophyllum* gr. *pustulatum, Peyssonnelia* sp.
1135-2-9D-10-15	8·36	Laminar *Lithothamnion, Spongites*? sp., *Peyssonnelia* sp.
1135-2-9D-80-87	9·06	Laminar *Lithothamnion, Lithophyllum* gr. *pustulatum, Peyssonnelia* sp.
1135-2-10C-55-57.1	9·76	*Lithothamnion prolifer* (heavily bored) in the nucleus, laminar *Lithothamnion, Peyssonnelia* sp.
1135-2-10C-55-57.2	9·76	Laminar *Lithothamnion, Lithophyllum* gr. *pustulatum, Peyssonnelia* sp.
1135-2-10C-95-100	10·16	Laminar *Lithothamnion, Peyssonnelia* sp.
1135-2-12A-22	11·44	Laminar *Lithothamnion, Lithophyllum acrocamptum, Lithoporella* sp.
1135-2-12A-94-100	12·16	Laminar *Lithothamnion, Lithophyllum acrocamptum*?, *Lithoporella* sp.

approximately 16 and 22 ka BP. Deposits below 9·16 mbsf were not dated, as fossils without encrustation were not found. Three intervals with distinct sedimentation rates can be differentiated: below 7 mbsf, the rate is around 0·6 mm year^{-1}, between 7 and 0·66 mbsf the values increase to 3·9 mm year^{-1} before dropping sharply to 0·04 mm year^{-1}.

Videograb sample SO236-51 is located at the toe of a 31 m high terrace which is positioned at a water depth of *ca* 97 m (Fig. 3A). The 132 m deep location is 500 m upslope of Site M74/4-1135. The sea floor at the sample locality and along the flank of Ari Atoll is an irregular rocky surface (Fig. 6A and B). The rock is a rudstone (Fig. 6C) with large benthic foraminifers (*Amphistegina*, *Heterostegina*, *Alveolinella*), encrusting and articulated red algae, encrusting foraminifers and bryozoa, *Halimeda* flakes, serpulids and planktonic foraminifers. The irregular rock surface, colonized by gorgonians, sponges and bryozoa, is characterized by holes and depressions with irregular shapes (Fig. 6B). For age dating of these

deposits, several *Amphistegina* specimens were isolated which yield an age of around 13·32 ± 0·11 ka cal BP.

INTERPRETATION AND DISCUSSION

Depositional geometry

The three seismic lines depicted in Fig. 2 cover the transition from the reef buildups of the atoll margins to the stratified succession of the fore-reef area and the basin. This transition is characterized by a change from discontinuous and chaotic reflections to more continuous reflections from the sediment wedges. The limit between both acoustic facies is placed at the toe of the 31 m high wall of the 97 m terrace, which has been attributed to reefs which drowned during Meltwater Pulse 1A (14·3 to 14·65 kyrs BP; Deschamps *et al.*, 2012) by Fürstenau *et al.* (2010). The successions of the different slopes analysed are arranged into an aggrading to slightly backstepping pattern (Fig. 2).

Fig. 6. (A) Underwater photograph of the irregular and rocky sea floor in the vicinity of the location of sample SO236-051. Numbers at the top of the photograph display the geographical location and the water depth. (B) Videograb sample SO236-051 on deck of RV SONNE just after recovery. The irregular block is overgrown by gorgonians (G), sponges (S) and bryozoa. Scale: 5 cm. (C) Thin section photograph of Videograb sample SO236-051 with a high interparticle porosity, abundant *Amphistegina*, bryozoa, planktic foraminifers, *Halimeda* debris and bioclasts. Scale bar: 0·5 mm. [Colour figure can be viewed at wileyonlinelibrary.com]

M74/4-1135, which shows that the succession was deposited between *ca* 22 and 16 kyrs BP (Table 1, Fig. 7). Applying the reconstruction for the sea-level position during and after the last glacial maximum (Deschamps *et al.*, 2012; Lambeck *et al.*, 2014), the site would have been located at water depths of *ca* 48 m at 22 kyrs BP, *ca* 55 m at 18 kyrs BP, *ca* 52 m at 16·9 kyrs BP and *ca* 57 m at 16 kyrs BP (Figs 7 and 8).

In general, Core M74/4 displays an upcore trend from a floatstone to rudstone to a packstone to grainstone texture. The red algal associations and the large benthic foraminifers provide two lines of evidence indicating that the deposits recovered in Core M74/4-1135 were formed *in situ* within this palaeobathymetric range, or are parautochthonous at most. The samples marked by grey shading in the red algal overview of Table 2 include components common in the 'intermediate water assemblage' of Webster *et al.* (2009), namely *Lithophyllum* gr. *acrocamptum* and *Lithothamnion prolifer*. In the Pacific Ocean, in the absence of shallow-water species, such as *Porolithon onkodes*, these coralline algae are typical of 20 to 60 m water depths; a similar depth range, however, can be expected for the Indian Ocean, where less data are available. This covers the water depth of *ca* 50 m for the site location applying the Deschamps *et al.* (2012) sea-level curve. In any case, this depth range obviously can change in extremely clear waters, which however, in the discussed case study is irrelevant, as the maximum water depth at the time of sedimentation is constrained by the sea-level position (Fig. 8).

The large benthic foraminifer *Operculina ammonoides*, which is frequent to abundant in the lower and middle part of the succession (Figs 4 and 7), lives in water depths of up to 70 m with an optimum around 40 m (Hohenegger, 2000). *Amphistegina lessoni* and *A. radiata* are also frequent to abundant and thrive in water depths of up to 70 m, with highest abundance around 40 m (Hohenegger, 2000). As is the case for the red algae, the palaeobathymetry indicated by the large benthic foraminifers in the succession falls within the water depth for the site (Deschamps *et al.*, 2012). Both large benthic foraminifer decrease in abundance as the water depth at Site M74/4-1135 deepened (Fig. 7). This trend is paralleled by

Lowstand sediment wedges

The sediment wedges were formed during the last glacial sea-level lowstand, as indicated by the dating of the core

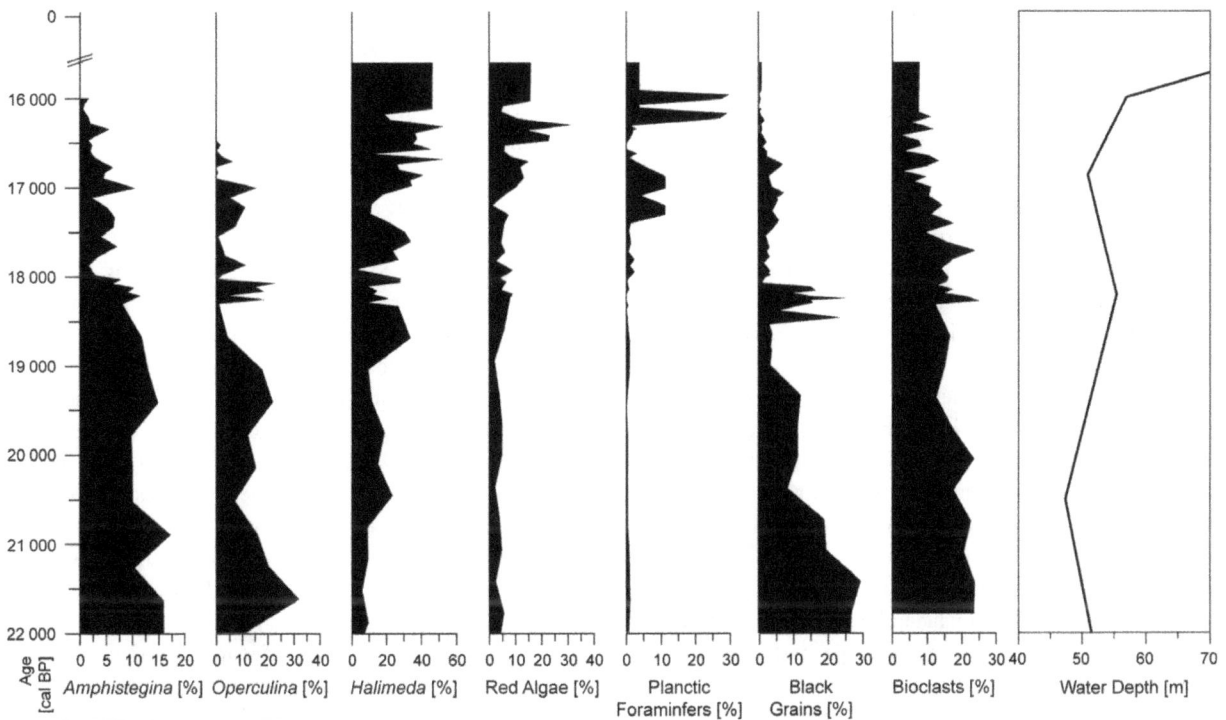

Fig. 7. Compositional variation of selected components in Core M74/4-1135 against age (position of samples used for age model is shown in Fig. 4). In the right-hand side of the figure, the variations in water depth at Site M74/4-1135 are shown, using the sea-level curve by Deschamps *et al.* (2012) and Lambeck *et al.* (2014).

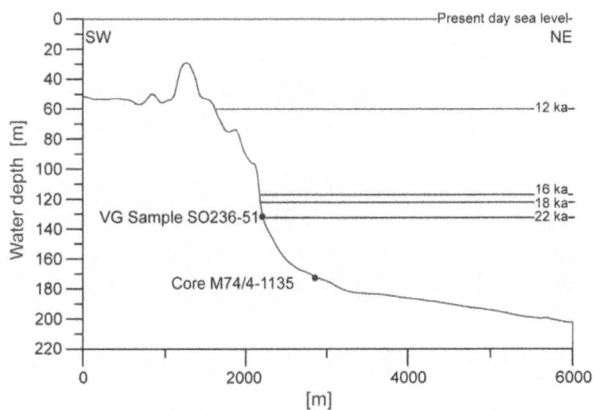

Fig. 8. SW–NE section through the northern flank of Ari Atoll along the position of seismic line 1 (see Fig. 1B for location) with sea-level stands between 22 ky BP and today.

increased amounts of *Halimeda* and red algae. In the Indo-Pacific, *Halimeda* and red algae may be more abundant at water depths exceeding those occupied by large benthic foraminifers (Wilson & Vecsei, 2005). Therefore, the upcore increase in algal abundance may well be a consequence of the deepening of Site M74/4-1135 during the post-glacial sea-level rise.

The dark-stained grains (Fig. 4) record a certain sediment influx from the exposed carbonate platform. In

carbonates, such grains blackened by organics and clays are known from intertidal and subtidal environments (Flügel, 2004). It is proposed that in Core M74/4-1135, these grains were shed from the emerging parts of the platform, which were just some hundreds of metres away from the site location. This export was reduced with the transgression due to the beginning of sea-level rise around 20·5 ka BP (Lambeck *et al.*, 2014) to recover a little bit around 18·5 ka BP with the temporary stop of sea-level rise at that time (Fig. 7). The base of this transgression correlates with the high-amplitude reflection in the parasound profile which crosses the core location (Fig. 3).

In summary, data indicate that the sediment wedge at position of Site M74/4-1135 is a lowstand wedge and that no or only very minor sediment is deposited in this area during the latest Pleistocene and Holocene. The parallelism in shape and seismic stratigraphy of this wedge compared with the other wedges as imaged in the seismic lines (Fig. 2) indicate that this also applies to other atoll flanks.

Intrinsic and extrinsic control on slope deposition

The seminal publication by Schlager *et al.* (1994) established the concept of the highstand shedding of Quaternary tropical flat-topped carbonate platforms, where the platforms export sediment into the adjacent basins when they are flooded and a shallow-water carbonate factory produces particles. Limitations were seen for carbonate ramps and carbonate platforms undergoing sea-level lowstands for extended periods of time. Applying the highstand shedding scenario, carbonate platform slope angles are seen as a function of grain size (Kenter, 1990), with coarser-grained sediments developing steeper slope angles than finer-grained ones. Both aspects are now widely accepted paradigms of carbonate sedimentology and stratigraphy applied when interpreting geological and subsurface geophysical data.

In the carbonate sedimentary succession of the Maldives, a differentiation of highstand and lowstand deposition has been demonstrated for the Pleistocene and Holocene drifts of the Inner Sea by Paul *et al.* (2012) and Betzler *et al.* (2013b). This is based on variations in the amount of fine-grained, platform-derived aragonite mud which at different basinal localities of the Maldives is most abundant in the highstand deposits. Therefore, it remains to be resolved why the slope and toe of slope deposits do not bear highstand sediments.

Sediment export by highstand shedding does not necessarily imply deposition at the slope of the carbonate platform, because other processes may outweigh gravitational transport processes. In carbonates, slope steepening behind the angle of repose has been suggested as an intrinsic mechanism allowing slope bypass of sediment exported from the platform into the basin (Schlager & Camber, 1986). The Bahamas escarpment has been reported as the type example to illustrate the case of slope sediment bypass. This, however, has to be revised because a number of recent studies show that along-slope contour currents winnow the slope and toe of slope (Mulder *et al.*, 2012; Betzler *et al.*, 2015; Jo *et al.*, 2015; Lüdmann *et al.*, 2016; Principaud *et al.*, 2016; Wunsch *et al.*, 2016). Consequently, it is reworked deposits that accumulate in periplatform drifts (Betzler *et al.*, 2015) and in detached drifts in the basin (Bergman *et al.*, 2010; Lüdmann *et al.*, 2016).

Large drifts in the Inner Sea together with submarine dunes show that currents also dictate sedimentation patterns in the Maldives (Fig. 1C and D) (Betzler *et al.*, 2009, 2013a,b, 2016; Lüdmann *et al.*, 2013). Current speeds in the channels between the atolls are typically in the range of 0·5 to 0·8 m s^{-1} and accelerate in the inter-faro channels and along the atoll flanks to 1·5 to 2·6 m s^{-1} (Preu & Engelbrecht, 1991; Owen *et al.*, 2011). These currents are the result of the interaction of ocean, tidal and wind-induced currents. As noted by Darwin (1842, p. 108), based on the reports of Capt. Moresby, 'The currents of the sea flow across these atolls (...) yet the currents sweep with greater force round their flanks'.

The effect of these currents is well-documented in the video survey performed along the flank of Ari Atoll around the location of Sample SO236-51, where the sea floor is rocky, with no sediment cover (Fig. 6A). A question which remains unanswered based on available data is if the current regime flowing around the atolls was different during the past sea-level lowstand thus allowing lowstand deposition.

The margin aggradation or backstepping (Fig. 2) in the Maldives can be traced back to the lower Pliocene (Betzler *et al.*, 2013a). Such a pattern elsewhere is interpreted to reflect carbonate platform growth under increasing accommodation, but the long-term platform trend discards sea-level changes as a trigger as it also applies to the rate of subsidence which has been calculated to be around 0·03 to 0·04 mm year^{-1} over the long term (Belopolsky & Droxler, 2004) and to 0·15 mm year^{-1} for the short term (Gischler *et al.*, 2008). The onset of the growth pattern correlates rather with the early Pliocene widening of the inter-atoll passages as mapped by Lüdmann *et al.* (2013). Inner Sea drift geometries indicate that the opening and widening of the passages allowed the current system that controls sedimentation nowadays to be established.

Summarizing, it is proposed that alongslope currents redistribute the sediment formed during the highstand shedding away from the atoll slopes, where the highest current speeds are to be expected, towards the more protected and tranquil areas of the Inner Sea. These are the zones of the Inner Sea, where large drift bodies accumulate (Fig. 1C and D). These drift sediments show a clear differentiation into aragonite-mud rich highstand deposits and aragonite poor lowstand deposits (Paul *et al.*, 2012; Betzler *et al.*, 2013b). Self-erosion of the slope (Schlager & Camber, 1986) is excluded because the seismics and parasound lines do not show the corresponding signatures (Fig. 2).

CONCLUSIONS

The slope successions of atolls in the Maldives archipelago consist of a series of basinward thinning wedges which were deposited during sea-level lowstands, whereas highstand deposits are condensed. Rather than intrinsic factors such as slope steepness, the vigorous currents around and in the atolls control this pattern. Highstand sediment is not accumulated along the slope but reworked and transported to be accumulated into drift deposits forming in current-protected areas. This observation confirms that contour currents are of equal importance for determining carbonate sedimentation patterns on carbonate platform slopes as is, for example, the windward–leeward orientation of the slope. This effect of contour currents is expected to be more common in the geological record of carbonate platform slopes than previously estimated.

ACKNOWLEDGEMENTS

The authors thank the officers and the crew of the RV METEOR during Cruise M74/4 and of RV SONNE during Cruise SO236 MALSTROM for their efficient assistance and the Shipboard Scientific Parties for the substantial support. Jörn Fürstenau and Hauke Petersen are thanked for their contributions within the project NEOMA. The German Federal Ministry of Education and Research is thanked for funding (03S0405, 03G0236A), the Ministry of Fisheries and Agriculture of the Maldives for granting the research permit for Maldivian waters. Thanks to the editor P. Swart and associate editor A. Klaus for their editorial support.

CONFLICT OF INTEREST

The authors declare that they have no conflict of interest.

References

Aubert, O. and Droxler, A.W. (1992) General Cenozoic evolution of the Maldives carbonate system (equatorial Indian Ocean). *Bull. Centres Rech. Explor.-Prod. Elf-Aquitaine*, **16**, 113–136.

Belopolsky, A.V. and Droxler, A.W. (2004) Seismic expressions and interpretations of carbonate sequences: the Maldives carbonate platform, equatorial Indian Ocean. *AAPG Stud. Geol.*, **49**, 57.

Bergman, K.L., Westphal, H., Janson, X., Poiriez, A. and Eberli, G.P. (2010) Controlling Parameters on Facies Geometries of the Bahamas, an Isolated Carbonate Platform Environment. In: *Carbonate Depositional Systems: Assessing Dimensions and Controlling Parameters* (Eds H. Westphal, B. Riegl and G.P. Eberli), pp. 5–80. Springer, Heidelberg.

Betzler, C., Hübscher, C., Lindhorst, S., Reijmer, J.J.G., Römer, M., Droxler, A.W., Fürstenau, J. and Lüdmann, T. (2009) Monsoonal-induced partial carbonate platform drowning (Maldives, Indian Ocean). *Geology*, **37**, 867–870.

Betzler, C., Fürstenau, J., Lüdmann, T., Hübscher, C., Lindhorst, S., Paul, A., Reijmer, J.J.G. and Droxler, A.W. (2013a) Sea-level and ocean-current control on carbonate-platform growth, Maldives, Indian Ocean. *Basin Res.*, **25**, 172–196.

Betzler, C., Lüdmann, T., Hübscher, C. and Fürstenau, J. (2013b) Current and sea-level signals in periplatform ooze (Neogene, Maldives, Indian Ocean). *Sed. Geol.*, **290**, 126–137.

Betzler, C., Lindhorst, S., Lüdmann, T., Weiss, B., Wunsch, M. and Braga, J.C. (2015) The leaking bucket of a Maldives atoll: implications for the understanding of carbonate platform drowning. *Mar. Geol.*, **366**, 16–33.

Betzler, C., Eberli, G.P., Kroon, D., Wright, J.D., Swart, P.K., Nath, B.N., Alvarez-Zarikian, C.A., Alonso-García, M., Bialik, O.M., Blättler, C.L., Guo, J.A., Haffen, S., Horozal, S., Inoue, M., Jovane, L., Lanci, L., Laya, J.C., Mee, A.L.H., Lüdmann, T., Nakakuni, M., Niino, K., Petruny, L.M., Pratiwi, S.D., Reijmer, J.J.G., Reolid, J., Slagle, A.L., Sloss, C.R., Su, X., Yao, Z. and Young, J.R. (2016) The abrupt onset of the modern South Asian Monsoon winds. *Sci. Rep.*, **6**, 29838.

Chabaud, L., Ducassou, E., Tournadour, E., Mulder, T., Reijmer, J.J.G., Conesa, G., Giraudeau, J., Hanquiez, V., Borgomano, J. and Ross, L. (2016) Sedimentary processes determining the modern carbonate periplatform drift of Little Bahama Bank. *Mar. Geol.*, **378**, 213–229.

Darwin, C. (1842) *Structure and Distribution of Coral Reefs*. Smith, Elder & Co., London, 214 pp.

Deschamps, P., Durand, N., Bard, E., Hamelin, B., Camoin, G., Thomas, A.L., Henderson, G.M., Okuno, J.I. and Yokoyama, Y. (2012) Ice-sheet collapse and sea-level rise at the Bolling warming 14,600 years ago. *Nature*, **483**, 559–564.

Droxler, A.W., Haddad, G.A., Mucciarone, D.A. and Cullen, J.L. (1990) Pliocene-Pleistocene aragonite cyclic variations in Holes 714A and 716B (the Maldives) compared with Hole 633A (the Bahamas): records of climate-induced $CaCO_3$ preservation at intermediate water depths. *Proc. ODP Sci. Results*, **115**, 539–577.

Eberli, G.P. and Ginsburg, R.N. (1987) Segmentation and coalescence of Cenozoic carbonate platforms, northwestern Great Bahama Bank. *Geology*, **15**, 75–79.

Flügel, E. (2004) *Microfacies of Carbonate Rocks*. Springer-Verlag, Berlin, 976 pp.

Fürstenau, J., Lindhorst, S., Betzler, C. and Hübscher, C. (2010) Submerged reef terraces of the Maldives (Indian Ocean). *Geo-Marine Lett.*, **30**, 511–515.

Gischler, E., Hudson, J.D. and Pisera, A. (2008) Late Quaternary reef growth and sea level in the Maldives (Indian Ocean). *Mar. Geol.*, **250**, 104–113.

Grammer, G.M. and Ginsburg, R.N. (1992) Highstand versus lowstand deposition on carbonate platform margins: insight from Quaternary foreslopes in the Bahamas. *Mar. Geol.*, **103**, 125–136.

Hohenegger, J. (2000) Coenoclines of larger foraminifera. *Micropaleontology*, **46**, 127–151.

Jo, A., Eberli, G.P. and Grasmueck, M. (2015) Margin collapse and slope failure along southwestern Great Bahama Bank. *Sed. Geol.*, **317**, 43–52.

Kenter, J.A.M. (1990) Carbonate platform flanks: slope angle and sediment fabric. *Sedimentology*, **37**, 777–794.

Lambeck, K., Rouby, H., Purcell, A., Sun, Y. and Sambridge, M. (2014) Sea level and global ice volumes from the Last Glacial Maximum to the Holocene. *Proc. Natl Acad. Sci. USA*, **111**, 15296–15303.

Lüdmann, T., Kalvelage, C., Betzler, C., Fürstenau, J. and Hübscher, C. (2013) The Maldives, a giant isolated carbonate platform dominated by bottom currents. *Mar. Pet. Geol.*, **43**, 326–340.

Lüdmann, T., Paulat, M., Betzler, C., Möbius, J., Lindhorst, S., Wunsch, M. and Eberli, G.P. (2016) Carbonate mounds in the Santaren Channel, Bahamas: a current-dominated periplatform depositional regime. *Mar. Geol.*, **376**, 69–85.

Mulder, T., Ducassou, E., Eberli, G.P., Hanquiez, V., Gonthier, E., Kindler, P., Principaud, M., Fournier, F., Léonide, P., Billeaud, I., Marsset, B., Reijmer, J.J.G., Bondu, C., Joussiaume, R. and Pakiades, M. (2012) New insights into the morphology and sedimentary processes along the western slope of Great Bahama Bank. *Geology*, **40**, 603–606.

Neumann, A.C. and Ball, M.M. (1970) Submersible observations in the Straits of Florida: geology and bottom currents. *Geol. Soc. Am. Bull.*, **81**, 2861–2874.

Owen, A., Kruijsen, J., Turner, N. and Wright, K. (2011) *Marine Energy in the Maldives*. Prefeasibility report on Scottish Support for Maldives Marine Energy Implementation, Main Report. Centre for Understanding Sustainable Practice, Robert Gordon University, Aberdeen, UK.

Paul, A., Reijmer, J.J.G., Fürstenau, J., Kinkel, H. and Betzler, C. (2012) Relationship between Late Pleistocene sea-level variations, carbonate platform morphology and aragonite production (Maldives, Indian Ocean). *Sedimentology*, **59**, 1540–1658.

Preu, C. and Engelbrecht, C. (1991) Patterns and processes shaping the present morphodynamics of coral reef islands. Case study from the North-Male atoll, Maldives (Indian Ocean). In: *From the North Sea to the Indian Ocean* (Eds H. Brückner and U. Radtke), pp. 209–220. Franz Steiner, Stuttgart.

Principaud, M., Ponte, J.P., Mulder, T., Gillet, H., Robin, C. and Borgomano, J. (2016) Slope-to-basin stratigraphic evolution of the northwestern Great Bahama Bank (Bahamas) during the Neogene to Quaternary: interactions between downslope and bottom currents deposits. *Basin Res.*. doi:10.1111/bre.12195.

Purdy, E.G. and Bertram, G.T. (1993) Carbonate concepts from the Maldives, Indian Ocean. *AAPG Stud. Geol.*, **34**, 56.

Reimer, P.J., Bard, E., Bayliss, A., Beck, J.W., Blackwell, P.G., Ramsey, C.B., Buck, C.E., Cheng, H., Edwards, R.L., Friedrich, M., Grootes, P.M., Guilderson, T.P., Haflidason, H., Hajdas, I., Hatté, C., Heaton, T.J., Hoffmann, D.L., Hogg, A.G., Hughen, K.A., Kaiser, K.F., Kromer, B., Manning, S.W., Niu, M., Reimer, R.W., Richards, D.A., Scott, E.M., Southon, J.R., Staff, R.A., Turney, C.S.M. and van der Plicht, J. (2013) IntCal13 and MARINE13 radiocarbon age calibration curves 0-50000 years cal BP. *Radiocarbon*, **55**, 1869–1887.

Rendle-Bühring, R.H. and Reijmer, J.J.G. (2005) Controls on grain-size patterns in periplatform carbonates: marginal setting versus glacio-eustacy. *Sed. Geol.*, **175**, 99–113.

Schlager, W. and Camber, O. (1986) Submarine slope angles, drowning unconformities and self-erosion of limestone escarpments. *Geology*, **14**, 762–765.

Schlager, W., Reijmer, J.J.G. and Droxler, A.W. (1994) Highstand shedding of carbonate platforms. *J. Sed. Res.*, **64**, 270–281.

Stuiver, M. and Reimer, P.J. (1993) Extended ^{14}C database and revised CALIB radiocarbon calibration program. *Radiocarbon*, **35**, 215–230.

Tomczak, M. and Godfrey, J.S. (2003) *Regional Oceanography: An Introduction*. Daya Publishing House, Delhi, 390 pp. Available at: www.es.flinders.edu.au/~mattom/regoc/pdfversion.html.

Tournadour, E., Mulder, T., Borgomano, J., Hanquiez, V., Ducassou, E. and Gillet, H. (2015) Origin and architecture of a Mass Transport Complex on the northwest slope of Little Bahama Bank (Bahamas): relations between off-bank transport, bottom current sedimentation and submarine landslides. *Sed. Geol.*, **317**, 9–26.

Webster, J.M., Braga, J.C., Clague, D.A., Gallup, C., Hein, J.R., Potts, D.C., Renema, W., Riding, R., Riker-Coleman, K., Silver, E. and Wallace, L.M. (2009) Coral reef evolution on rapidly subsiding margins. *Global Planet. Change*, **66**, 129–148.

Wilson, M.E.J. and Vecsei, A. (2005) The apparent paradox of abundant foramol facies in low latitudes: their environmental significance and effect on platform development. *Earth Sci. Rev.*, **69**, 133–168.

Wunsch, M., Betzler, C., Lindhorst, S., Lüdmann, T. and Eberli, G.P. (2016) Sedimentary dynamics along carbonate slopes (Bahamas archipelago). *Sedimentology*. doi:10.1111/sed.12317.

2

Examining the interplay of climate and low amplitude sea-level change on the distribution and volume of massive dolomitization: Zebbag Formation, Cretaceous, Southern Tunisia

RICHARD NEWPORT*, CATHY HOLLIS* (iD), STÉPHANE BODIN† and JONATHAN REDFERN*

*School of Earth and Environmental Sciences, Manchester University, Manchester M13 9PL, UK (E-mail: Cathy.Hollis@manchester.ac.uk)
†Department of Geoscience, Aarhus University, Høegh-Guldbergs Gade 2, 8000 Aarhus C, Denmark

Keywords
Cretaceous, dolomitization, mesosaline, Sahara platform.

ABSTRACT

During the Cretaceous, a humid global climate, calcitic seas, high relative sea-level and low amplitude changes in relative sea-level largely prevented large-scale dolomitization in many carbonate successions. However, the well-exposed shallow-water carbonate sediments of the Upper Albian–Lower Turonian Zebbag Formation on the Jeffara Escarpment, southern Tunisia, are pervasively dolomitized. This study considers why dolomitization was so widespread in this region during a period of Earth history when platform-scale dolomitization is rare. Marine conditions were established in the Upper Albian, evidenced by stacked upward-shallowing packages of shallow subtidal to peritidal carbonate sediments in the basal Rhadouane Member. A gradual increase in the volume of subtidal sediments in the Cenomanian Kerker Member, culminated in deposition of laterally extensive marls, during maximum flooding of the platform in the Lower Turonian. The overlying Gattar Member was then deposited in shallower water as relative sea-level fell. The entire Zebbag Formation is pervasively replaced by stratabound, fabric-retentive, dolomite, except within the marl at the top of the Kerker Member, which is only partially dolomitized. Petrographic textures indicate dolomitization largely post-dated marine cementation and platform emergence but pre-dated chemical compaction. Slightly more positive oxygen isotope signatures, slightly elevated concentrations of Sr and a near-absence of evaporites are consistent with dolomitization by reflux of mesohaline sea water. An upward-decrease in major element concentrations and higher $^{87}Sr/^{86}Sr$ compared to Upper Cretaceous sea water suggest that basal, Albian siliciclastic beds acted as aquifers facilitating dolomitization by fluxing fluids offshore. Dolomitization is interpreted to have resulted from multiple fluxes of sea water over periods of 0·5 to 2·5 Ma. The unusually high volume of dolostone for a platform of this age most probably reflects deposition within an arid climate belt, where an efficient reflux system was facilitated by basal, permeable siliciclastic strata.

INTRODUCTION

Despite numerous studies of the mechanisms, fluid sources, temporal and geographical distribution of dolomite in the Phanerozoic, there is still active discussion as to the primary controls on its distribution, which varies significantly in time and space. Several studies have related temporal variations in dolomite abundance to changes in Mg/Ca ratio, glacio-eustacy and atmospheric CO_2 (Given & Wilkinson, 1987; Mackenzie & Morse, 1992; Sun, 1994; Burns et al., 2000; Wright & Oren, 2005). Greenhouse periods, characterized by warm climates, are interpreted to have a high abundance of dolomite, probably reflecting favourable conditions for evaporite deposition and dolomitization via hypersaline reflux (Warren, 2000). While this is certainly true for the Afro-Arabian Plate during the Permo-Triassic and Jurassic, there was a more humid climate during the Cretaceous within the circum-Tethys region, and a transition to calcitic seas (Sandberg, 1983). This change in climate was coupled with the highest relative sea-level of the Phanerozoic (Haq *et al.*, 1987), with atmospheric pCO_2 3 to 12 times higher than present day (Kuypers *et al.*, 1999). This led to an ice-free, hot and stable greenhouse climate by the middle Cretaceous (MacLeod *et al.*, 2013; Bodin *et al.*, 2015). The Upper Cretaceous has a low global abundance of dolomite (Given & Wilkinson, 1987; Sun, 1994), although massive dolomitization is observed in North Africa (Abdallah, 2003; Touir *et al.*, 2009; Bodin *et al.*, 2010) and southern Europe (Korbar *et al.*, 2001; Husinec & Sokač, 2006; Benito & Mas, 2007; Iannace *et al.*, 2013). This study considers why dolomitization was so widespread in this region, despite calcitic seas, low amplitude changes in relative sea-level and a humid global climate – none of which are conducive to dolomitization.

The Upper Albian to Middle Turonian Zebbag Formation is pervasively dolomitized and provides an excellent opportunity to study the processes and controls on massive dolomitization of marine carbonates on the northern margin of the Afro-Arabian Plate during the middle Cretaceous. Strata crop out along the laterally continuous Jeffara Escarpment located in the central part of southern Tunisia. The escarpment records an Upper Permian to Upper Cretaceous succession, with excellent pseudo-3D exposure. It extends for over 200 km, from central southern Tunisia into north-western Libya separating the Ghadames Basin and Dahar uplift in the south-west from the Jeffara coastal plain in the north-east (Fig. 1). Dolomitization is observed along the length of the escarpment (Badalini *et al.*, 2002; Bodin *et al.*, 2010). The excellent exposure of the Zebbag Formation provides a high-quality outcrop analogue for hydrocarbon reservoirs in the region, including the Miskar gas field offshore Tunisia (Zappaterra, 1995) and the Arous Al-Bahar gas field within the offshore Sirt Basin (Belopolsky *et al.*, 2012).

Previous studies document the diagenesis and dolomitization of Albian to Turonian strata in north-western Libya (Koehler, 1982; Chaabani *et al.*, 2003; El-Bakai *et al.*, 2010) and central Tunisia (M'rabet, 1981; Abdallah, 2003; Touir *et al.*, 2009), but no previous work has been conducted on the Zebbag Formation of southern Tunisia. The origin of dolomitization within laterally equivalent strata has been variously attributed to reflux of hypersaline brines (M'rabet, 1981; Abdallah, 2003; Touir *et al.*, 2009), normal marine fluids (Al-Aasm, 2005; Touir *et al.*, 2009), mixed marine-meteoric fluids (Koehler, 1982;

Abdallah, 2003; Al-Aasm, 2005) and deep phreatic fluids of continental origin (M'rabet, 1981). This study will present a multidisciplinary and multi-scale evaluation, using petrographical, geochemical and field evidence in order to fully evaluate the controls on dolomitization.

GEOLOGICAL SETTING

During the Late Cretaceous, southern Tunisia was located on the Saharan Platform, a passive margin. It was located around 12°N within a hot, arid climate belt (Fig. 2) (Chumakov *et al.*, 1995; Scotese, 2001; Sellwood & Valdes, 2006). Prevailing trade winds moved in a south-westerly direction (Fig. 2; Fabre & Mainguet, 1991; Poulsen *et al.*, 1998). Sea water temperature in the southern part of the Tethys is estimated to have been between 21°C and 36°C (Kolodny & Raab, 1988; Pearson *et al.*, 2001; Schouten *et al.*, 2003; Steuber *et al.*, 2005; Amiot *et al.*, 2010; Linnert *et al.*, 2014).

During the late Albian, regional marine transgression established a broad and shallow, carbonate platform across much of the Saharan Platform of North Africa (Benton *et al.*, 2000; Wood *et al.*, 2014). The rise in relative sea-level is marked by a transition from siliciclastic sediments of the 'Continental Intercalaire' Ain El Guettar Formation (De Lapparent & Gorce, 1960) to the overlying marine carbonate sequence, which in the Jeffara Escarpment is represented by the Zebbag Formation (Lefranc & Guiraud, 1990; Bodin *et al.*, 2010; Fig. 1). The Formation is divided into the Charenn, Rhadouane, Kerker and the Gattar Members (Fig. 1). The Charenn and Radhouane Members are only found in the northernmost part of the Jeffara Escarpment, thinning towards the south (Bodin *et al.*, 2010). The Charenn Member is a coarse-grained siliciclastic deposit which contains marine fauna (bryozoans and bivalves) and locally shows herringbone crossbedding, indicating a tidal component. The Rhadouane Member conformably overlies the Charenn Member or sits unconformably above the Lower Albian fluvial to marginal marine Ain El Guettar Formation (Bodin *et al.*, 2010). It is characterized by bivalve and gastropod floatstones with abundant microbialites (Koehler, 1982; Bodin *et al.*, 2010) and has been dated as Late Albian, based on the presence of the ammonite *Knemiceras* sp. (Abdallah *et al.*, 1995).

Continued sea-level rise over a vast and flat area in the Cenomanian led to deposition of shallow-water peritidal and shallow subtidal facies with low diversity, salinity-tolerant fauna such as gastropod, and miliolid, wackestones to mudstones and a single evaporite horizon (Bodin *et al.*, 2010). This suggests variable salinity levels with minimal circulation during deposition (Touir *et al.*, 2009). The boundary between the Rhadouane and Kerker

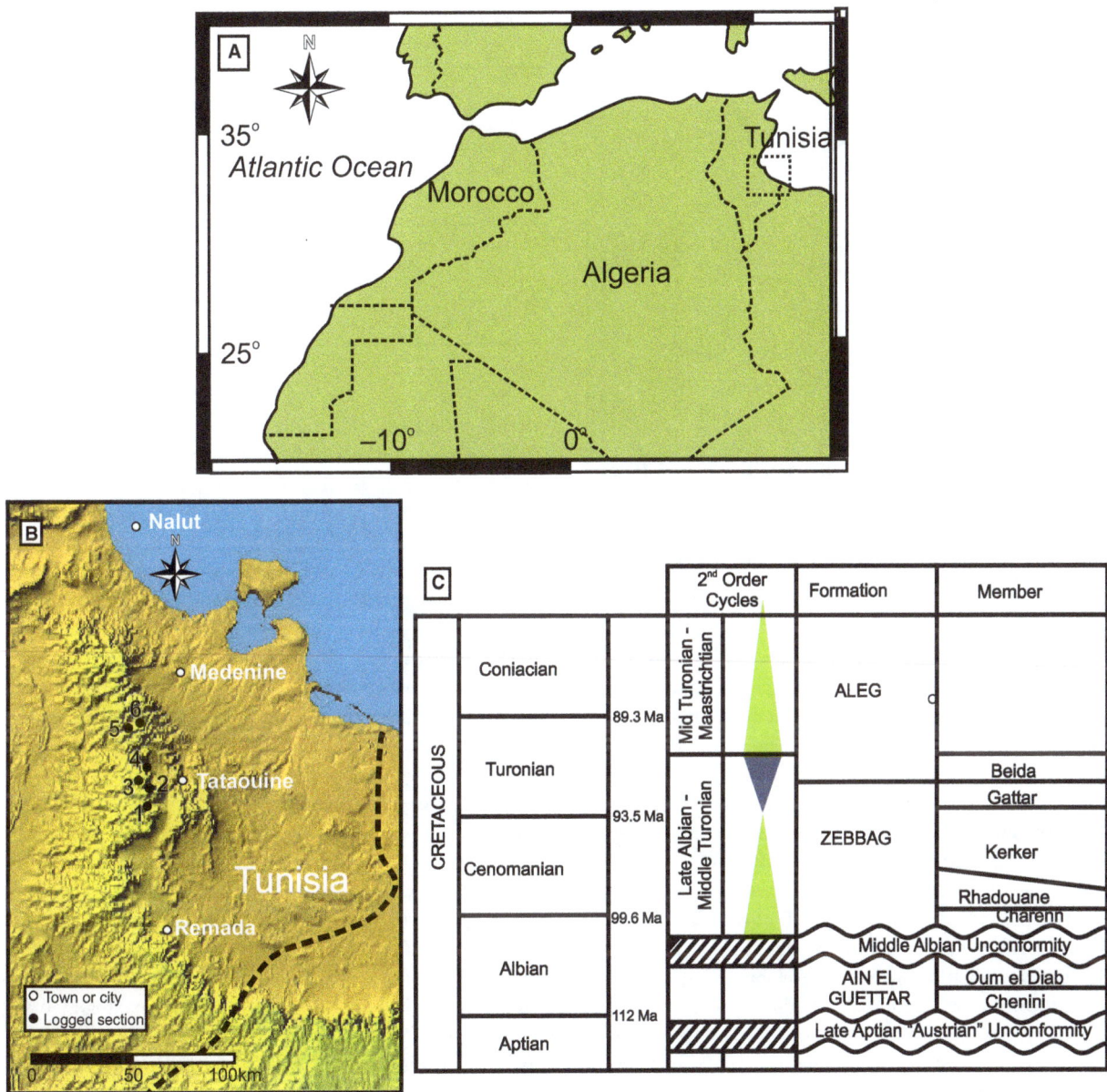

Fig. 1. (A) The location of the field area. (B) A digital elevation map of the Jeffara Escarpment and location of logged sections. Localities referred to in text are 1: Douiret Road; 2: Douiret Ancien; 3 Chenini to Diouret Road (i) and (ii); 4: Chenini; 5: K'Hil and 6: Ghomrassen. (C) Stratigraphic column showing the various formations and members from the Aptian to Coniacian of the area around Tataouine. Data from Bodin et al. (2010).

Members is gradual and defined by a transition from clay-free to clay rich marl-dominated facies (Bodin et al., 2010). The Kerker Formation resembles the Rhadouane Member, and the laterally equivalent Yifran Formation (Bodin et al., 2010), being composed of stacked peritidal cycles of gastropod and bivalve wackestones capped by microbialite beds (El-Bakai, 1997). The presence of bird tracks, mammal tracks and abundant tepee structures within the central part of the Kerker Member south of the study area (Contessi & Fanti, 2012a,b; Contessi, 2013)

suggests periods of subaerial exposure, whereas lateral-equivalent strata in the study area are dominated by sub-tidal facies.

The Cenomanian–Turonian boundary occurs within the uppermost part of the Kerker Member, with maximum flooding at ca 92·2 Ma (Lüning et al., 2004). It forms part of the second-order transgressive–regressive cycle of Marie et al. (1984). Other authors have correlated the Cenomanian–Turonian boundary to the third-order UZA 2·5 global sea-level cycle of Haq et al. (1987) (Touir

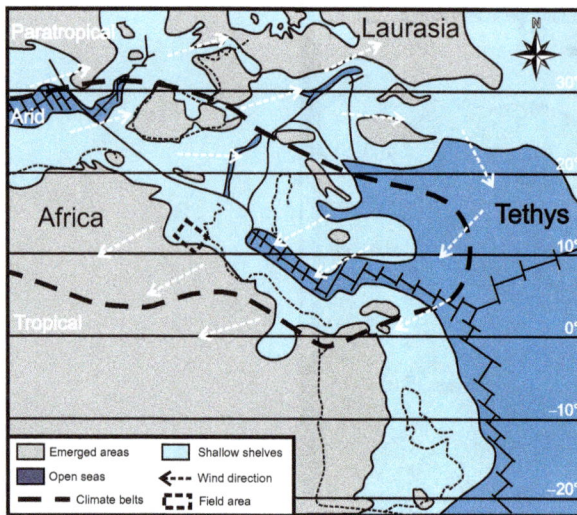

Fig. 2. A palaeogeographical map of the Tethys during the early Aptian adapted from Bodin *et al.* (2010). Climate belts based on those of Scotese (2001) and wind direction taken from Poulsen *et al.* (1998).

Table 1. Table showing GPD coordinates of logged sections of the Zebbag Formation in southern Tunisia

Section name	Location	
	Latitude	Longitude
Douiret Road	N32°50495	E10°18041
Douiret Ancien	N32°52264	E10°17541
Chenini to Douiret Road (i)	N32°53671	E10°14084
Chenini to Douiret Road (ii)	N32°53920	E10°14106
Chenini	N32°54899	E10°16252
K'Hil	N33°05035	E10°13101
Ghomrassen	N33°07190	E10°15007

& Soussi, 2003). Cenomanian sediments in this study contain abundant echinoid debris, pelagic foraminifera and oysters indicating a continued rise in relative sea-level and open marine conditions (Grosheny *et al.*, 2013). The uppermost beds are the lateral equivalent of the upper Cenomanian to lower Turonian Bahloul Formation of Northern and Central Tunisia (Abdallah *et al.*, 1995; Grosheny *et al.*, 2013). As the rate of relative sea-level rise slowed, large rudist build-ups of the Gattar Member became established and prograded northwards. A fully upward-shallowing succession is seen within the Gattar Member, culminating with the development of tidal flats and deposition of evaporites within troughs in central Tunisia (Camoin, 1991; Abdallah, 2003). The Annaba marls, which directly overlie the Gattar Formation in northern central Tunisia, are not seen in the Jeffara Escarpment where the Gattar Formation is covered by the Beida Anhydrites (Touir *et al.*, 2009).

METHODS

Six sections from around the town of Tataouine in Tunisia were logged using a Jacobs staff and traditional field methods. The carbonate rock textures were described using the Dunham (1962) classification. The GPS coordinates of logged sections are given in Table 1 and locations are shown on Fig. 1. Logged sections were systematically sampled to cover the full range of facies and textures observed within each member of the Zebbag Formation. Collected samples were prepared as 30 μm covered thin sections, impregnated with blue resin to highlight

porosity. Sections were also stained with Alizarin Red S and potassium ferricyanide to determine carbonate mineralogy and iron content (Dickson, 1966). Petrographic description led to construction of a paragenetic sequence using transmitted light and cross-polarized light techniques. Dolomite textures were classified according to Sibley & Gregg (1987). Point counting of thin sections for mineralogy and porosity was undertaken using Petrog© point counting software and stepper stage. For each section, 250 points were counted using an approximate stepping distance of 1·15 mm in the × direction and 1·14 mm in the y direction. A subset of samples was prepared as 30 μm uncovered, polished sections for cathodoluminescence, conducted using a Citl 8200 mark 2 luminoscope with an accelerating voltage of 6 to 8 kV, a vacuum of *ca* 0·2 mbar and a cathode current of between 310 to 335 μA.

The mineralogy of samples was analysed using X-ray diffraction (XRD) at the University of Manchester. Bulk samples were ground using a pestle and mortar and analysed using a Bruker D8 advanced instrument and a Cu $K_{\alpha 1}$ radiation. Data were collected over a range of 5° 2θ to 70° 2θ with a step size of 0·02° 2θ. Standards of known composition were run prior and post data collection to ensure accuracy. Quantification was determined using Siroquant software, which uses an area under the peak method of quantification.

Stable oxygen and carbon isotope (39 samples), major element and major element analysis (39 samples) were conducted at the Ruhr University, Bochum. Bulk rock powders were extracted from selected samples using a dentist drill. For stable isotope analysis, 0·30 mg ± 0·04 mg was weighed out and analysed using a Thermo 253 mass spectrometer attached to a Gasbench II and a PAL auto sampler. All values are reported relative to the Vienna-Pee Dee Formation belemnite (V-PDB) and standard deviations are 0·05 ‰ for carbon and 0·11 ‰ for oxygen. Where appropriate $\delta^{18}O$ values calculated for fluids are reported relative to Standard Mean Ocean

Water (SMOW). For major element analysis, *ca* 0·15 mg of sample was weighed and dissolved in 3 M HNO₃. The solution was then diluted with 2 ml of deionized water (>18·2 MΩ cm⁻¹). The concentration of the elements Ca, Mg, Fe, Mn and Sr were analysed using a Thermo Scientific iCAP 6500 DUO inductively coupled plasma optical emission spectrometer. Eight reference samples were also analysed (BSC-CRM-512, dolomite and BSC-CRM-513, limestone) to ensure accuracy. The relative standard deviation, which is the ratio of the standard deviation to the mean, for all elements and all samples was <5%. All major element and isotope data presented are bulk samples with >80% dolomite based on XRD and point counting techniques.

The Sr isotope measurements (six samples) were carried out at the SGIker-Geochronology and Isotopic Geochemistry facility of the University of the Basque Country UPV/EHU (Spain). The procedure for sample treatment and extraction of Sr was carried out according to the method of Pin & Bassin (1992) and Pin *et al.* (1994). The ⁸⁷Sr/⁸⁶Sr ratios were measured by MC-ICP-MS using a high-resolution Thermo Fisher Scientific Neptune instrument in static multicollection mode. Values were corrected for mass fractionation by normalizing to ⁸⁸Sr/⁸⁶Sr = 8·375209 (Steiger & Jäger, 1977). The

uncertainty for individual measurements of ⁸⁷Sr/⁸⁶Sr isotopes and average ratio under the same conditions for NBS-987 standard over the period of analyses was 0·710269 ± 0·000015 (2 SD). The full analytical details can be found in Newport (2014).

RESULTS

Sedimentology and petrography

A composite log of the Zebbag Formation and field photographs are shown in Figs 3 and 4–6, respectively. The detailed sedimentology of each member is described in detail below.

Rhadouane member

The average thickness of the Rhadouane Member in the field area is 13 m (Fig. 3) but reaches a maximum of 35 m in the northernmost part of the escarpment (Bodin *et al.*, 2010). The base of the Rhadouane Member is defined by a thin, yellow marl bed which rests unconformably above the Ain el Guettar Formation. The transition between the Rhadouane and Kerker Members is gradual and defined by the progressive loss of siliciclastic

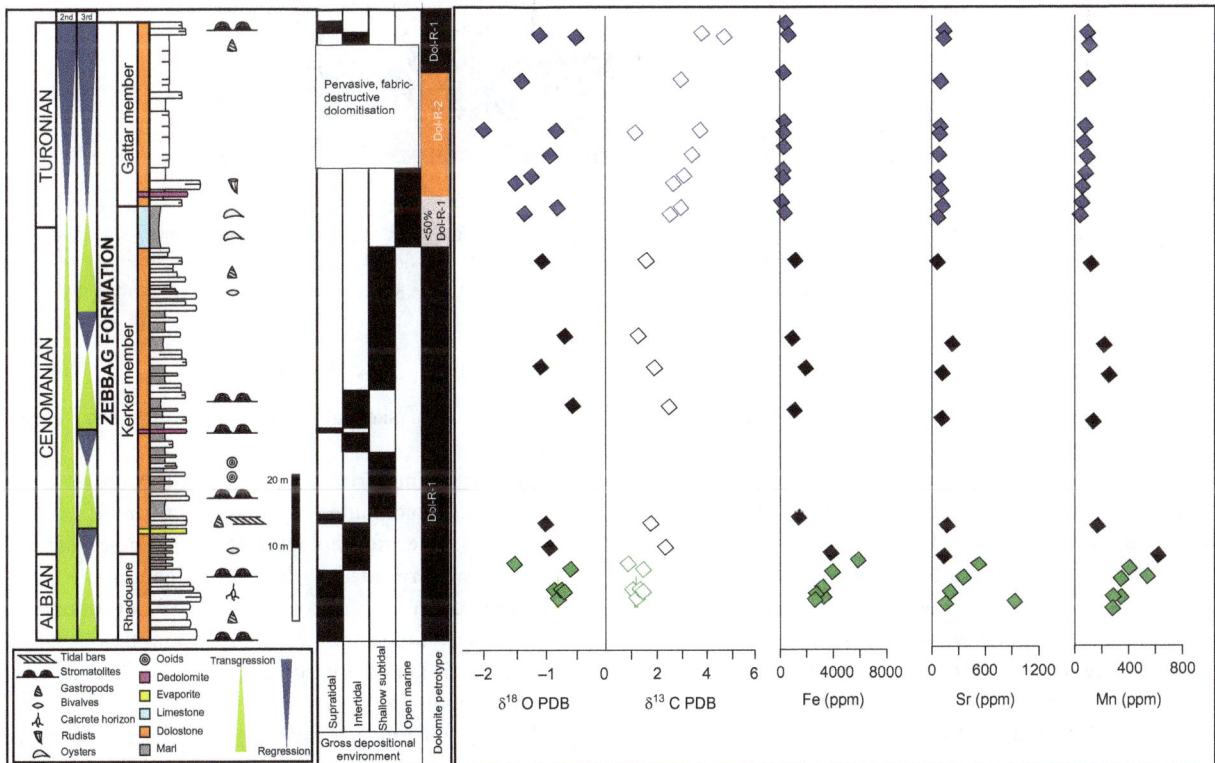

Fig. 3. Composite log through the Zebbag Formation in southern Tunisia showing interpreted transgressive and regressive cycles, gross depositional environments, dolomite petrotype distribution and all stable isotope and trace element data by depth.

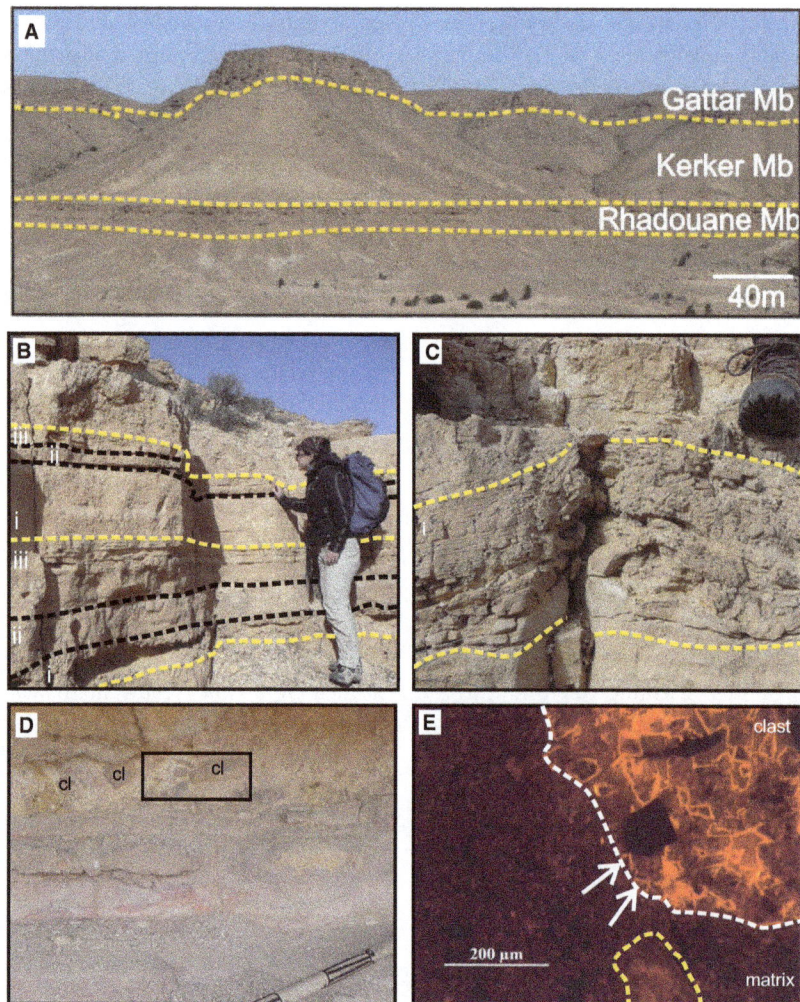

Fig. 4. (A) A field panorama showing the three studied members (Rhadouane, Kerker and Gattar) of the Zebbag Formation. (B) Stacked peritidal cycles with three separate units (described in text). Red lines separate units and yellow lines represent top/base of different cycles. (C) Stromatolites with tepee structures. (D) Laterally discontinuous breccias with dolomite clasts (cl) and calcrete horizons (10 cm increments on Jacob staff) *ca* 2 m above base of Rhadouane Member. (E) CL image of breccia clast and matrix within Rhadouane Member. Note truncated rhombs at edge of clast (arrowed).

sand and an increase in the thickness and frequency of carbonate marl units (Fig. 3). The Rhadouane Member is characterized by an upward-decrease in clay content and pervasively dolomitized, metre-scale, upward-shallowing successions, as follows (Fig. 4B):

1 Basal unit of highly bioturbated (by *Thalassonoides*) beds of skeletal (gastropod and benthic foraminiferal) peloidal wackestones to packstones with abundant rounded to sub-rounded, moderately sorted detrital quartz grains. No orig-inal shell material is present and skeletal composition is only recorded by the presence of biomouldic porosity or ghosts of grains. The presence of marine fauna and rela-tively muddy textures suggest deposition in a moderate-energy to low-energy lagoonal environment.

2 Central unit of skeletal peloidal wackestone to grain-stone.

3 Uppermost planar-crinkly algal laminites with rare, low amplitude stromatolites with *ca* 5 cm relief and a crinkly to ridge morphology (Fig. 4C). These beds are commonly overprinted by laterally discontinuous brec-cia and calcrete horizons, composed of abundant dolomitized intraclasts, within a dolomitized matrix, cutting into underlying beds (Fig. 4D). The matrix-replacive dolomite is very finely crystalline (*ca* 15 μm) with very dull orange luminescence, whereas clasts are composed of dolomite which is often coarser (*ca* 50 μm) than the matrix, with a dull orange core and brighter orange rim under CL. Truncated dolomite

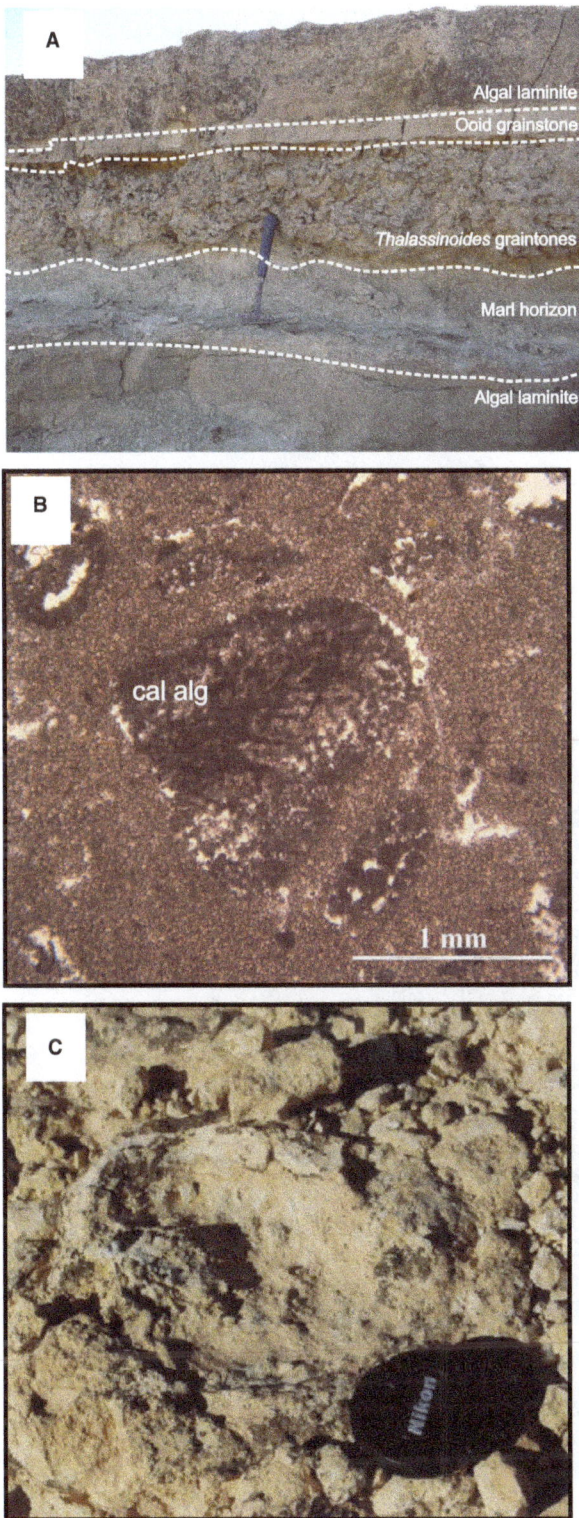

Fig. 5. (A) An example of stacked peritidal cycles in the Kerker Member, (B) calcareous algal mudstone typical of marl beds separating upward-shallowing cycles within Kerker Member, and (C) oysters in the uppermost part of the Kerker Member.

rhombs within clasts and abraded detrital dolomite rhombs are also observed (Fig. 4E).

Kerker member

The thickness of the Kerker Member in the field area is 56 m (Fig. 3) but can reach a total thickness of 140 m further to the north (Bodin *et al.*, 2010). The Kerker Member shows similar stacked, upward-shallowing facies to that observed in the Rhadouane Member, albeit separated by decimetre-scale marl beds. Bioturbated skeletal peloidal pack/wackestones are overlain by peloidal wackestones to grainstones that, in the middle part of the Kerker Member, are in part substituted by lenticular beds of cross-bedded oolitic or skeletal grainstones up to 30 cm thick (Fig. 5A). These are capped by planar-crinkly algal laminites and common domal stromatolites with *ca* 2 to 10 cm relief. Other important differences compared to the Rhadouane Member include:

1 In the lower Kerker Member, there is a prominent gypsum horizon, the only evidence for evaporite deposition within the Zebbag Formation in the study area.
2 The middle part of the Kerker Member has a thin (*ca* 50 cm) dedolomitized, brecciated algal laminite bed that is highly deformed. Dedolomitization of this bed is laterally continuous over the length of the outcrops at several locations.
3 In the upper part of the Kerker Member, there is a gradual loss in the number and thickness of algal laminites, and in the uppermost *ca* 20 m, none are observed. This is coupled with an upwards-increase in the limestone–marl ratio and thickness of gastropod and shelly wackestones to grainstones units.
4 The uppermost Kerker Member is defined by a *ca* 6 m thick marl horizon which contains open marine fauna including bivalves, echinoids and oysters, showing that this member gradually becomes more subtidal, upwards (Fig. 5C). This marl unit is only partially dolomitized.

Gattar member

The Gattar Member is 30 m thick in the field area (Figs 3 and 6A) and has a relatively constant thickness across the escarpment (Bodin *et al.*, 2010). Due to intense dolomitization in the study area, detailed facies analysis of the Gattar Member is challenging; however, some general trends and facies are noted. The lowermost part of the Gattar Member is characterized by poorly defined low-relief (*ca* 5 m) rudist colonies measuring up to 40 m in diameter (Fig. 6A). Rudists are rarely preserved (Fig. 6B) and generally only recognizable from calcite-cemented biomoulds, in some rare cases, ghost rudist grainstone fabrics are seen in proximity to the rudist

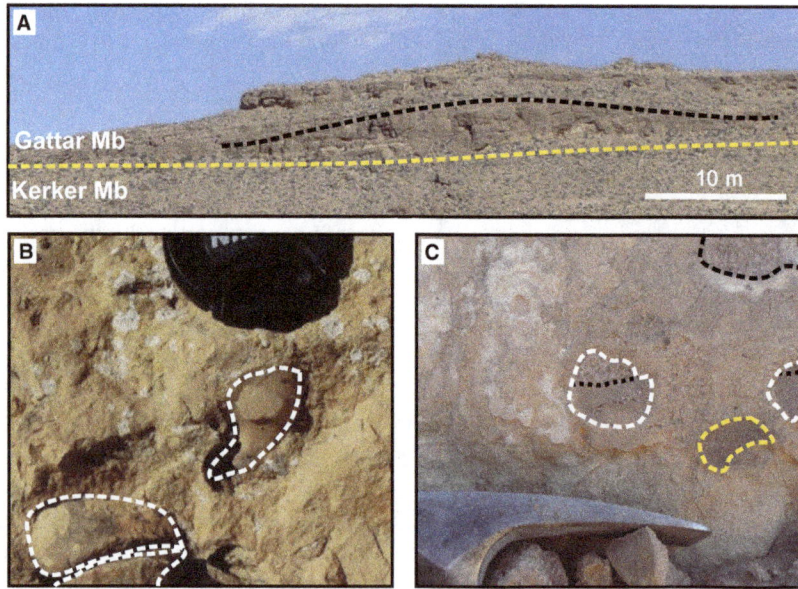

Fig. 6. Field photographs from the Gattar Member showing (A) low-relief rudist build-ups seen at the base of the Gattar Member, (B) rarely preserved rudist moulds and casts, (C) geopetal structures filled with sediment (red outline), cement and sediments (white outline with boundary shown by black line) and cement (yellow outline) seen within the upper parts of the Gattar Member.

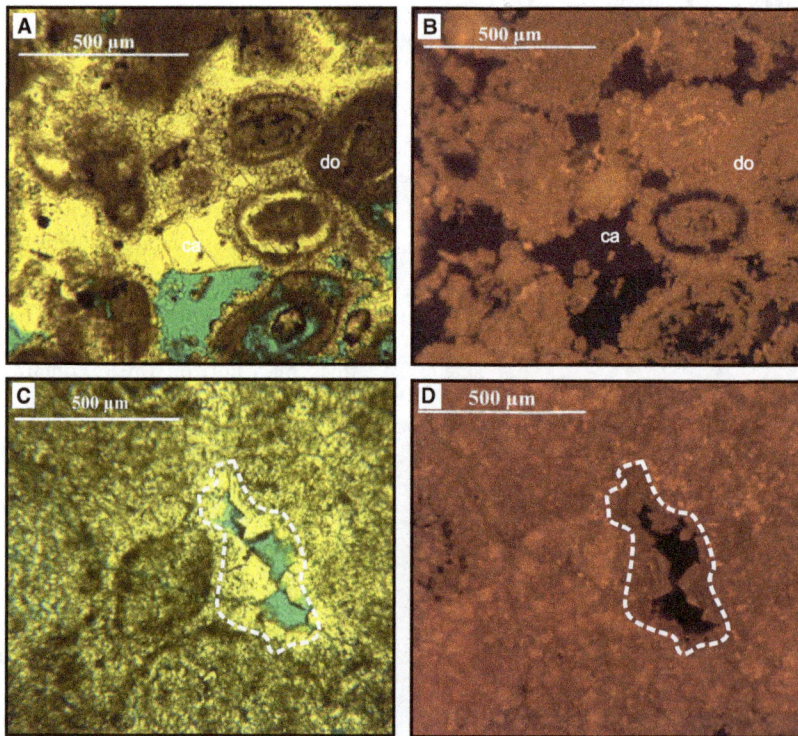

Fig. 7. Dolomite petrotypes in the Zebbag Formation. (A) PPL photomicrograph of Dol-R-1 replacing ooids and isopachous fringing cements. (B) Same image as in (A) but taken under CL. (C) PPL photomicrograph of zoned dolomite cement lining intergranular porosity. (D) Same image as (C) but taken under CL. do, dolomite; ca, calcite cement.

moulds. Rarely, towards the base of the Gattar Member, beds contain ghosts of peloids and ooids. Dolomitization is mostly fabric-destructive and so these beds cannot be traced laterally or correlated between outcrops. In the upper parts of the Gattar Member, algal laminated beds and stromatolites, with *ca* 2 m radii and 50 cm of relief become common, and geopetal structures are observed (Fig. 6C). The Gattar Member is therefore interpreted to have been deposited under dominantly subtidal conditions in its lower part with more common peritidal facies higher in the succession. This member is overlain by an evaporitic succession of the Beida Formation, further to the north of the study area; however, evaporites were not noted in these outcrops. Fractures lined by calcite and subsequently filled with sediments cut through the Gattar Member.

Petrography of diagenetic phases

There are two main types of dolomite, Dol-R-1 and Dol-R-2, and their distribution varies throughout the studied succession.

The entire Rhadouane, the lower-middle parts of the Kerker Member and the upper Gattar Member are completely replaced by finely crystalline (20 μm to 50 μm diameter), partially fabric-preserving dolomite with planar-e to planar-s textures (Fig. 7A and B). Isopachous marine cements coating ooids within oolitic grainstones are also dolomitized (Fig. 7A). This phase is classified as Dol-R-1 and is the most common dolomite type within the Zebbag Formation. Under CL, Dol-R-1 has a homogeneous dull orange–yellow luminescence (Fig. 7B). Rarely, bright to dull luminescent, finely zoned dolomite cements line intergranular porosity (Fig. 7C and D). Residual porosity is often occluded by non-luminescent, non-ferroan calcite cement. Dol-R-1 is cross-cut by both calcite-cemented fractures and very weakly formed bed-parallel dissolution seams and stylolites.

The lower parts of the Gattar Member are pervasively dolomitized by fabric-destructive, coarsely crystalline dolomite (ca 250 μm) with planar-e to planar-s textures and a cloudy core-clear rim morphology (Dol-R-2). Some crystals show sutured contacts with adjacent rhombs. Only rare

ghosts of rudist fragments, undifferentiated molluscs and ooids are preserved. On the boundary between the cloudy core and the clear rim, there is often partial dissolution and cementation by non-luminescent calcite, while there is common etching and partial dissolution within the clear rim (Fig. 8A and B). Under CL, cloudy (inclusion rich) cores are dull orange luminescent with slight mottling, while the clear rims exhibit slightly duller orange luminescence with no mottling (Fig. 8B). Rhombs of Dol-R-2 typically exhibit serrated contacts with other rhombs, but no pressure dissolution is noted in intercrystalline calcite cements (Fig. 8A). There is no spatial variability in the dolomite phases observed either along strike or along depositional dip.

Calcite cement

Throughout the Zebbag Formation, intercrystalline, mouldic, fracture, intragranular and vuggy porosity is lined, and usually filled, by calcite cement. Two different luminescence patterns were observed under CL; (ii) non-luminescent calcite and (ii) fine, sharp oscillatory zonation alternating between bright, dull and non-luminescent (Fig. 9A and B).

Dedolomite

Stratabound dedolomitization is observed at the base of the Gattar Member, and ca 15 m above the base of the Kerker Member. These horizons have ghost dolomite textures, are non-luminescent under CL, excepting minor, localized, dull orange luminescence that probably reflect the presence of minor inclusions of dolomite (Fig. 9C and D).

Geochemistry

Stable isotope analysis, major element concentrations and strontium isotope analysis were used to constrain the

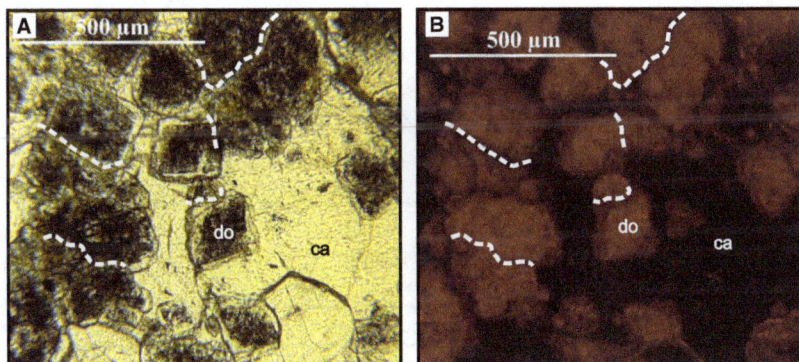

Fig. 8. (A) Image showing Dol-R-2 cloudy core, clear rim morphology. Intercrystalline porosity now filled with calcite cement. (B) Same image as in (A) but taken under CL. Note slight difference in CL intensity of clear rim and cloudy core. White lines show convex–concave contacts between dolomite rhombs. do, dolomite; ca, calcite cement.

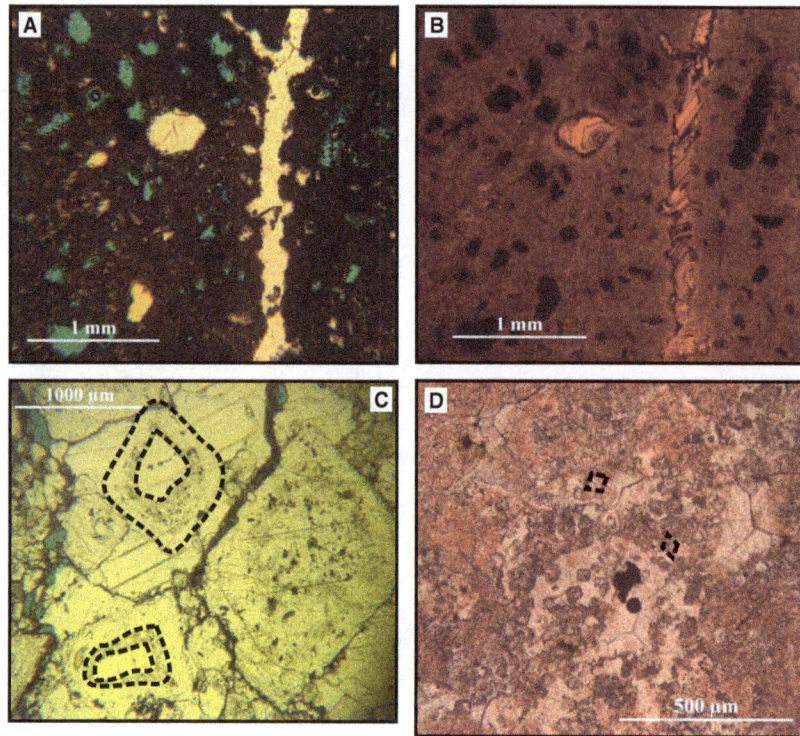

Fig. 9. Photomicrographs of (A) sharply zoned calcite cement filling fracture and mouldic pore space, (B) same photograph as in A but taken under CL, (C) coarse dedolomite crystals with minor inclusions of dolomite (black outline) and (D) dedolomite with remnant and ghost dolomite rhombs (black outline).

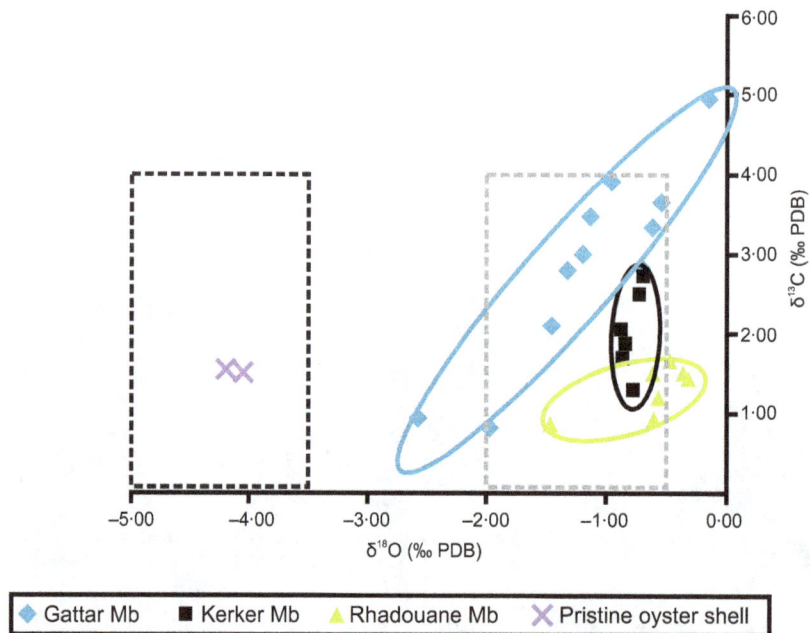

Fig. 10. Oxygen and carbon stable isotope cross plot of dolomite samples from the Zebbag Formation. Black box represents marine limestone signature and grey box represents expected marine dolomite signature based on an +3 per mil difference between co-precipitated limestone and dolomite (based on data from Budd (1997).

source of dolomitizing fluids and data are summarized in Fig. 3.

Carbon, oxygen and strontium isotopes

Isotopic analysis of pristine oyster shells from the upper parts of the Kerker Member have $\delta^{13}C = 1.54\%_0$ and $\delta^{18}O = -4.13\%_0$ (Fig. 10). There is a clear distinction in the oxygen and carbon isotopic signature between each formation as show in Fig. 3. The $\delta^{18}O$ signature of dolomite within the Rhadouane Member has a mean value of $-0.63\%_0$ (-1.47 to $-0.33\%_0$) and $\delta^{13}C$ values of $1.30\%_0$ (0.87 to $1.66\%_0$). The oxygen isotope signature of dolomite within the Kerker Member is slightly more negative compared to the Rhadouane Member, with a much narrower range (mean = $-0.80\%_0$, -0.88 to $-0.70\%_0$). Carbon isotope ratios of dolomite in the Kerker Member average $2.02\%_0$ (1.29 to $2.72\%_0$). The Gattar Member shows the most negative and the widest ranging oxygen isotope signature for the whole of the Zebbag Formation (mean = $-1.15\%_0$, -2.57 to $-0.15\%_0$) but slightly more positive $\delta^{13}C$ values (mean = $2.89\%_0$, 0.83 to $4.95\%_0$) (Fig. 10).

Given the presence of calcite cement and the difficulty in obtaining samples of pure dolomite for isotope analysis, an 'endmember' isotopic composition was estimated. This was carried out by measuring the C and O isotopic composition of mixtures of calcite cement and dolomite. The proportions of each phase in each sample were measured by point counting and mineralogy verified using XRD (Newport, 2014). Figure 11 shows the results of this analysis, and demonstrates a linear relationship between samples with *ca* 100% calcite and samples with *ca* 100% dolomite. Based on these data, an 'endmember' isotopic composition can be interpreted using simple linear regression (Fig. 11 and Table 2). The $\delta^{18}O$ value of precipitating fluids for each member was calculated (1.1 to $2.0\%_0$; Table 2) based on a sea surface temperature of 35°C and the dolomite fractionation factor of Matthews & Katz (1977). This fractionation factor was chosen in line with the recommendation of Murray & Swart (2017).

Dolomite from the Rhadouane and Kerker members has a mean $^{87}Sr/^{86}Sr$ of 0.707739 (0.707685 to 0.707785; Fig. 12), while dolomite in the Gattar Member shows a much wider range of $^{87}Sr/^{86}Sr$ values (mean = 0.707676, 0.707538 to 0.707908; Fig. 12).

Major element concentrations

Overall, there is a clear decrease in the concentration of major elements from top to base of the Zebbag Formation (Fig. 3), with the dolomitized limestone of each member of the Zebbag Formation showing distinctive elemental concentrations. Iron concentrations in the Rhadouane Member are the highest for the entire Zebbag Formation, with mean = 4205 p.p.m. (2757 to 6772 p.p.m.) compared to mean values of 1042 p.p.m. in the Kerker Member (1042 to 4261 p.p.m.) and 271 p.p.m. in the Gattar Member (76 to 696 p.p.m.) (Fig. 3). The Rhadoaune Member also has the highest Mn (mean = 363 p.p.m., 308 to 441 p.p.m.) and Sr concentration for the whole Zebbag Formation (mean = 383 p.p.m., 170 to 1011 p.p.m.) compared to the Kerker Member (mean = 274 p.p.m., 103 to

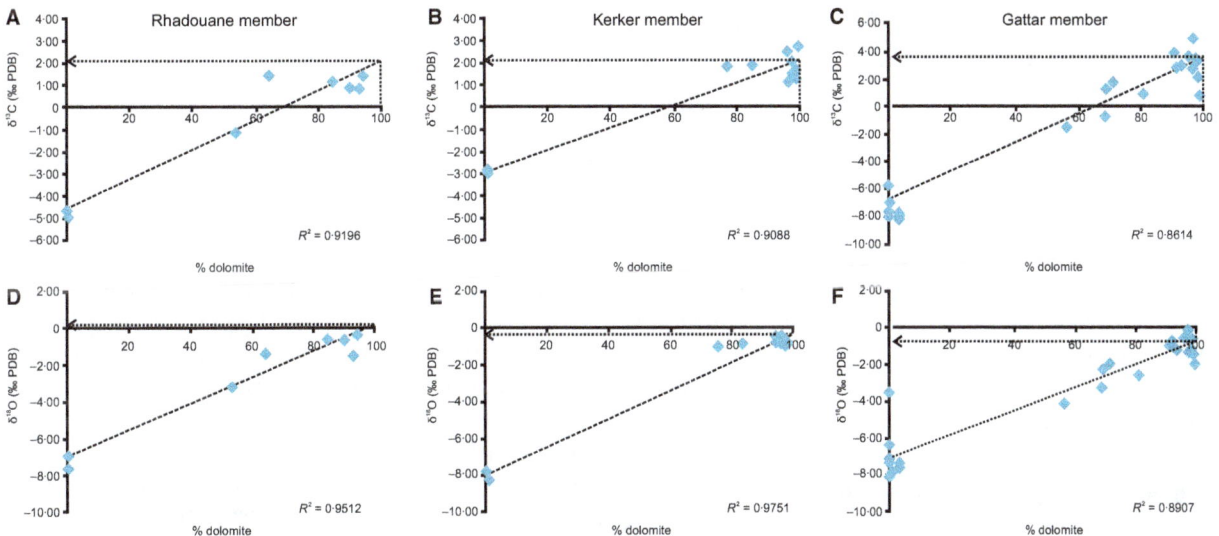

Fig. 11. Isotopic signature of known mixtures of calcite and dolomite from the Zebbag Formation to determine endmember isotopic signatures. Top row shows values of ^{13}C and bottom row shows values of ^{18}O. R^2 values are also shown. See text for details.

Table 2. Table of average major element composition and endmember stable isotope signatures for the Zebbag Formation, arranged in stratigraphic order. Isotopic signatures are calculated using a reverse linear regression. The $\delta^{18}O$ is calculated for water at 35°C using the equation by Matthews & Katz (1977)

Member	Sr (p.p.m.)	Fe (p.p.m.)	Mn (p.p.m.)	$\delta^{13}C‰$	$\delta^{18}O‰$	$\delta^{18}O‰$
Gattar	93	271	55	3·80	−0·80	1·1
Kerker	132	1876	274	2·10	−0·40	1·5
Rhadouane	383	4205	363	2·10	0·10	2·0

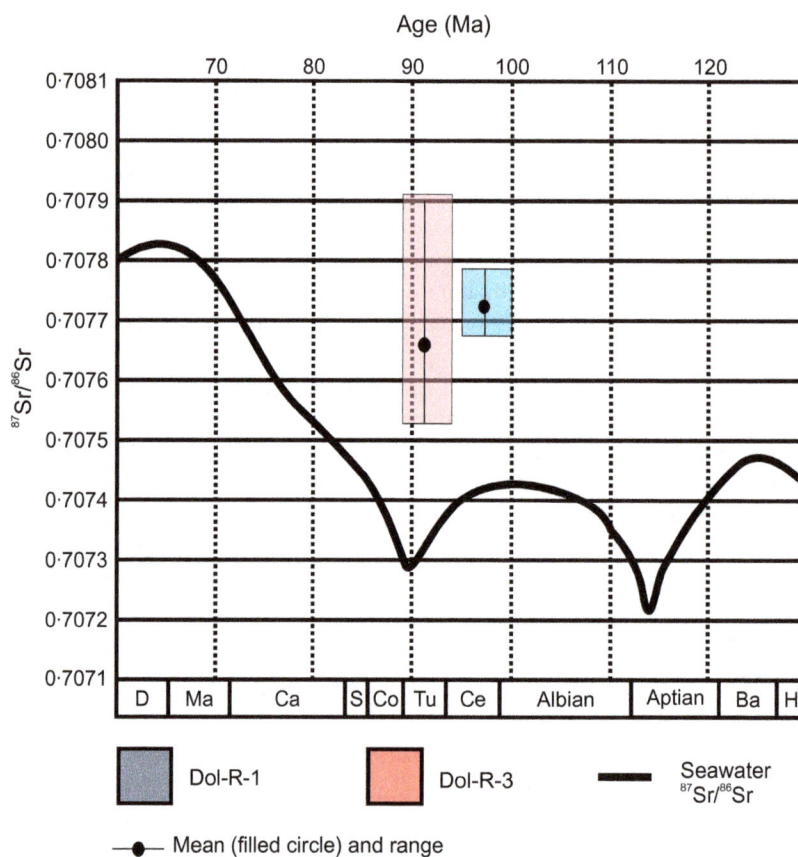

Fig. 12. A chart showing the $^{87}Sr/^{86}Sr$ signature for dolomite types Dol-R-1 and Dol-R-3 compared to sea water signature for the Upper Cretaceous. Grid and sea water curve adapted from McAthur *et al.* (2001). H, Hauterivian; Ba, Barremian; Ce, Cenomanian; Tu, Turonian; Co, Coniacian; S, Santonian; Ca, Campanian; Ma, Maastrichtian; D, Danian.

714 p.p.m.; Sr mean = 132 p.p.m., 81 to 244 p.p.m.) and Gattar Member (mean = 55 p.p.m., 30 to 93 p.p.m.; Sr mean = 93 p.p.m., 67 to 148 p.p.m.) (Fig. 3).

DISCUSSION

Sedimentology and stratigraphy

The progressive transition from clastic to carbonate deposition during deposition of the Rhadouane Member is consistent with global sea-level rise and associated global climate change, influencing drainage patterns and limiting siliciclastic sediment supply. The dominance of stacked, upward-cleaning cycles is suggestive of sedimentation within a subtidal lagoonal to peritidal setting bounded by oolitic sand bars. The presence of ooids, the single horizon of gypsum and the microbial facies that occur towards the top of cycles implies stressed, saline conditions, while the low relief of the crinkly laminites and stromatolites indicates water depths were very shallow. Nevertheless, the near-absence of evaporites suggests that salinity rarely reached gypsum saturation.

Discontinuous calcrete and breccia horizons indicate that the top of upward-shallowing cycles became emergent during deposition of the Rhadouane Formation. These are not interpreted as evaporite-removal breccias as there is no clear indication of collapse structures, which would indicate the removal of evaporites, nor any petrographical evidence of evaporite crystal moulds, evaporite pseudomorphs or ghosts. The identification of higher relief, domal stromatolites up-section suggests a gradual increase in water depth, culminating with deposition of a thick marl with an open marine skeletal assemblage in the uppermost Kerker Member. These observations are in line with regional interpretations of rising sea-level to a maximum flooding event in the lowermost Turonian (Lüning et al., 2004; Fig. 3).

The base of the Gattar Member in the study area is dominated by rudist bioherms and ooid shoals, marking a return to relatively shallower water and higher energy conditions. It is possible that these facies define a low-relief shelf break, but pervasive dolomitization has hindered identification of detailed stratal relationships. Within the uppermost part of the member, an increase in abundance and frequency of algal laminite beds and stromatolites suggests upward-shallowing and progradation of peritidal facies. A continued basinward shift in facies led to the development of supratidal conditions and deposition of the overlying Beida Anhydrite as part of the Abiod Formation, which are not seen in the field area, and eventually exposure of the platform (Camoin, 1991; Touir et al., 2009).

Timing of dolomitization

The replacement of marine isopachous fringing cements by dolomite indicates that dolomitization occurred after marine cementation. Dolomite Dol-R-1 is cross-cut by stylolites, however, and there is a lack of convex–concave grain contacts in ooid grainstones, indicating that dolomitization pre-dated compaction. Calcrete horizons within the Rhadouane Member are dolomitized and reworked as clasts (Fig. 4C), and truncated dolomite rhombs occur within brecciated clasts that form above calcretes (Fig. 4D), while single dolomite rhombs occur within the matrix. This suggests that dolomitization occurred immediately prior to, and immediately following platform emergence. The subtle difference in texture and luminescence of dolomite between clasts and matrix in breccias at exposure surfaces (Fig. 4D) also implies that the Rhadouane Member was affected by multiple pulses of dolomitizing fluids.

Dolomite phase Dol-R-2 shows sutured crystal contacts between adjacent rhombs but lacks clear, well-formed stylolites. As well-developed stylolites need at least ca 500 m of overburden to form in dolomitized sediments

(Mountjoy et al., 1999), replacement by Dol-R-2 must have occurred prior to significant burial. We interpret Dol-R-2 to have formed during shallow burial, prior to any significant chemical compaction.

Origin of dolomite

Stratigraphic observations and dolomite texture

Dolomitization in the Zebbag Formation occurs as stratabound dolomite bodies, replacing all limestone within a single escarpment that extends over a distance of 200 km (Bodin et al., 2010). This stratabound geometry and spatial extent of dolomitization rules out the possibility of dolomitization via mixed meteoric-marine fluids as proposed by previous authors (Koehler, 1982; Abdallah, 2003), a conceptual model that is anyway widely questioned (Warren, 2000). As well as its wide extent, the lack of spatial variability in dolomite texture and geochemistry within each member of the Zebbag Formation is notable. Overall, the planar texture of dolomite phases Dol-R-1 and Dol-R-2 imply dolomitization at temperatures below the crystal roughening temperature of 50°C (Sibley & Gregg, 1987). However, texture does vary stratigraphically. Within the Rhadouane, Kerker and upper parts of the Gattar members, which are dominated by peritidal facies, dolomite is finely crystalline and partially fabric-preserving (Dol-R-1). Here, muddy sediment in packstones and wackestones provided a high reactive surface area for dolomitization and nucleation rates outpaced growth rates (Sibley & Gregg, 1987). In the lower parts of the Gattar Member, replacive dolomite is coarsely crystalline and exhibits a cloudy core, clear rim morphology (Dol-R-2). Coarse sediments in rudist grainstones have a lower reactive surface area leading to growth of coarser dolomite rhombs. This strongly suggests that dolomite texture is, at least in part, facies controlled.

Fluid chemistry

The carbon and oxygen isotopic composition of carbonate precipitated from Cretaceous sea water has been estimated to vary between $\delta^{13}C = 0.00‰$ and $4.00‰$ and $\delta^{18}O = -5.0$ to $-3.5‰$ (Scholle & Arthur, 1980; Bogoch et al., 1994). The composition of pristine oyster shells ($\delta^{13}C = 1.54‰$, $\delta^{18}O = -4.1‰$) from the Kerker member is consistent with this. All three Members of the Zebbag formation also show normal marine $\delta^{13}C$ signatures ($\delta^{13}C = 1.29$ to $4.95‰$), probably due to buffering of the carbon isotopic composition by Upper Cretaceous limestone during recrystallization (Land, 1980; Rott & Qing, 2013). Marine dolomites are expected to have more positive $\delta^{18}O$ values ($+3.00‰$) (Land, 1980; Budd, 1997) compared to stratigraphically age-equivalent limestone, such that $\delta^{18}O$ values between $-2\cdot00‰$ and $0\cdot50‰$ would be anticipated for the Zebbag formation if dolomitization was from penecontemporaneous sea water. The endmember measured $\delta^{18}O$ value, based on known mixtures of calcite and dolomite for the Rhadouane member is slightly more positive ($\delta^{18}O = -1\cdot47$ to $-0\cdot33‰$). This apparent increase is less marked for the Kerker ($\delta^{18}O = -0\cdot88$ to $-0\cdot70‰$) and Gattar members ($\delta^{18}O = -2\cdot57$ to $-0\cdot15‰$).

Based on measured isotopic values, and a sea surface temperature of 35°C, the composition of the dolomitizing fluid has been calculated using the equation of Matthews & Katz (1977) as $\delta^{18}O_{water} = 2‰$, $1\cdot5‰$ and $1‰$, respectively, for the Rhadouane, Kerker and Gattar members (Table 2). These values are all more positive compared to Cenomanian sea water ($\delta^{18}O_{water} = -1‰$) (Wilson et al., 2002; Voigt et al., 2004), but more negative values would be calculated at lower temperatures. It is highly probable that the oxygen isotopic composition of sea water became more positive as a result of evaporation, although rarely did fluids reach gypsum saturation; only one gypsum bed was identified within the study area. This suggests that although evaporation was only temporarily high enough to reach gypsum saturation, brine pools were extensively developed across the peritidal zone. This is supported by the presence of peritidal facies, with a low diversity faunal assemblage, within the Rhadouane and lower Kerker members and the uppermost Gattar Member. The slight increase in the $\delta^{18}O$ values of brines in the Rhadouane Member compared to the Kerker and Gattar members is consistent with deposition of these facies within very shallow water, implying a closer spatial proximity to the brine pool. The presence of the Beida Anhydrites (Touir et al., 2009) above the Gattar Member suggests that hypersaline fluids would have been available to the platform later in the Turonian.

Further evidence for elevated fluid salinity can be interpreted from concentrations of Sr (e.g. M'rabet, 1981; Touir et al., 2009). The K_D of Sr with respect to water and gypsum is <1 (Veizer, 1983), so strontium will be preferentially concentrated into the fluid phase, and removal of Ca by gypsum will increase Sr/Ca ratios (Swart et al., 2005). Stoichiometric dolomite precipitated from fluids with typical sea water ratios of Sr/Ca should have Sr concentrations of ca 100 p.p.m. (Vahrenkamp & Swart, 1990). The upper Rhadouane Member shows the highest concentration of Sr, below the evaporite marker horizon (mean = 384 p.p.m.), suggesting that dolomitizing fluids had elevated Sr/Ca ratios and absolute concentrations of this element. The Kerker and Gattar members have average Sr concentrations of 133 p.p.m. and 94 p.p.m., respectively, more typical of normal sea water. In conclusion, oxygen isotope ratios and facies analysis

suggests that dolomite precipitated from sea water that was slightly above normal salinity, but a lack of evidence for thick evaporate beds and an absence of Sr-enrichment suggests fluids rarely reached gypsum saturation.

The concentration of Fe and Mn within dolomite can be used to interpret the redox state of precipitating fluids; the K_D of Fe and Mn is >1 and so both of these elements are preferentially incorporated into the dolomite crystal under reducing conditions (Machel, 1988). Holocene dolomites associated with marine fluids and evaporites have Fe values ranging between 10 p.p.m. and 2000 p.p.m. and Mn values ranging between 5 p.p.m. and 275 p.p.m. (Gregg et al., 1992; Montanez & Read, 1992b). The measured values of Fe (76 to 2338 p.p.m.) and Mn (36 to 276 p.p.m.) for the Gattar Member and for the middle and upper Kerker Member are consistent with these values. In all cases, this implies precipitation under at least moderately reducing conditions, consistent with the dull luminescence under CL.

The lowermost Kerker and Rhadouane members have much higher concentrations of Fe (2757 to 6772 p.p.m.) and Mn (308 to 714 p.p.m.) than the overlying strata. This could in part be due to leaching of Fe from Fe-oxides that are commonly associated with exposure surfaces (Montanez & Read, 1992b; Christ et al., 2012; Vandeginste & John, 2012), which are common in the lower Rhadouane Member. It cannot, however, explain the increased Fe concentration in the upper Rhadouane and lower Kerker members, where there is little evidence for emergence. While the presence of peritidal facies suggests there could be short-lived exposure events, we propose an alternative interpretation.

The nearest reservoir of Fe and Mn would be the underlying Ain el Guettar Formation, which comprises marginal marine siliciclastic sediments. If dolomitizing fluids had used this sandstone as an aquifer, then water–rock interaction could have enriched fluids in Fe and Mn. Further support for this interpretation comes from measured $^{87}Sr/^{86}Sr$. Carbonates precipitated from Cenomanian sea water should have $^{87}Sr/^{86}Sr$ values of $0\cdot707400$ to $0\cdot707420$, while measured values for the Rhadouane and Kerker members are enriched (mean $0\cdot707739$; Fig. 12). Such enrichment could have also occurred by fluid–rock interaction during brine migration (Fig. 13A). As such, it is possible that the enriched Sr concentrations and oxygen isotopic ratios ascribed previously to increased brine salinity could also reflect fluid interaction with siliciclastic sediments. Such a model has significant implications for the distribution of dolomitization, as migration of refluxed brines offshore within the basal sandstone aquifer provides a mechanism for fluid transport over tens of kilometres to distal locations on the carbonate platform. Given that the peritidal sediments that dominate the Rhadouane and Kerker members would have a low initial permeability, such a model

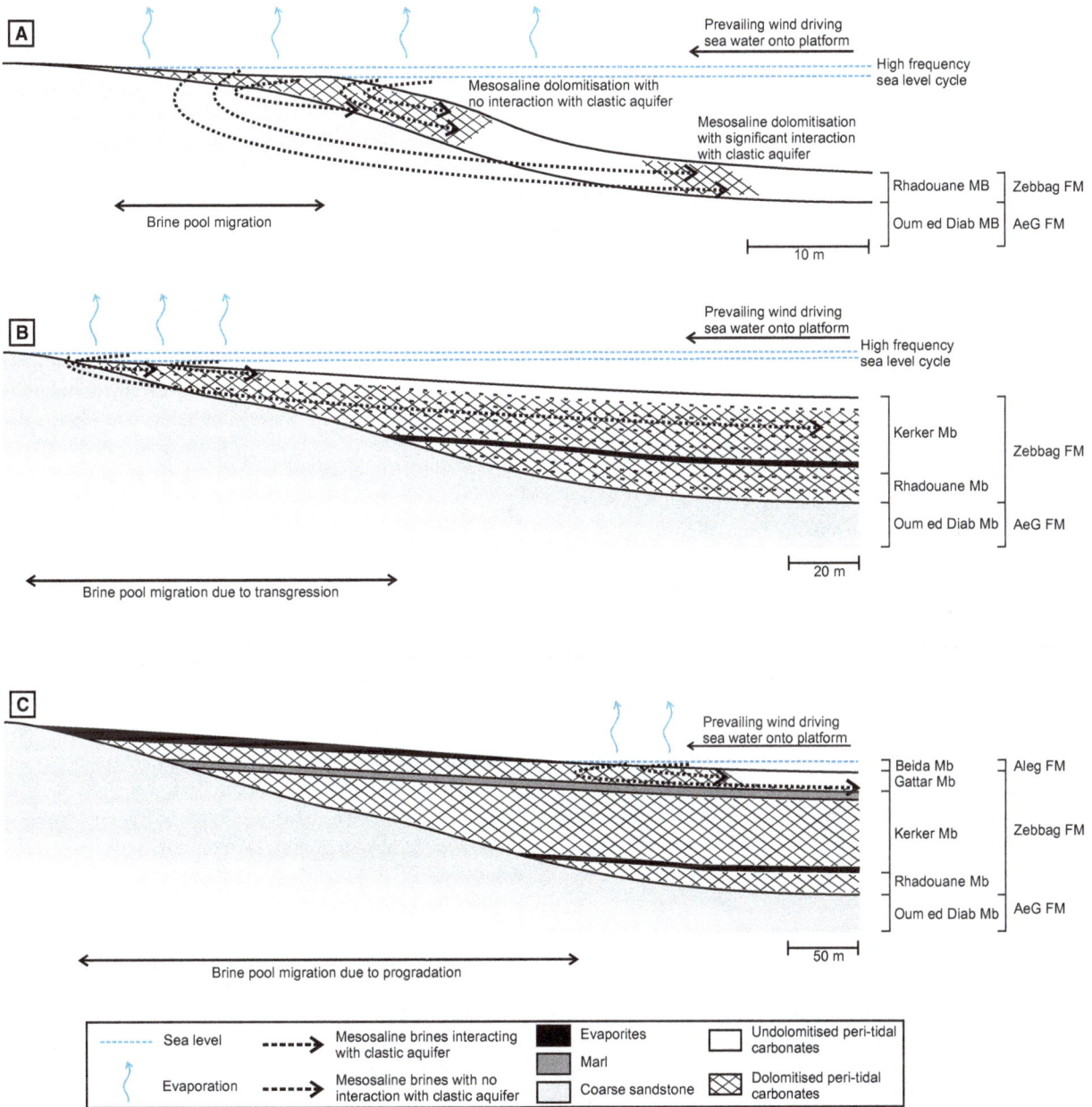

Fig. 13. A conceptual model of dolomitization for the Zebbag Formation in southern Tunisia. (A) Rhadouane Member, sea water ponded within brine pools on the platform top and fluxed seawards by density-driven brine reflux. Lateral, basinwards flux was facilitated by fluid flow along the underlying Oum ed Diab Member. (B) Kerker Member: brine reflux continued, but became less effective as sea-level rose towards the Cenomanian Turonian boundary. Eventually sea-level rise isolated the study area from proximal brine pools. (C) Gattar Member, a fall in relative sea-level in the Turonian re-established shallow-water carbonate productivity. Under increasingly arid conditions, brine pools formed in the south, supplying dolomitizing fluids that were channelled laterally above the marl horizon in the upper Kerker Member. Under falling sea-level, northwards progradation of the platform ensured laterally extensive dolomitization of the Jeffara Arch. Note the change in scale between figures.

provides an efficient mechanism for dolomitizing the platform to a wide extent.

Turonian sea water should precipitate carbonate with $^{87}Sr/^{86}Sr$ between 0·707290 and 0·707350 (Fig. 12; McAthur et al., 2001), and the Gattar Member shows an enrichment above this range (mean 0·707676). This is less easy to explain hydrogeologically by interaction with the Ain el Guettar Formation. An alternative explanation is that recrystallization of the Gattar Member took place during interaction with younger fluids, responsible for precipitation of calcite cements in fractures. These fluids could have interacted with the overlying Miocene Beglia

Formation, which is now observed as a fracture-filling phase in the Gattar Member. Further support for this assertion comes from the pervasive dedolomitization and calcite cementation observed within the basal Gattar Member. The presence of Miocene-aged sandstones filling fractures that are lined by meteoric calcite and associated with dedolomitization supports a model for dedolomitization in the Miocene (Al-Aasm, 2005).

MODEL OF DOLOMITIZATION

Overall, dolomite texture, distribution and geochemistry is consistent with dolomitization from low temperature ($<50°C$), slightly reduced, mesosaline fluids. Although conceptual models of reflux dolomitization commonly invoke penesaline fluids, close to gypsum saturation, there is good theoretical evidence that pervasive dolomitization can occur from mesosaline brines, given sufficient time for reflux and large enough fluid volumes (Jones & Xiao, 2005).

Using the method presented by Frazer et al. (2014), the total number of moles of Mg required for dolomitization can be calculated. The total volume of dolomite is estimated to be ca 1,800,000 km^3 based on dimensions of 200 km × 125 km × 0·1 km (Bodin et al., 2010) and an estimated initial porosity of 40%. This gives a total mass of $5·14 \times 10^{18}$ kg of dolomite based on a mineral density of 2·86 g cm^{-3}. Using a molar mass of dolomite of 0·184 kg mol^{-1}, a total of $2·80 \times 10^{19}$ mols of Mg would be needed to completely dolomitize the Zebbag Formation.

Fluid inclusion analysis of marine halite suggests a Cretaceous sea water concentration of Mg of 0·034 mol L^{-1} (Timofeeff et al., 2006), so a total volume of $8·24 \times 10^{20}$ l of Cretaceous sea water would have been required to completely dolomitize the Zebbag Formation assuming 100% efficiency. More realistically, the system would not have been 100% efficient, so this provides a minimum required volume. Biostratigraphic data and the age of sequence boundaries observed at the base of the Rhadouane Member and top of the Gattar Member, indicates that dolomitization must have occurred during a 9 Myr period. This requires at least $91·5 \times 10^{12}$ l of sea water to flow through the Zebbag Formation per year. Montanez & Read (1992a) estimated that a flow rate of ca 5×10^{12} l could be fluxed through a single peritidal cycle each year in the Devonian Knox Group using measured flow rates through modern day sabkhas (McKenzie et al., 1980). Both the Zebbag Formation and the Knox Group have similar areas (both estimated to be 25,000 km^2), and consist of cycles of upward-shallowing peritidal sediments of comparable thickness. This estimates brine flux to be ca 3·7 m year^{-1}, a reasonable estimate given the modelling work of Garcia-Fresca et al. (2012) and Garcia-Fresca & Jones (2011).

Climate within the study area during the Cenomanian–Turonian at has been modelled as hot and arid (Fig. 2). Sea surface temperatures are predicted to have been between 30°C and 36°C (Schouten et al., 2003; Steuber et al., 2005); an optimum temperature for dolomitization to take place, without precipitation of anhydrite cements. Above this temperature, the amount of anhydrite that is precipitated, even from mesohaline brines increases, potentially creating significant barriers to dolomitizing fluid flux (Jones & Xiao, 2005; Al-Helal et al., 2012). Further support for appropriate climate conditions comes from global circulation modelling, which indicates that wind direction was towards the south-west, forcing waves and currents onto the Saharan shield, thereby maintaining a constant supply of sea water (Poulsen et al., 1998; Ufnar et al., 2008).

Rhadouane and Kerker members

The Rhadouane and Kerker members comprise stacked cycles capped by peritidal facies, and the upward-increase in ooids and microbialites until the middle of the Kerker Member suggests that salinity was increasing until sea-level began to rise during deposition of the uppermost Kerker Member. Texturally distinct dolomitized clasts and dolomitized matrix within breccias formed at emergent surfaces suggests dolomitization took place prior to and immediately following exposure. Each cycle is therefore interpreted to have been dolomitized as a result of individual pulses of dolomitizing fluids supplied during deposition of the shallowest part of small-scale upward-shallowing cycles (Fig. 13A). Refluxing brines can cause dolomitization of the uppermost part of upward-shallowing cycles (Montanez & Read, 1992a) as a result of sinking of brines through peritidal cycles, although recent work has also shown that the more porous and permeable beds within the lowermost part of cycles can become preferentially dolomitized while the upper parts of cycles act as permeability barriers and remain undolomitized (Iannace et al., 2013). In the case of the Rhadouane and Kerker members, repeated fluxing of brines is interpreted to have led to complete dolomitization of the sediment stack as shown by numerical models (Garcia-Fresca & Jones, 2011; Garcia-Fresca et al., 2012). The very shallow, broad and very low angle dip of the carbonate platform, with laterally extensive facies belts, probably meant that brine pools formed and underwent evaporation over a vast peritidal area. Combined with migration of the brine pool during changes in relative sea-level, this may in part explain the lack of spatial variation in dolomite fabric and geochemistry.

In the Rhadouane Member, there are few permeability barriers separating individual cycles. Dolomitization is therefore interpreted to have occurred as dense brines sank through the platform causing 'top-down' dolomitization. Each pulse of dolomitizing fluids would probably penetrate to depths greater than the previous sedimentary cycle, particularly as most cycles are less than *ca* 1 m thick, anhydrite cements are absent and dolomitization would modify porosity in such a way that permeability would be increased, up to a porosity optimum (Saller & Henderson, 1998). In the Kerker Member, the high vertical layering of the succession, imparted by marls at the base of upward-shallowing cycles, could have imparted some lateral flow on dolomitizing fluids. Multiple pulses of dolomitizing fluid would need to occur for pervasive dolomitization to occur; hydrogeologic and reactive transport models have shown that repeated pulses of dolomitizing fluids can cause pervasive dolomitization of a sediment stack given sufficient time, porosity and permeability (Garcia-Fresca & Jones, 2011; Garcia-Fresca et al., 2012). The fine-grained texture of many facies in the Rhadouane and Kerker members would have permitted rapid nucleation and dolomitization (Gabellone & Whitaker, 2016), resulting in very finely crystalline dolomite textures. One dimensional modelling of stacked peritidal cycles indicate that pervasive dolomitization can occur within 1·5 kyr as a result of multiple reflux fronts, although the modelled brines were evaporated to 4× the concentration of sea water (Garcia-Fresca & Jones, 2011). The estimated duration of between 48 and 207 kyr for metre-scale peritidal cycles (Montanez & Read, 1992a; García et al., 1996) suggests that there was sufficient time for brine reflux to cause pervasive dolomitization of the Zebbag Formation. Furthermore, the repeated stacking of similar facies throughout the Rhadouane and Kerker members suggests low amplitude relative sea-level changes, consistent with the greenhouse climate and tectonic stability of the Arabian-Saharan Shield. Many authors have proposed that dolomitization is associated with relative sea-level fall (Mutti & Simo, 1994; Touir et al., 2009). This study clearly shows that within a greenhouse climate, when amplitude of sea-level change is relatively small, dolomitization can still occur during sea-level rise, terminating only during periods of maximum flood.

It has been argued that elevated concentrations of Fe, Mn and enriched [87]Sr/[86]Sr signatures within the Rhadouane Member derive from the interaction of dolomitizing fluids with underlying clastic strata. In this model, downward refluxing dolomitizing fluids would sink through carbonate strata and into the underlying Ain el Guettar Formation. Prominent clay horizons capping the Chennini Member of the Ain el Guettar Formation could have facilitated the lateral flux of brines, and provided a source of Fe and Mn. The decrease in Fe and Mn concentrations in the Kerker Member could reflect a decrease in the concentration of Fe and Mn in the fluid as the flow path extended laterally, as a result of relative sea-level rise, and with distance above the base Zebbag Formation. However, such a model would need to be tested by hydrogeological modelling.

The relatively close proximity of cycles to the brine source and the fine-grained, high reactive surface area of finer sediments that characterize the Rhadouane and Kerker members means sediments could dolomitize quickly (Jones & Xiao, 2005; Al-Helal et al., 2012; Gabellone & Whitaker, 2016). Two-dimensional modelling of penesaline reflux and subsurface studies has shown that the amount of dolomitization decreases with distance away from brine source (Saller & Henderson, 1998; Jones & Xiao, 2005; Al-Helal et al., 2012), and so duration of flux must be sufficient to cause dolomitization over a geographically wide area. Reactive transport modelling of mesohaline brines has shown that *ca* 50% of calcium carbonate can be replaced over distances of up to 6 km and to a depth of *ca* 500 m over 1 Myr (Jones & Xiao, 2005). Given the estimated duration of *ca* 2·4 Myr during deposition of the Rhadouane and Kerker members, dolomitization would be expected to extend further than 6 km and deeper than 500 m. Within each upward-shallowing succession, the reflux zone would have prograded northwards, such that the total lateral extent of dolomitization would be significantly more than 6 km. Brine reflux continued until the platform top became completely flooded in the Uppermost Cenomanian. At this point, evaporation was insufficient to form refluxing fluids and so dolomitization via active reflux was shut off (Fig. 13B). This agrees with modelling work conducted on the San Andres Formation (Garcia-Fresca et al., 2012).

Gattar member

In the Lower Turonian, rudist bioherms of the Gattar Member were deposited above the Kerker Member, with facies passing southwards into tidal flat deposits (Camoin, 1991; Touir & Soussi, 2003). This platform geometry would have restarted the flux of dolomitizing fluids, from south to north along depositional dip. The lack of evaporites within the basal Gattar Member, normal marine dolomite major element signatures and only slightly more positive oxygen isotope values, suggests that dolomitization occurred from mesosaline fluids. The uppermost marl bed in the Kerker Member probably acted as a permeability barrier to dolomitizing fluids, focusing fluids laterally and downdip. Furthermore, the grainer sediments at the base of the Gattar Member would have excellent permeability, facilitating the flux of dolomitizing fluids offshore (Fig. 13C). Layered sequences of grainy and muddy layers have been shown to provide an effective mechanism for driving dolomitizing fluids to a greater distance from source (Al-Helal *et al.*, 2012) despite them becoming less saturated with respect to Mg as a result of dolomite precipitation along the flow path. This, coupled with the coarse-grained sediment texture, resulted in fewer nucleation sites and growth rates of crystals effectively outpaced nucleation rates forming coarsely crystalline Dol-R-2.

Facies progradation within the Gattar Member suggests that the source of dolomitizing fluid moved basinward through time, leading to pervasive dolomitization of the entire Gattar Member (Fig. 13C). A similar situation has been modelled in the San Andres Formation of Texas (Garcia-Fresca *et al.*, 2012). The presence of the Beida Anhydrite facies above the Gattar Member, to the north of the study area, indicates that sea water salinity gradually increased, potentially leading to more effective dolomitization of the platform. Given the poor resolution of biostratigraphy in the Gattar Member, it is difficult to estimate how long reflux was maintained. However, using the maximum flooding event at 92·2 Ma (Lüning *et al.*, 2004) and sequence boundary at 91·7 Ma (Camoin, 1991), a period of 500 kyr is estimated. This prolonged period of fluid flow, coupled with facies progradation, would have been sufficient for massive dolomitization over the distance observed, particularly given the hypothesized lateral component to the flow. This is supported by the sharp contact between pervasively dolomitized strata in the Gattar Member and partial dolomitization in the top Kerker Member.

CONCLUSIONS AND IMPLICATIONS

Integration of field, petrographical and geochemical data of the Upper Cretaceous Zebbag Formation in southern Tunisia have shown that dolomitization occurred as a result of near surface mesosaline reflux of dolomitizing fluids over a period of less than 9 Myr, during the Cenomanian and Turonian. Despite an apparent global scarcity of platform-scale dolomitization during the Upper Cretaceous, the Jeffara Escarpment of southern Tunisia shows extensive and pervasive dolomitization. This is interpreted to primarily reflect the regional climate. The studied interval of the Zebbag Formation was located within a hot and arid climate belt that resulted in sufficient evaporation of sea water to create mesosaline brines. This occurred along a discrete climate zone north of the hot, humid equatorial conditions that presided across the Arabian Plate, and explains the abundance of dolomitized platforms restricted to the circum-Mediterranean region at this time.

The Saharan Platform was tectonically stable during the Cenomanian–Turonian such that reflux was maintained

for prolonged periods of time allowing strata to become pervasively dolomitized. This probably occurred by numerous passes of penecontemporaneous sea water that was slightly evaporated, but rarely to the point of gypsum saturation. Laterally extensive dolomitization was probably further facilitated during the Upper Albian and Cenomanian (Rhadouane and Kerker members) by fluid flux along an underlying Albian sandstone aquifer. Dolomitization was much less pervasive at the Cenomanian–Turonian boundary, when sea-level rise associated with a second-order maximum flooding event resulted in deposition of relatively deep water marls that were probably located far from the source of dolomitizing fluids. Subsequently, during the Turonian, growth and progradation of the carbonate platform and basinward migration of brine pools ensured pervasive dolomitization over distances of several hundred kilometres was re-established. Deposition of the Beida Anhydrites in the uppermost Turonian, which are not seen in the study area, indicates the penesaline brines may have led to more efficient dolomitization of the uppermost Gattar Member.

ACKNOWLEDGEMENTS

This forms part of the PhD thesis of Richard Newport, work was funded by the North Africa Research Group at University of Manchester, sponsored by Hess, Anadarko, BG, Wintershall, Repsol, RWE, Dana, BP, Cairn, Chevron, Kosmos and Conoco-Phillips. Strontium isotope analysis was carried out at the University of the Basque Country. Carbon and oxygen isotopic analysis and major, minor and trace element geochemistry was conducted at the Ruhr University, Bochum and all other geochemical analysis was undertaken at the Williamson Resource Centre, University of Manchester. The authors thank Beatriz Garcia-Fresca and Dave Cantrell for their careful reviews of this paper.

References

Abdallah, H. (2003). Genesis and diagenesis of the Gattar carbonate platform, Lower Turonian, northern southern Tunisia. In: North African Cretaceous Carbonate Platform Systems (Eds E. Gili, N. El Hédi and P. Skelton), *NATO Science Series (Series IV: Earth and Environmental Sciences)*, **28**, pp. 31–51. Springer, Dordrecht.

Abdallah, H., Memmi, L., Damotte, R., Rat, P. and Magniez-Jannin, F. (1995) Le Crétacé de la chaîne nord des Chotts (Tunisie du centre-sud): biostratigraphie et comparaison avec les régions voisines. *Cretac. Res.*, **16**, 487–538.

Al-Aasm, I.S. (2005) *Dolomitization and Dedolomitization of the Lower Turonian Gattar Carbonates, Southern Tunisia: Paleogeographic, Geochemical and Sequence Stratigraphy Controls.* GAC-MAC-CSPG-CSSS Joint Meeting, Halifax.

Al-Helal, A.B., Whitaker, F.F. and Xiao, Y. (2012) Reactive transport modeling of brine reflux: dolomitization, anhydrite precipitation, and porosity evolution. *J. Sed. Res.*, **82**, 196–215.

Amiot, R., Wang, X., Lécuyer, C., Buffetaut, E., Boudad, L., Cavin, L., Ding, Z., Fluteau, F., Kellner, A.W. and Tong, H. (2010) Oxygen and carbon isotope compositions of middle Cretaceous vertebrates from North Africa and Brazil: ecological and environmental significance. *Palaeogeogr. Palaeoclimatol. Palaeoecol.*, **297**, 439–451.

Badalini, G., Redfern, J. and Carr, I. (2002) A synthesis of current understanding of the structural evolution of North Africa. *J. Pet. Geol*, **25**, 249–258.

Belopolsky, A., Tari, G., Craig, J. and Iliffe, J. (2012) New and emerging plays in the Eastern Mediterranean: an introduction. *Petrol. Geosci.*, **18**, 371–372.

Benito, M. and Mas, R. (2007) Origin of Late Cretaceous dolomites at the southern margin of the Central System, Madrid Province, Spain. *J. Iberian Geol.*, **33**, 41–54.

Benton, M.J., Bouaziz, S., Buffetaut, E., Martill, D., Ouaja, M., Soussi, M. and Trueman, C. (2000) Dinosaurs and other fossil vertebrates from fluvial deposits in the Lower Cretaceous of southern Tunisia. *Palaeogeogr. Palaeoclimatol. Palaeoecol.*, **157**, 227–246.

Bodin, S., Petitpierre, L., Wood, J., Elkanouni, I. and Redfern, J. (2010) Timing of early to mid-cretaceous tectonic phases along North Africa: new insights from the Jeffara escarpment (Libya–Tunisia). *J. Afr. Earth Sc.*, **58**, 489–506.

Bodin, S., Meissner, P., Janssen, N.M.M., Steuber, T. and Mutterlose, J. (2015) Large igneous provinces and organic carbon burial: controls on global temperature and continental weathering during the Early Cretaceous. *Global Planet. Change*, **133**, 238–253.

Bogoch, R., Buchbinder, B. and Magaritz, M. (1994) Sedimentology and geochemistry of lowstand peritidal lithofacies at the Cenomanian-Turonian boundary in the Cretaceous carbonate platform of Israel. *J. Sed. Res.*, **64**, 733–740.

Budd, D.A. (1997) Cenozoic dolomites of carbonate islands: their attributes and origin. *Earth-Sci. Rev.*, **42**, 1–47.

Burns, S.J., Mckenzie, J.A. and Vasconcelos, C. (2000) Dolomite formation and biogeochemical cycles in the Phanerozoic. *Sedimentology*, **47**, 49–61.

Camoin, G.F. (1991) Sedimentologic and paleotectonic evolution of carbonate platforms on a segmented continental margin: example of the African Tethyan margin during Turonian and Early Senonian times. *Palaeogeogr. Palaeoclimatol. Palaeoecol.*, **87**, 29–52.

Chaabani, F., Manaai, M., Souissi, F., Souissi, R. and Sassi, S. (2003). *Dolomitization and Calcitization Stages in Rocks of the Nalut Formation, Jabal Nafüsah, NW Libya. Geology of North West Libya*. Second Symposium on the Sedimentary Basins of Libya, pp. 171–182.

Christ, N., Immenhauser, A., Amour, F., Mutti, M., Preston, R., Whitaker, F.F., Peterhänsel, A., Egenhoff, S.O., Dunn, P.A. and Agar, S.M. (2012) Triassic Latemar cycle tops—subaerial exposure of platform carbonates under tropical arid climate. *Sed. Geol.*, **265**, 1–29.

Chumakov, N., Zharkov, M., Herman, A., Doludenko, M., Kalandadze, N., Lebedev, E., Ponomarenko, A. and Rautian, A. (1995) Climatic belts of the mid Cretaceous time. *Stratigr. Geol. Correl.*, **3**, 241–260.

Contessi, M. (2013) First report of mammal-like tracks from the Cretaceous of North Africa (Tunisia). *Cretac. Res.*, **42**, 48–54.

Contessi, M. and Fanti, F. (2012a) First record of bird tracks in the Late Cretaceous (Cenomanian) of Tunisia. *Palaios*, **27**, 455–464.

Contessi, M. and Fanti, F. (2012b) Vertebrate Tracksites in the Middle Jurassic-Upper Cretaceous of South Tunisia. *Ichnos*, **19**, 211–227.

De Lapparent, A.F. and Gorce, F. (1960) Les dinosauriens du" Continental intercalaire" du Sahara central. *Mere. Soc. Geol. Fr.*, **88A**, 1–57.

Dickson, J.A.D. (1966) Carbonate identification and genesis as revealed by staining. *J. Sed. Res.*, **36**, 491–505.

Dunham, R.J. (1962) Classification of carbonate rocks according to depositional texture. In: *Classification of Carbonate Rocks* (Ed. W.E. Ham). *Memoir of AAPG*, **1**, 108–121.

El-Bakai, M. (1997) Petrography and palaeoenvironment of the Sidi as Sid Formation in Northwest Libya. *Petrol. Res. J.*, **9**, 9–26.

El-Bakai, M., Idris, M. and Sghair, A. (2010). Petrography, geochemistry and stable isotopes constraints on the origin of the Cretaceous dolomite (Ain Tobi Member) in NW Libya. *Sedimentary Events and Hydrocarbon Systems – CSPG-SEPM Joint Convention: Program with Abstracts*, 89.

Fabre, J. and Mainguet, M. (1991) Continental sedimentation and palaeoclimates in Africa during the Gondwanian Era (Cambrian to Lower Cretaceous): the importance of wind action. *J. Afr. Earth Sci.*, **12**, 107–115.

Frazer, M., Whitaker, F. and Hollis, C. (2014) Fluid expulsion from overpressured basins: implications for Pb–Zn mineralisation and dolomitisation of the East Midlands platform, northern England. *Mar. Pet. Geol.*, **55**, 68–86.

Gabellone, T. and Whitaker, F. (2016) Secular variations in seawater chemistry controlling dolomitization in shallow reflux systems: insights from reactive transport modelling. *Sedimentology*, **63**, 1233–1259.

García, A., Segura, M. and García-Hidalgo, J. (1996) Sequences, cycles and hiatuses in the Upper Albian-Cenomanian of the Iberian Ranges (Spain): a cyclostratigraphic approach. *Sed. Geol.*, **103**, 175–200.

Garcia-Fresca, B. and Jones, G.D. (2011) Apparent stratigraphic concordance of reflux dolomite: new insight from high-frequency cycle scale synsedimentary reactive transport models. In: *Carbonate Geochemsitry: Reactions and Process in Aquifers and Reservoirs* (Eds A. Summers Engel, S. Engel, P. Moore and H. DuChene), *Karst Waters Inst. Spec. Publ.*, **16**, 27–30.

Garcia-Fresca, B., Lucia, F.J., Sharp, J.M. and Kerans, C. (2012) Outcrop-constrained hydrogeological simulations of brine reflux and early dolomitization of the Permian San Andreas Formation. *AAPG Bull.*, **96**, 1757–1781.

Given, R.K. and Wilkinson, B.H. (1987) Dolomite abundance and stratigraphic age; constraints on rates and mechanisms of Phanerozoic dolostone formation. *J. Sed. Res.*, **57**, 1068–1078.

Gregg, J.M., Howard, S.A. and Mazzullo, S. (1992) Early diagenetic recrystallization of Holocene (<3000 years old) peritidal dolomites, Ambergris Cay, Belize. *Sedimentology*, **39**, 143–160.

Grosheny, D., Ferry, S., Jati, M., Ouaja, M., Bensalah, M., Atrops, F., Chikhi-Aouimeur, F., Benkerouf-Kechid, F., Negra, H. and Aït Salem, H. (2013) The Cenomanian-Turonian boundary on the Saharan Platform (Tunisia and Algeria). *Cretac. Res.*, **42**, 66–84.

Haq, B.U., Hardenbol, J. and Vail, P.R. (1987) Chronology of fluctuating sea levels since the Triassic. *Science*, **235**, 1156–1167.

Husinec, A. and Sokač, B. (2006) Early Cretaceous benthic associations (foraminifera and calcareous algae) of a shallow tropical-water platform environment (Mljet Island, southern Croatia). *Cretac. Res.*, **27**, 418–441.

Iannace, A., Frijia, G., Galluccio, L. and Parente, M. (2013) Facies and early dolomitization in Upper Albian shallow-water carbonates of the southern Apennines (Italy): paleotectonic and paleoclimatic implications. *Facies*, 1–26.

Jones, G.D. and Xiao, Y. (2005) Dolomitization, anhydrite cementation, and porosity evolution in a reflux system: insights from reactive transport models. *AAPG Bull.*, **89**, 577–601.

Koehler, R.P. (1982) *Sedimentary Environment and Petrology of the Ain Tobi Formation, Tripolitania, Libya.* Unpublished thesis, Rice University, USA, 572 pp.

Kolodny, Y. and Raab, M. (1988) Oxygen isotopes in phosphatic fish remains from Israel: paleothermometry of tropical Cretaceous and Tertiary shelf waters. *Palaeogeogr. Palaeoclimatol. Palaeoecol.*, **64**, 59–67.

Korbar, T., Fuček, L., Husinec, A., Vlahović, I., Oštrić, N., Matičec, D. and Jelaska, V. (2001) Cenomanian carbonate facies and rudists along shallow intraplatform basin margin-the island of Cres (Adriatic Sea, Croatia). *Facies*, **45**, 39–58.

Kuypers, M.M., Pancost, R.D. and Damsté, J.S.S. (1999) A large and abrupt fall in atmospheric CO_2 concentration during Cretaceous times. *Nature*, **399**, 342–345.

Land, L.S. (1980) The isotopic and trace element geochemistry of dolomite: the state of the art. Concepts and models of dolomitization. *Soc. Econ. Paleontol. Mineral. Spec. Publ.*, **28**, 87–110.

Lefranc, J.P. and Guiraud, R. (1990) The Continental Intercalaire of northwestern Sahara and its equivalents in the neighbouring regions. *J. Afr. Earth Sci.*, **10**, 27–77.

Linnert, C., Robinson, S.A., Lees, J.A., Bown, P.R., Pérez-Rodríguez, I., Petrizzo, M.R., Falzoni, F., Littler, K., Arz, J.A. and Russell, E.E. (2014) Evidence for global cooling in the Late Cretaceous. *Nat. Commun.*, **5**, p. 4194.

Lüning, S., Kolonic, S., Belhadj, E., Belhadj, Z., Cota, L., Barić, G. and Wagner, T. (2004) Integrated depositional model for the Cenomanian-Turonian organic-rich strata in North Africa. *Earth Sci. Rev.*, **64**, 51–117.

Machel, H.G. (1988) Fluid flow direction during dolomite formation as deduced from trace-element trends. Sedimentology and geochemistry of dolostones. *SEPM Spec. Publ.*, **43**, 115–125.

Mackenzie, F.T. and Morse, J.W. (1992) Sedimentary carbonates through Phanerozoic time. *Geochim. Cosmochim. Acta*, **56**, 3281–3295.

MacLeod, K.G., Huber, B.T., Berrocoso, Á.J. and Wendler, I. (2013) A stable and hot Turonian without glacial $\delta^{18}O$ excursions is indicated by exquisitely preserved Tanzanian foraminifera. *Geology*, **41**, 1083–1086.

Marie, J., Trouve, P., Desforges, G. and Dufaure, P. (1984). Nouveuax elements du Paleogeographie du Cretace du Tunisie. *Notes Mem. Total*, **19**, 37.

Matthews, A. and Katz, A. (1977) Oxygen isotope fractionation during the dolomitization of calcium carbonate. *Geochim. Cosmochim. Acta*, **41**, 1431–1438.

McAthur, J., Howarth, R. and Bailey, T. (2001) Strontium isotope stratigraphy: LOWESS version 3: best fit to the marine Sr-isotope curve for 0–509 Ma and accompanying look-up table for deriving numerical age. *J. Geol.*, **109**, 155–170.

McKenzie, J.A., Hsu, K.J. and Schneider, J.F. (1980) Movement of subsurface waters under the sabkha, Abu Dhabi, UAE, and its relation to evaporative dolomite genesis. In: *Concepts and Models of Dolomitization* (Eds D. Zenger, J. Dunham and R. Ethington), *Soc. Econ. Paleon. Miner.*, **28**, 11–30.

Montanez, I.P. and Read, J.F. (1992a) Eustatic control on early dolomitization of cyclic peritidal carbonates: evidence from the Early Ordovician Upper Knox Group, Appalachians. *Geol. Soc. Am. Bull.*, **104**, 872–886.

Montanez, I.P. and Read, J.F. (1992b) Fluid-rock interaction history during stabilization of early dolomites, upper Knox Group (Lower Ordovician), US Appalachians. *J. Sed. Res.*, **62**, 753–778.

Mountjoy, E.W., Machel, H.G., Green, D., Duggan, J. and Williams-Jones, A.E. (1999) Devonian matrix dolomites and deep burial carbonate cements: a comparison between the Rimbey-Meadowbrook reef trend and the deep basin of west-central Alberta. *Bull. Can. Pet. Geol.*, **47**, 487–509.

M'rabet, A. (1981) Differentiation of environments of dolomite formation, Lower Cretaceous of Central Tunisia. *Sedimentology*, **28**, 331–352.

Murray, S.T. and Swart, P.K. (2017) Evaluating formation fluid models and calibrations using clumped isotope paleothermometry on bahamian dolomites. *Geochim. Cosmochim. Acta*, **206**, 73–93.

Mutti, M. and Simo, T. (1994) Distribution, petrography and geochemistry of early dolomite in cyclic shelf facies, Yates Formation (Guadalupian), Capitan Reef Complex, USA. *Dolomites: A Volume in Honour of Dolomieu* (Eds B. Purser, M. Tucker and D. Zenger), *In. Assoc. Sedimentol. Spec. Publ.*, **51**, 91–109.

Newport, R.J. (2014) *Controls on Dolomitisation of Upper Cretaceous Strata of North Africa and Western Mediterranean*. PhD Thesis, University of Manchester, Manchester, UK, 284 pp.

Pearson, P.N., Ditchfield, P.W., Singano, J., Harcourt-Brown, K.G., Nicholas, C.J., Olsson, R.K., Shackleton, N.J. and Hall, M.A. (2001) Warm tropical sea surface temperatures in the Late Cretaceous and Eocene epochs. *Nature*, **413**, 481–487.

Pin, C. and Bassin, C. (1992) Evaluation of a strontium-specific extraction chromatographic method for isotopic analysis in geological materials. *Anal. Chim. Acta*, **269**, 249–255.

Pin, C., Briot, D., Bassin, C. and Poitrasson, F. (1994) Concomitant separation of strontium and samarium-neodymium for isotopic analysis in silicate samples, based on specific extraction chromatography. *Anal. Chim. Acta*, **298**, 209–217.

Poulsen, C.J., Seidov, D., Barron, E.J. and Peterson, W.H. (1998) The impact of paleogeographic evolution on the surface oceanic circulation and the marine environment within the Mid-Cretaceous tethys. *Paleoceanography*, **13**, 546–559.

Rott, C.M. and Qing, H. (2013) Early dolomitization and recrystallization in Shallow Marine Carbonates, Mississippian Alida Beds, Williston Basin (Canada): evidence from petrography and isotope geochemistry. *J. Sed. Res.*, **83**, 928–941.

Saller, A.H. and Henderson, N. (1998) Distribution of porosity and permeability in platform dolomites; insight from the Permian of West Texas. *AAPG Bull.*, **82**, 1528–1550.

Sandberg, P.A. (1983) An oscillating trend in Phanerozoic non-skeletal carbonate mineralogy. *Nature*, 19–22.

Scholle, P.A. and Arthur, M.A. (1980) Carbon isotope fluctuations in Cretaceous pelagic limestones: potential stratigraphic and petroleum exploration tool. *AAPG Bull.*, **64**, 67–87.

Schouten, S., Hopmans, E.C., Forster, A., Van Breugel, Y., Kuypers, M.M. and Damsté, J.S.S. (2003) Extremely high sea-surface temperatures at low latitudes during the middle Cretaceous as revealed by archaeal membrane lipids. *Geology*, **31**, 1069–1072.

Scotese, C.R. (2001). Paleomap project. Available at: http://www.scotese.com/.

Sellwood, B.W. and Valdes, P.J. (2006) Mesozoic climates: general circulation models and the rock record. *Sed. Geol.*, **190**, 269–287.

Sibley, D.F. and Gregg, J.M. (1987) Classification of Dolomite Rock Textures. *J. Sed. Petrol.*, **57**, 967–975.

Steiger, R.H. and Jäger, E. (1977) Subcommission on geochronology: convention on the use of decay constants in geo-and cosmochronology. *Earth Planet. Sci. Lett.*, **36**, 359–362.

Steuber, T., Rauch, M., Masse, J.-P., Graaf, J. and Malkoč, M. (2005) Low-latitude seasonality of Cretaceous temperatures in warm and cold episodes. *Nature*, **437**, 1341–1344.

Sun, S.Q. (1994) A reappraisal of dolomite abundance and occurrence in the Phanerozoic. *J. Sed. Res.*, **64**, 396–404.

Swart, P.K., Cantrell, D.L., Westphal, H., Handford, C.R. and Kendall, C.G. (2005) Origin of dolomite in the Arab-D reservoir from the Ghawar Field, Saudi Arabia: evidence from petrographic and geochemical constraints. *J. Sed. Res.*, **75**, 476–491.

Timofeeff, M.N., Lowenstein, T.K., Da Silva, M.A.M. and Harris, N.B. (2006) Secular variation in the major-ion chemistry of seawater: evidence from fluid inclusions in Cretaceous halites. *Geochim. Cosmochim. Acta*, **70**, 1977–1994.

Touir, J. and Soussi, M. (2003). The growth and migration of two turonian rudist-bearing carbonate platforms in Central Tunisia. Eustatic and tectonic controls. *North African Cretaceous Carbonate Platform Systems* (Eds E. Gili, N. El Hédi and P. Skelton), *NATO Science Series (Series IV: Earth and Environmental Sciences)*, **28**, pp. 53–81. Springer, Dordrecht.

Touir, J., Soussi, M. and Troudi, H. (2009) Polyphased dolomitization of a shoal-rimmed carbonate platform: example from the middle Turonian Bireno dolomites of central Tunisia. *Cretac. Res.*, **30**, 785–804.

Ufnar, D.F., Ludvigson, G.A., González, L. and Gröcke, D.R. (2008) Precipitation rates and atmospheric heat transport during the Cenomanian greenhouse warming in North America: Estimates from a stable isotope mass-balance model. *Palaeogeogr. Palaeoclimatol. Palaeoecol.*, **266**, 28–38.

Vahrenkamp, V.C. and Swart, P.K. (1990) New distribution coefficient for the incorporation of strontium into dolomite and its implications for the formation of ancient dolomites. *Geology*, **18**, 387–391.

Vandeginste, V. and John, C.M. (2012) Influence of climate and dolomite composition on dedolomitization: insights from a multi-proxy study in the central Oman Mountains. *J. Sed. Res.*, **82**, 177–195.

Veizer, J. (1983) Chemical diagenesis of carbonates: theory and application of trace element technique. Stable isotopes in sedimentary geology. *SEPM Short Course*, **10**, 1–100.

Voigt, S., Gale, A.S. and **Flögel, S.** (2004) Midlatitude shelf seas in the Cenomanian-Turonian greenhouse world: temperature evolution and North Atlantic circulation. *Paleoceanography*, **19**, PA4020.

Warren, J. (2000) Dolomite: occurrence, evolution and economically important associations. *Earth Sci. Rev.*, **52**, 1–81.

Wilson, P.A., Norris, R.D. and **Cooper, M.J.** (2002) Testing the Cretaceous greenhouse hypothesis using glassy foraminiferal calcite from the core of the Turonian tropics on Demerara Rise. *Geology*, **30**, 607–610.

Wood, J., Bodin, S., Redfern, J. and **Thomas, M.** (2014) Controls on facies evolution in low accommodation, continental-scale fluvio-paralic systems (Messak Fm, SW Libya). *Sed. Geol.*, **303**, 49–69.

Wright, D.T. and **Oren, A.** (2005) Nonphotosynthetic bacteria and the formation of carbonates and evaporites through time. *Geomicrobiol J.*, **22**, 27–53.

Zappaterra, E. (1995) The Pelagian Block (central Mediterranean): exploration and new opportunities. AAPG Search and Discovery Article #90956, AAPG International Convention and Exposition Meeting, Nice, France, 10–13 September 1995.

3

Sedimentology and Geochemistry of the 2930 Ma Red Lake–Wallace Lake Carbonate Platform, Western Superior Province, Canada

TIMOTHY MCINTYRE[1] and PHILIP FRALICK

Department of Geology, Lakehead University, Thunder Bay, Ontario, Canada P7B 5E1 (E-mails: tmcintyre@lakeheadu.ca; tm3@ualberta.ca)

Keywords

Archean sedimentology, carbonate geochemistry, carbonate platform, Mesoarchean ocean chemistry.

[1]Present address: Department of Earth and Atmospheric Sciences, University of Alberta, Edmonton, Alberta, Canada T6G 2R3

ABSTRACT

Chemical sedimentary rocks in the Red Lake–Wallace Lake area of the Canadian Shield form the Earth's oldest known carbonate platform and, as such, provide a unique opportunity to explore the floor of a 2930 Myr old warm, shallow sea. Peritidal depositional features dominate the platform top, ranging from colloform crusts, teepee structures, and evaporate pseudomorphs in the supratidal; pseudomorph crystal fans and laterally linked domal stromatolites, associated with stromatactis-like structures, sheet cracks and liminoid fenestral fabrics extending from the lower supratidal through the intertidal; and herringbone cross-stratification, isolated domal stromatolites and herring-bone calcite cement in tidal channels. These lithofacies indicate a low energy, restricted evaporitic environment. Limited subtidal platform top deposits are characterized by laterally linked domal stromatolites and pseudomorph crystal fans on a larger scale than those found in the intertidal areas. A transitional lithofacies to deeper water were deposited further offshore. It consists of ribbon rock (mixed laminae and beds of carbonate, slate and iron oxide sediments), slump structures composed of intraclastic carbonate lithoclasts in a marl matrix and carbonate-associated iron formation. Basinal deposits consist of chert and chert-oxide facies iron formation. Peritidal lithofacies are composed of ferroan dolomite, whereas deep subtidal to upper slope lithofacies are composed of calcite. The dolomites have O and Sr isotopic ratios which were probably reset during dolomitization, whereas only O is altered in the limestones. The $\delta^{13}C$ values for all the carbonate samples were $0 \pm 1\cdot1\%_{\text{V-PDB}}$, with samples formed in deeper water having lighter $\delta^{13}C$ values, suggesting the deeper open ocean had a lighter $\delta^{13}C$ budget than water in the platform interior. Post Archean Australian Shale normalized rare earth element patterns for the carbonate samples have positive La and Eu anomalies, suprachondritic Y/Ho ratios and slight heavy rare earth element enrichment in most samples. Basin lithofacies are characterized by heavy rare earth element enrichment and positive Eu anomalies. One sample had a significant negative Ce anomaly. These data probably indicate restricted circulation in the *supra* and intertidal areas and possibly the development of spatially limited areas where oxygen production could move the redox boundary out into the water column.

INTRODUCTION

Carbonate strata deposited in the shallow Archean ocean provide an archive of information on palaeogeography, biologic productivity, the general chemistry of the water from which they formed, and diagenetic pathways (Grotzinger & James, 2000; Webb & Kamber, 2000; Kamber & Webb, 2001). Extensive study of the latest Archean Campbellrand-Malmani carbonate platform of South Africa (Sumner, 1996; Sumner & Grotzinger, 1996, 2004;

Wright & Altermann, 2000; Kamber & Webb, 2001; Rouxel *et al.*, 2005; Schneiderhan *et al.*, 2006; Scott *et al.*, 2008; Fischer *et al.*, 2009; Knoll & Beukes, 2009; Ono *et al.*, 2009; Waldbauer *et al.*, 2009; Heimann *et al.*, 2010; Voegelin *et al.*, 2010) and the Hamersley carbonate platform in Australia (Becker & Clayton, 1972; Kaufman *et al.*, 1990; Veizer *et al.*, 1990; Simonson *et al.*, 1993; Eigenbrode & Freeman, 2006; Czaja *et al.*, 2010) has improved our understanding of their architecture, sedimentation processes, biologic activity and the chemistry of the overlying water. There is still much to be learnt about the shallow sea floor during the latest Archean Campbellrand time interval, such as the factors controlling the formation of the giant domes composed of sea floor crystal fans and whether oxygen was able to sporadically accumulate in aerially limited restricted environments, but, because of the wealth of previous work, researchers have a much better vantage point from which to ask further questions.

Moving deeper in time the 2·8 Ga Steep Rock carbonate platform has not been studied as extensively, but has provided evidence that photosynthesis had begun before its formation (Riding *et al.*, 2014; Fralick & Riding, 2015). It also furnishes a glimpse at the organization of the shallow carbonate sea floor at that time (Wilks & Nisbet, 1985, 1988; Wilks, 1986; Fralick & Riding, 2015), which exhibited both similarities to, and differences from the younger South African and Australian platforms, and was a foreign world compared to today's carbonate build-ups.

Here, we reach a dead-end in the study of thick carbonate platforms. Some thin carbonate assemblages have been extensively investigated, such as the *ca* 20 m thick carbonate units in the 3·4 Ga Strelley Pool Formation (Allwood *et al.*, 2010; and references therein), but the relative lack of carbonate platforms prior to 2·8 Ga limits our ability to investigate platform architecture and its relationship to water chemistry during this interval. However, a 200 m thick 2·93 Ga carbonate platform does exist, although it has received very little attention. The Red Lake–Wallace Lake carbonate platform in Superior Province of north-western Ontario has only been the subject of an overview by Hofmann *et al.* (1985). It is the oldest known thick carbonate platform and is exposed in wave washed outcrops on the shores of Red and Wallace Lakes (Fig. 1).

REGIONAL GEOLOGY

The Red Lake and Wallace Lake Greenstone Belts are located in Superior Province on the southern margin of the Uchi Subprovince within the North Caribou Terrane (NCT) (Fig. 1A and B) and are characterized by lithostrati-

graphically and chronostratigraphically similar volcano-sedimentary sequences (Sanborn-Barrie *et al.*, 2001; Sasseville *et al.*, 2006). Correlative chemical precipitates that formed the Red Lake–Wallace Lake carbonate platform are found within the Wallace Lake Assemblage of the Wallace Lake Greenstone Belt and Ball Assemblage of the Red Lake Greenstone Belt. The Wallace Lake Assemblage is comprised of a transgressive siliciclastic succession, the Conley Formation, overlain by chemical sediments of the East Bay Formation, which in turn is conformably overlain by the dominantly volcanic Overload Bay Formation (Sasseville *et al.*, 2006; Fralick *et al.*, 2008). The base of the Conley Formation consists of a conglomeratic braided fluvial system upwardly gradational into an overlying distributary mouth bar – distal bar assemblage (Fralick *et al.*, 2008). Approximately 800 m of prodelta turbidites cap the formation. These become thinly bedded and finer grained near the top and at the transition to the East Bay Formation interlayer with thinly bedded chert and magnetite. This is overlain by iron formation with chlorite-rich, millimetre-thick layers or, in other areas, 50 m of dolomitic carbonates (Fralick *et al.*, 2008). These chemical sediments define the western portion of the Red Lake–Wallace Lake carbonate platform at Wallace Lake and are overlain by chloritic sandstones and siltstones, and volcanic rocks of the mafic-ultramafic Overload Bay Formation.

Detrital zircons found within sandstone near the base of the Conley Formation provide a maximum age constraint of 2997 to 2999 Ma (Davis, 1994) and a lower limit is provided by a 2921 Ma crosscutting felsic dyke (Davis, 1994). The Overload Bay Formation is cut by a 2921 ± 2 Ma dacitic dike and intruded by a 2921 ± 1 Ma tonalitic batholith (Tomlinson *et al.*, 2001; Sasseville *et al.*, 2006). The volcano-sedimentary sequence is interpreted as an ocean plateau (Hollings *et al.*, 1999) where plume induced shallowing created sub-aerial conditions that subsided to shallow marine during an interval of restricted volcanism and reduced heat-flow. The thermal subsidence allowed first shallow water siliciclastic deposition, and, as the source area flooded, siliciclastic starved carbonate sedimentation, and finally deeper water chert and iron formation (Fralick *et al.*, 2008). The Overload Bay Formation was deposited during renewed plume volcanism (Hollings *et al.*, 1999).

The Ball Assemblage of the Red Lake Greenstone Belt is comprised of a dominantly calc-alkaline mafic to felsic volcanic sequence with minor komatiite, siliciclastics and stromatolitic carbonates (Sanborn-Barrie *et al.*, 2001). These 200 m thick stromatolitic carbonates are the eastern extension of the Red Lake–Wallace Lake carbonate platform in the Red Lake area. Their age is constrained by overlying and underlying felsic volcanic rocks to

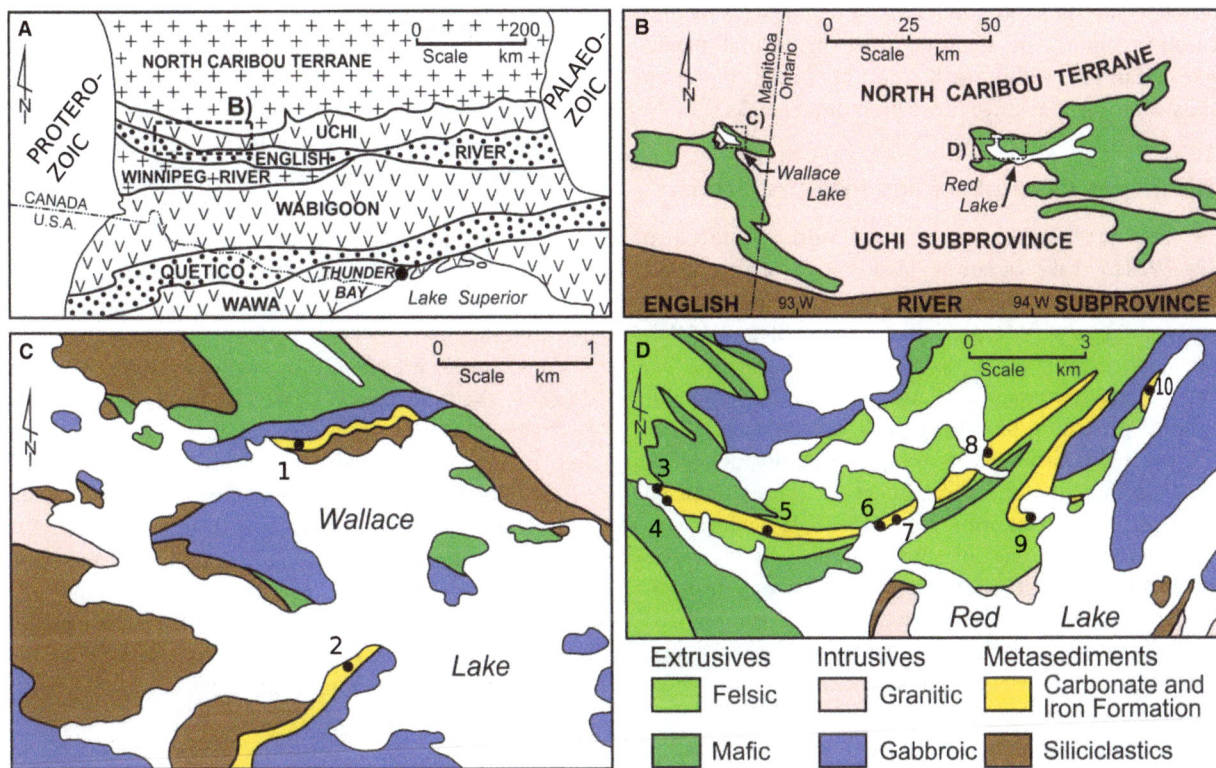

Fig. 1. (A) Regional geology of north-western Ontario. (B) Locations of the Wallace Lake Greenstone Belt and Red Lake Greenstone Belt. Insets are the field locations for both greenstone belts. (C) Wallace Lake field locations (modified from Sasseville *et al.*, 2006). (D) Red Lake field locations (modified from Hofmann *et al.*, 1985). Points 1 to 10 in C and D are the locations of the studied outcrops and are so numbered in the text.

between 2940 ± 2 Ma and 2925 ± 3 Ma (Corfu & Wallace, 1986). Like the Wallace Lake Assemblage, the Ball Assemblage was formed by initial plume-related volcanism and tectonic activity, followed by a hiatus in volcanic activity and a period of siliciclastic starved sedimentation during which time chemical precipitates were deposited. Renewed volcanism occurred at *ca* 2925 Ma (Sanborn-Barrie *et al.*, 2001).

METHODS

A total of 131 carbonate, iron formation, marl and siliciclastic samples were collected from the 10 locations shown in Fig. 1C and D over three separate field seasons during the summers of 1998, 2012 and 2013. The best exposures of these carbonates were along the shore at both Red Lake and Wallace Lake. Samples that showed well-preserved primary features were preferentially selected for sampling. Much of the shoreline exposure was neomorphosed and altered carbonate lacking any discernable primary sedimentary structures. Where well-preserved primary sedimentary structures were identified, the structures were measured and logged.

Many of the collected samples were cut, slabbed and polished. Thin sections were produced from selected samples and characterized by petrographic microscope. Of the 131 samples, 38 were selected for whole rock major and trace element geochemical analysis and 20 of these were analysed for carbon and oxygen isotopes. Samples that contained primary sedimentary features were selected for geochemical analysis while any obvious zones of alteration were excluded. The samples were coarsely crushed using a tungsten carbide mallet and plate and powdered in an agate mill. Carbon and oxygen isotopes were measured from the whole rock powders at G.G. Hatch Isotope Laboratories, University of Ottawa, by methods described in Revez *et al.* (2001). Major elements and trace elements were measured by inductively coupled plasma (ICP) atomic emission spectroscopy (AES) and ICP mass spectrometry (MS), respectively, at Lakehead University Instrument Laboratories. For major and trace element analysis, 0·500 g aliquots of each sample were digested by open vessel methods with three treatments of nitric and hydrofluoric acid. Final dilution of the samples was to 200 times for AES and 2000 times for MS. Duplicate samples and reference materials were used to evaluate accuracy and precision in each run and were within acceptable limits. In addition, three procedural blanks were measured within each run. Minimum detection limits were taken as three times procedural blanks.

Lithofacies description and interpretation

The only previous sedimentological work on the Red Lake–Wallace Lake carbonate platform was a descriptive study of the stromatolites and atikokania (pseudomorphed fanning sea floor precipitates; terminology of Walcott, 1912) present in the Red Lake area (Hofmann et al., 1985). Here, stratigraphic relationships of the stromatolites and pseudomorph fans, as well as their associations with other lithofacies across the platform, are described. Many of the lithofacies are similar to those found in other carbonate platforms from the Palaeoarchean to Neoproterozoic. Some of the outcrops have several metres of carbonate stratigraphy that have groups of lithofacies, and for these, the stratigraphy is summarized in Fig. 2. Along the shoreline, alteration has obscured some areas of stratigraphy and where this has happened, the lithofacies have been inferred from the nearest identifiable structures. Stratigraphic units were grouped by lithofacies and detailed descriptions of these are given in this section.

Wavy Laminae

The wavy laminae lithofacies are characterized by flat, wavy and crinkly laminae and massive orange weathering carbonate separated, at somewhat regular intervals, by millimetre-scale, light coloured chert laminae (Fig. 3). Sheet cracks, stromatactis-like cavities, roll-ups, crinkly laminae, irregular folding, teepee structures, probable desiccation cracks, small millimetre-scale fibrous colloform cement crusts, pseudomorphs after gypsum and colleniella occur locally.

Sheet cracks are found in some ancient carbonate deposits (Kennedy et al., 2001; Aubrecht et al., 2009; Hoffman & McDonald, 2010; Fralick & Riding, 2015) and, in the present study, consist of irregular, bedding parallel elongate, spar filled voids ranging in length from 2 to 10 cm and commonly 1 cm in height (Fig. 3B). The spar crystals are perpendicular to the void walls and often overlain by dark micritic sediment which outlines the crystal terminations. The remaining void is sometimes filled with blocky megaquartz. The characteristics of the spar are reminiscent of the void filling cement, radiaxial fibrous calcite (Kendall & Tucker, 1973; Kendall, 1985; Aubrecht et al., 2009). The sheet cracks are similar to those found in Maronian cap dolostones (Kennedy et al., 2001; Hoffman & McDonald, 2010).

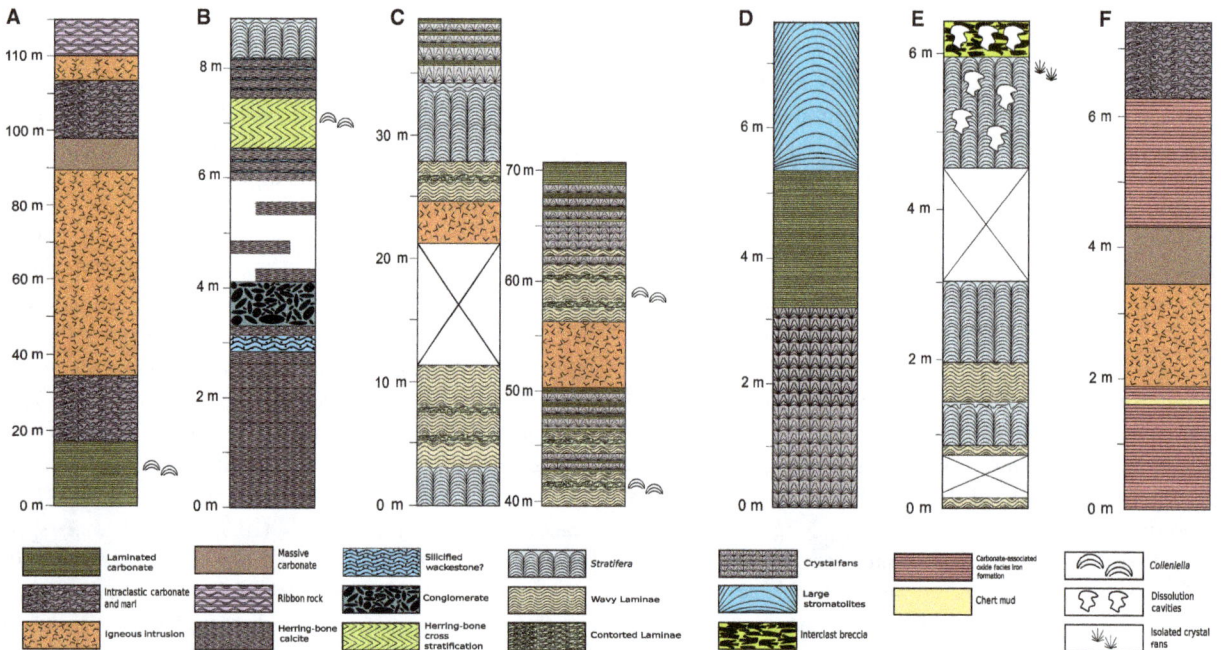

Fig. 2. Stratigraphic sections corresponding to outcrop locations described in Fig. 1C and D: A is outcrop 1, B is outcrop 5, C is outcrop 6, D is outcrop 7, E is outcrop 8 and F is outcrop 9. Massive carbonate, silicified wackestone and contorted laminae are not mentioned in text as they are poorly preserved and positive identification was putative.

Fig. 3. Some characteristic features of the wavy laminae lithofacies. (A) Partially silicified wavy laminae interbedded with massive carbonate. The scale bar is 10 cm. (B) Bedding parallel sheet cracks within wavy laminae. The arrows indicate the sheet cracks. The yellowish light coloured areas have been altered to chert. (C) A stromatolitic laminae and overlying mafic volcanics (dark areas above the stromatolitic laminae). The lower section of the clast is characteristic of the wavy laminae lithofacies and, in this instance, is gradational to *stratifera*. The arrows indicate stromatactis-like cavities in both wavy laminae and *stratifera*. The scale card is 8·5 cm long. (D) Swallow-tail gypsum pseudomorphs, now preserved as carbonate, in wavy laminae. The coin in B and E is 1·9 cm in diameter.

The stromatactis-like cavities are similar to those found in the stromatolite lithofacies, but confined between uneven laminae rather than stromatolitic convexity (Fig. 3C). The irregular folds and roll-ups are similar to structures found in the Carawine Dolomite in the Hemersley Basin by Simonson & Carney (1999). Crinkly mat also occurs locally and is interbedded with irregular and wavy laminae. Silicified carbonate close to chert beds tends to preserve fenestral fabrics and millimetre-scale vuggy porosity similar to birds eye structures. The teepee structures are small, *ca* 4 cm across and *ca* 2 cm in height. The colloform cement crusts are composed of densely packed upwardly diverging fibrous crystals, about 5 to 15 mm in height, which nucleate on underlying laminae. These cements are similar to sub-parallel fibrous crystals found in the Kotuikan Formation of Northern Siberia (Bartley et al., 2000). The pseudomorphs after gypsum are now quartz, consisting of small (*ca* 5 mm) pseudomorphs with swallow tailed terminations that appear to have grown at random orientations along bedding planes (Fig. 3D). There are possible erosional surfaces occurring within the lithofacies defined by irregularly shaped silicified horizons. Lithofacies similar to that of the wavy laminae lithofacies are common in many Archean and Proterozoic carbonate occurrences (e.g. Clough & Goldhammer, 2000; Sumner

& Grotzinger, 2004; Tice & Lowe, 2006; Allwood *et al.*, 2007).

Stromatolites

The stromatolites of the Red Lake–Wallace Lake carbonate platform commonly occur as laterally linked domal (LLD) stromatolites, isolated domal (ID) stromatolites and large low relief domes. The LLD and ID stromatolites of the Red Lake area have been termed *statifera* and *colleniella*, respectively, by Hofmann *et al.* (1985). The stromatolites are composed of coarse to fine laminoid fenestral fabric (terminology of Logan (1974)) defined by translucent white chert interbedded with orange weathering carbonate. Silicified areas of the stromatolites commonly preserve 1 to 3 mm birds eye vugs. Modern examples of birds eye fabric can be found in Shinn (1968, 1983), Flugel (2010), and references therein. *Statifera* are generally associated with wavy laminae, laminated carbonate and pseudomorph fans. In addition, herring-bone cross-strata are found in transitional facies between *statifera* and *colleniella*. Stromatolites, like those found in the Red Lake and Wallace Lake Carbonate Platform, are widely represented in the literature in larger carbonate platforms that date from the Neoarchean to the Proterozoic (Wilks & Nisbet, 1985; Grotzinger, 1986; Demicco &

Hardie, 1994; Clough & Goldhammer, 2000; Sumner & Grotzinger, 2004; Hofmann *et al.*, 2004; Schröder *et al.*, 2009).

Statifera are the most common stromatolitic facies found in the Red Lake–Wallace Lake carbonate platform and can be defined by irregularly stacked laterally linked domes creating biostromes up to 3 m in thickness (Fig. 4A, B and C). The stromatolites of these biostromes have broadly varying convexities on the order of 25 to 50 cm in diameter and 5 to 30 cm of topographic relief. Single stromatolites can have inheritance of up to 1 m. Layers of stromatolites within the *statifera* biostromes are sometimes truncated by planar or irregular surfaces and sharply overlain by the next generation of stromatolites or wavy laminae lithofacies. In many cases, *statifera* are gradational to wavy laminae lithofacies (i.e. stromatolites lose relief up section and become flat to wavy laminated carbonate beds). *Statifera* commonly have cherty reticulate stromatactis-like structures that conform to the curvature of the upper portions of the concave surfaces of stromatolitic layering (Fig. 4A, B and C). These stromatactis-like structures are like those found in the Doushantuo Cap Carbonate by Jiang *et al.* (2006) in that they have a cement rim, flat upper surfaces, irregular lower surfaces, and are filled with what appears to be sediment similar to that which surrounds the voids, in this case ferroan dolomite. In addition, *statifera* host dissolution features that occur as cavities in stromatolitic laminae in areas proximal to intraclastic breccia. Similar dissolution cavities to those present in the stromatolites have also been reported by Skotnicki & Knauth (2007) for the Mescal Limestone of the middle Proterozoic Apache Group.

Colleniella have similar dimensions and characteristics (laminoid fenestral fabric, birds eye vugs and reticulate stromatactis-like structures) to, and occur in close association with, *stratifera* (Fig. 4D). They differ from *statifera* in that they are isolated domal stromatolites hosted in other sedimentary facies rather than forming stromatolite biostromes. This stromatolite type is best preserved at outcrop 5 where it is overlain by herring-bone crossbedding composed of intraformational mud chips. The herring-bone crossbedding can be seen in the dip direction of the elongate rip-up clasts, where the clasts are ferroan dolomite 2 to 3 mm in thickness and 2 to 10 mm in length within a siliceous matrix. Individual herring-bone cross-stratified beds are *ca* 5 cm in thickness. Other *colleniella* associations in this area are herring-bone calcite (HBC) and *statifera*. The other occurrences of *colleniella* are poorly preserved but occur with laminated carbonate, small crystal fans and *statifera*.

Herring-bone calcite, that is associated with *colleniella*, is a cement that occurs as 1 to 3 cm thick laterally continuous layers (Fig. 5) and rarely 1 to 3 cm thick coatings on probable intraclastic breccia. Herring-bone calcite is characterized by irregular banding composed of serrated, sub-millimetre-scale layers of fibrous calcite or dolomite cement with crystals elongate and perpendicular to banding. In outcrop, banding is dominantly produced by preferential weathering of some layers creating an alternating high and low banded topography. In thin section, dolomitization and recrystallization are prolific leaving behind equant crystals and patches of fibrous spar that only faintly retain the banded extinction which is characteristic of HBC. Herring-bone calcite is a common cement found in large carbonate platforms throughout the Archean and Proterozoic (Ricketts & Donaldson, 1981; Sumner & Grotzinger, 1996). It occurs in some Phanerozoic carbonates, although only in restricted pore spaces (Sumner & Grotzinger, 1996; de Wet *et al.*, 2004).

Large low relief dome-type stromatolites form mounds on the order of 2 m in diameter and 1 m of topographic relief. The layering of these domes is more regular and smooth than those of *statifera* and *colleniella*, but in general they have similar characteristics to these other stromatolites.

Crystal fans

Crystal fans occur in the Red Lake–Wallace Lake carbonate platform in varying sizes and lithofacies associations. Some of these have been termed atikokania by Hofmann *et al.* (1985), based on the similarity of the fans with atikokania of the Mosher Carbonate (e.g. Walcott, 1912; Sumner & Grotzinger, 2000; Fralick & Riding, 2015), and here we present new occurrences and evidence for the widespread pervasiveness of these fans in the Red Lake–Wallace Lake carbonate platform. In all areas in which the crystal pseudomorphs occur, they have an acicular radiating habit and are preserved by preferential silicification of the needle-like crystals in a fine-grained ferroan dolomite or calcite matrix. Similar structures found in Archean carbonates are usually interpreted as pseudomorphs after aragonite (Sumner & Grotzinger, 2000, 2004; Fralick & Riding, 2015). The difference between fans in different occurrences is in their lithofacies associations, density of packing and size of individual fans.

The most abundant occurrence of these fans is as vertically stacked radiating needle-like crystals *ca* 2 to 7 cm in height (Fig. 6A and B). The proximity of nucleation points of individual fans is such that adjacent fans truncate each other imparting a more vertical orientation to the crystals rather than a botryoidal morphology. This density of packing has been previously described for similar crystal pseudomorphs from other Archean carbonate platforms (Sumner & Grotzinger, 2000, 2004; Fralick & Riding, 2015). These stacked fan beds generally range

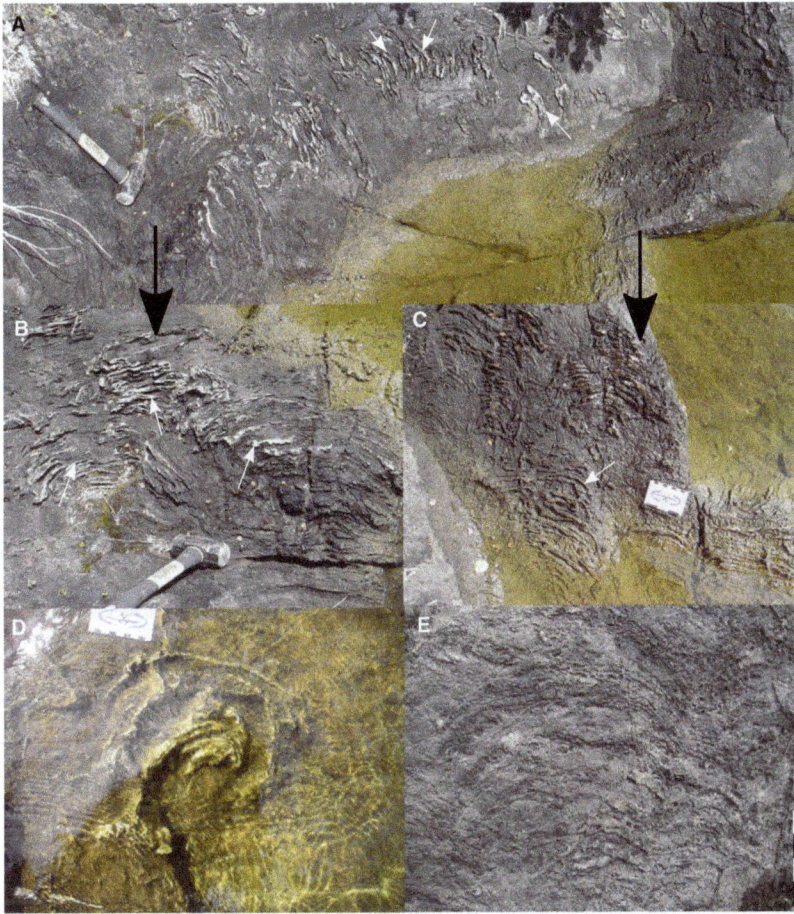

Fig. 4. Some stromatolitic outcrops of the Wallace Lake–Red Lake Carbonate Platform. (A) Large metre-scale *stratifera* bioherm. The black arrows indicate the areas from which images B and C were taken. In A, B and C, examples of well-defined and partially silicified stromatactis-like structures that can be seen to conform to the stromatolitic layering are pointed out by white arrows. These structures are like those in Fig. 2C. (D) *Colleniella*. To the left of this stromatolite, there are herring-bone cross-strata. (E) Wavy stromatolitic laminae with abundant bird's eye vugs. The hammer in A and B is 39 cm, the scale card in C and D is 8.5 cm long and the scale bar in E is 4 cm.

Fig. 5. The serrated banding seen in this picture is characteristic of herring-bone cement. The coin is 2.4 cm in diameter.

from 5 to 10 cm in thickness, but can form beds up to a few metres thick. The beds of stacked fans alternate with beds 2 to 10 cm thick of flat to wavy carbonate laminae like those found in the wavy laminae lithofacies. This type of crystal fan occurrence is associated with *stratifera* and wavy laminae lithofacies. The interbedding of these fans with laminated carbonate and their stromatolite association is common in the *ca* 2.54 Ga Campbelrand-Malmani Platform (Sumner & Grotzinger, 2004) and the *ca* 2.6 Ga Cheshire Formation (Sumner & Grotzinger, 2000). For the purposes of referencing the occurrence of these fans with laminated carbonate, their density of packing, and association with other lithofacies (with *stratifera* and wavy and irregular laminae facies) these fans are termed type 1 fans.

Another fan type (type 2) also occurs as more broadly fanning pseudomorphs with laminea that conform to the

Fig. 6. (A and B) Type 1 fans interbedded with crinkly laminae. Preferential silicification has preserved portions of the fans, but in general they are not well preserved. (C) Atikokania forming a large dome. The scale card is 8·5 cm long. (D) Close-up of C showing nucleation points of one of the fan layers close to the coin. The coin in A, B and D is 1·9 cm in diameter.

botryoidal shape of the fans. These fans are rare in the Red Lake–Wallace Lake carbonate platform and poorly preserved. They are also similar to those found in the subtidal lithofacies assemblage of the Campberland Malmani platform (Sumner & Grotzinger, 2000, 2004; Riding, 2008).

Atikokania are larger, with fans ranging from 5 to 30 cm in height and are densely packed and stacked (Fig. 6C and D). The stacked fan beds produce a dome-like structure about 2 m in diameter and 1·5 m in height.

Laminated carbonate

Laminated carbonate consists of flat layered orange weathering carbonate laminae. In polished sections, laminated carbonate is characterized by light coloured chert and dark grey millimetre-scale limestone bands. This type of occurrence is associated with large low relief domes and crystal fans.

Ribbon rock

Ribbon rock consists of wavy and laminated, fine- to medium-grained (0·05 to 0·5 mm), light and dark coloured carbonate couplets, occurring with irregular chert laminae, carbonate-shale couplets and rare magnetite laminae (Fig. 7A). Laminated light and dark carbonate couplets consist of light coloured, medium-grained

(0·2 to 0·5 mm), carbonate laminae with millimetre-scale thicknesses separated by fine-grained (<0·2 mm), dark, carbonate laminae with sub-millimetre-scale thicknesses. Carbonate and shale couplets alternate on a centimetre- to decimetre-scale. Magnetite laminae are rare but occur in locally abundant areas. Irregular folding occurs in some areas and is similar to folded structures described by McIlearth & James (1978). The ribbon rock lithofacies is similar to ribbon and parted limestone and carbonaceous rhythmite of the Victor Bay Formation (Sherman et al., 2000). In addition to areas where ribbon rock is preserved in stratigraphic section (Fig. 2A), these lithofacies also characterize outcrop 4 (Fig. 1D).

Intraclastic carbonate lithoclasts and marl lithofacies

In the Red Lake–Wallace Lake carbonate platform, these lithofacies occur as clasts of fine- to medium-grained (0·05 to 0·5 mm) calcite in a marl or carbonate matrix (Fig. 7B). The clasts of the disaggregated calcite beds are variable in morphology, but are generally elongate and deformed clasts up to 8 cm long and 0·2 to 3 cm thick. These are hosted in a marl matrix of fine-grained (0·01 to 0·2 mm) calcite and biotite in variable proportions and rare magnetite. The constituents of this lithofacies have the appearance of disaggregated ribbon rock facies. In addition to the occurrences of this lithofacies in stratigraphic sections in

Fig. 7. (A) Ribbon rock. The thick light coloured layer at the top and similar layers throughout the image are chert. Recessively weathered sections are light and dark carbonate. Dark areas with positive relief are coarsely crystalline carbonate. The coin is 1·9 cm in diameter. (B) Intraclastic carbonate lithoclasts and marl. The light areas are lithoclasts of carbonate and the dark layers are marl. (C) Carbonate-associated oxide facies iron formation (OF-IF). The carbonate weathers recessively relative to the iron oxide. The white arrow shows a contact with intraclastic carbonate lithoclasts and marl to the right of the image. The scale card in B and C is 8·5 cm long.

Fig. 2A and F, this intraclastic carbonate lithoclast and marl lithofacies occur in outcrop 3 (Fig. 1D) where it overlie interbedded decimetre-scale chert and shale beds.

Carbonate-associated oxide facies iron formation

Carbonate-associated oxide facies iron formation (OF-IF) is characterized by yellowish carbonate laminae interbedded with dark magnetic laminae (Fig. 7C). The laminae are planar to wavy, occasionally irregularly folded, and have light coloured carbonate disks throughout. The carbonate laminae of sub-millimetre thickness occur in packages separated by thin layers of dark, fine-grained material. These carbonate packages are up to 1 cm in thickness and have disseminated magnetite grains throughout. The dark magnetic laminae are composed of euhedral magnetite grains in a matrix of carbonate. In some instances, individual carbonate laminae grade into magnetite and are sharply overlain by carbonate producing chemically graded rhythmites. Thin, millimetre-scale, slate beds occur locally.

Chert-oxide facies iron formation

The chert-OF-IF is characterized by fine-grained, millimetre-scale laminations of light coloured chert and magnetite. Laminations are usually even and continuous, but deformation sometimes causes folding of the laminae. These lithofacies gradationally overlie the Conley Formation of the Wallace Lake area at outcrop 1. The gradation from the Conely Formation to the chert-OF-IF is characterized by a change from chloritic sandstones to siltstone at the top of the Conley Formation, which is transitional to siltstone, chert and oxide laminae, and overlain by chert-OF-IF. At Red Lake, chert-OF-IF occurs as layers 2 to 7 cm thick interbedded with ferroan dolomite layers of similar thicknesses.

GEOCHEMISTRY: RESULTS AND INTERPRETATION

The geochemical results are given in Tables S1–S3. They show that most of the samples are dolomite with only the atikokania crystal fans, laminated carbonate below them, ribbon rock and one sample of herring-bone cement composed of calcite (Table 1). The intraclastic carbonate lithoclasts forming the slump deposits and marl lithofacies associated with iron formation contain both dolomite and calcite portions. Strontium values are generally higher in the calcite atikokania fans, with amounts 10 times greater than the dolomite type 1 fans (Fig. 8). The calcite also contains larger amounts of Ba and Pb. This is expected as Sr, Ba and Pb will substitute for calcium, especially in aragonite (McIntyre, 1963; Kretz, 1982), whereas elements with smaller radii, such as Fe, Mn, Ni, Cu and Co, should have greater concentrations in the dolomite (Swart, 2015). The Red Lake–Wallace Lake samples do not show the latter trend, with insignificant differences in Ni, Cu and Co concentrations between the dolomite and calcite (Table 1). As the amounts in both minerals are very small, this may reflect low concentrations of these elements in the 2·93 Ga ocean. The Mn and Fe are lowest in the atikokania, but are highest in the calcite layers directly below it. This is probably related to a local variation in water chemistry rather than a mineralogical difference.

There is also a difference in the $\delta^{18}O$ values of the dolomite and calcite, with dolomite values of $-15\cdot2 \pm 1\cdot2\text{‰}_{\text{V-PDB}}$, and calcite values of $-11\cdot7 \pm 1\cdot2\text{‰}$ (Fig. 9). The $^{87}Sr/^{86}Sr$ ratios are also dependent on mineralogy, as the five dolomite samples analysed were $0\cdot7032 \pm 0\cdot0002$, whereas three calcite samples were $0\cdot7017$, $0\cdot7018$ and $0\cdot7020$ (Satkoski et al., 2017). The values for calcite are very close to the hypothetical value for sea water at 2·9 Ga (Veizer & Jansen, 1979; Taylor & McLennan, 1985; Veizer et al., 1989a,b; Godderis & Veizer, 2000; Veizer, 2003; Satkoski et al., 2017). As all samples had very low levels of Rb, and it was adjusted for, the Sr isotopic results strongly indicate that the $^{87}Sr/^{86}Sr$ for the calcite samples reflect sea water composition and have not been significantly reset, whereas the dolomite samples have had their Sr isotopic ratios altered, probably during dolomitization, the only known diagenetic or metamorphic event that affected the dolostones but not the limestones. The higher $^{87}Sr/^{86}Sr$ ratios for the dolomite were likely caused by interaction with meteoric fluids that had been involved in weathering reactions with older crustal rocks. As higher water/rock ratios are commonly needed to alter the carbon isotopic signature than the strontium isotopic signature (Veizer et al., 1989b; Jacobsen & Kaufman, 1994; Veizer, 2003), it is reasonable to assume that $\delta^{13}C$ in the calcite samples reflects the original composition of the water-mass from which the calcium carbonate precipitated. As the dolomite samples have similar $\delta^{13}C$ values to the calcite (Fig. 9A), it is also reasonable to assume that dolomitization did not alter the C isotopic values. Conversely, oxygen isotopic values commonly are reset at lower water/rock ratios than those necessary to alter the Sr isotopic values. Therefore, it is likely that the $\delta^{18}O$ values for dolomite have been reset, but it is unknown whether or not the $\delta^{18}O$ for calcite are primary (Fig. 9).

Stromatolitic units in the 2·7 Ga Cheshire and Manjeri Formations of Zimbabwe have $\delta^{18}O$ values (Abell et al., 1985a) similar to the calcite samples from Red Lake and Wallace Lake, although their $\delta^{13}C$ values are mostly slightly negative as opposed to those from Red Lake–Wallace Lake calcite, which are slightly positive. The $\delta^{18}O$ and $\delta^{13}C$ values for the Red Lake–Wallace Lake dolomite are similar to Archean Mushandike stromatolites from Zimbabwe (Abell et al., 1985b). Abell et al. (1985b) believed that this difference between the Cheshire and Mushandike was caused by progressively higher metamorphic grades. However, this explanation cannot be applied to the Red Lake samples as both the dolomite and calcite have experienced the same grade of metamorphism. Furthermore, Abell et al. (1985b) believed that the C and O isotopic values for the Cheshire Formation represented those of the water from which the limestone precipitated, with little later modification. The similar $\delta^{18}O$ values in the Red Lake–Wallace Lake calcite samples may, by comparison, also be close to their starting compositions.

Carbon isotope values obtained from the 2·55 Ga (Altermann & Nelson, 1998) Campbellrand carbonate platform of South Africa tend to be slightly more negative than the Red Lake–Wallace Lake samples with an average of $-0\cdot5\text{‰}$ (Fischer et al., 2009). The Campbellrand $\delta^{18}O$ values are heavier than those from the Red Lake–Wallace Lake samples with samples from two drill cores having $\delta^{18}O$ values averaging $-7\cdot6\text{‰}_{\text{V-PDB}}$ and $-8\cdot5\text{‰}_{\text{V-PDB}}$ with upper limits on both holes of $-5\cdot5\text{‰}$ and $-7\cdot6\text{‰}$ and lower limits of $-16\cdot7\text{‰}$ and $-10\cdot8\text{‰}$, respectively (Fischer et al., 2009). The Hamersley carbonate platform of Australia is similar with $\delta^{13}C$ values of $0 \pm 1\cdot2\text{‰}$ and $\delta^{18}O$ values ranging between -5 to -15‰_{PDB} (Becker & Clayton, 1972; Veizer et al., 1990). The older 2·8 Ga Steep

Table 1. Concentrations of select elements in dolomite and calcite samples

	Dolomite		Calcite	
		Type 1		Laminae below
	Stromatolites	fans	Atikokania	atikokania
Element	($n = 6$)	($n = 4$)	($n = 4$)	($n = 1$)
Sr p.p.m.	20	23	226	231
Ba p.p.m.	2·1	1·1	5·4	12·1
Pb p.p.m.	0·92	0·93	2·2	1·4
Fe_2O_3 %	2·26	2·57	1·27	6·77
MnO%	0·53	0·70	0·38	0·99
Ni p.p.m.	6·8	7·7	7·7	9·8
Cu p.p.m.	3·0	1·9	3·1	3·7
Co p.p.m.	3·1	4·1	4·3	3·4
V p.p.m.	1·0	1·3	1·0	5·0
Cr p.p.m.	B.D. (2)	B.D.	B.D.	4·0
Mo p.p.m.	B.D.	0·20	0·15	0·30
U p.p.m.	0·17	0·05	0·13	0·45

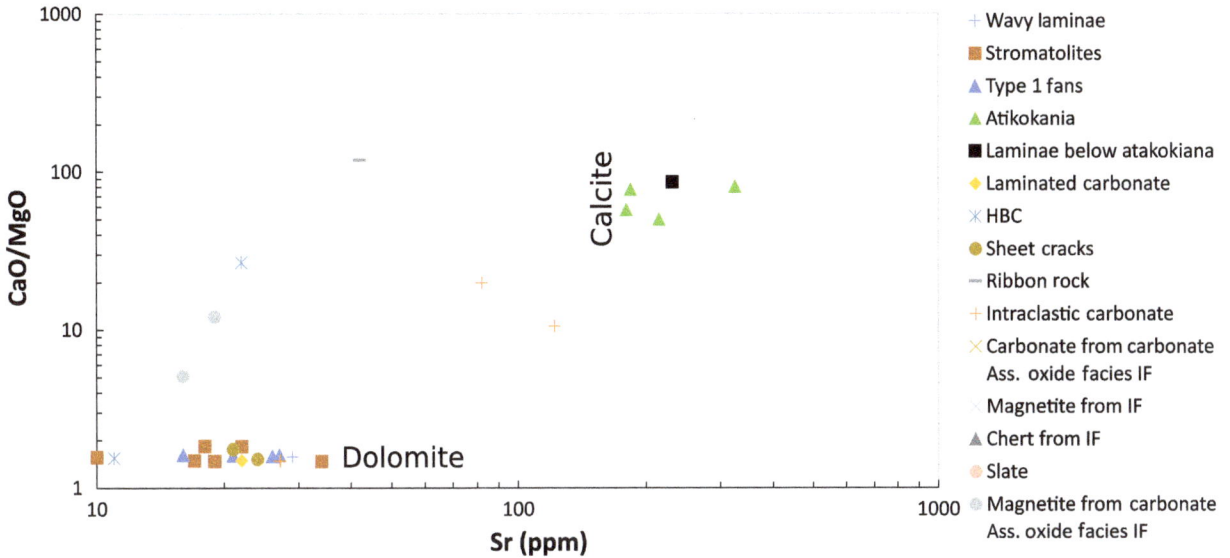

Fig. 8. The atikokania samples and laminated carbonate directly below an atikokania mound are calcite and contain considerably more Sr than the dolomite samples. The samples of slumped intraclastic material are mixtures of high Sr calcite and dolomite, whereas ribbon rock and herring-bone cement samples are low Sr calcite.

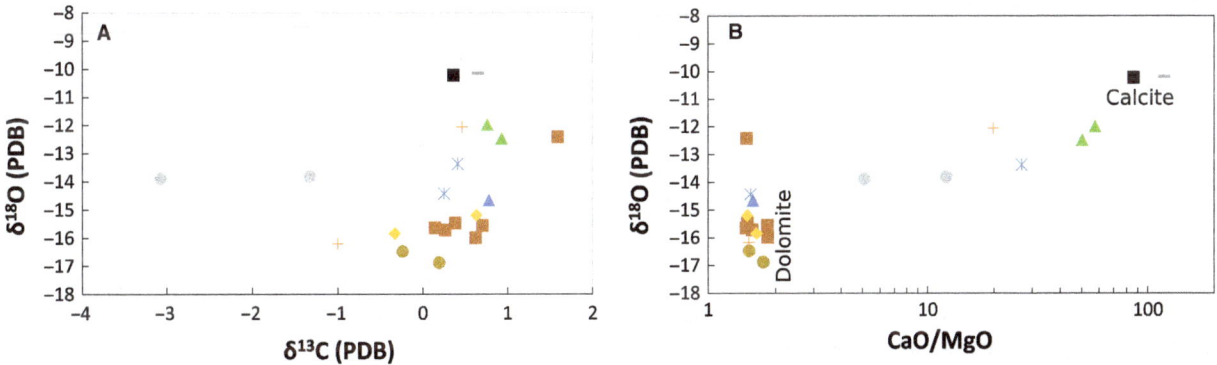

Fig. 9. (A) Cross-plot of O and C isotopic ratios. The carbonate associated with oxide facies iron formation (OF-IF) and a slump deposit have lower $\delta^{13}C$ values than the shallower water deposits, with the majority of shallow water deposits of both calcite and dolomite having similar $\delta^{13}C$ values. (B) Most calcite samples have higher $\delta^{18}O$ values than the dolomite samples. Symbols are the same as in Fig. 8.

Rock Carbonate Platform shows greater similarity to Red Lake–Wallace Lake calcite samples with $\delta^{13}C$ values of $1\cdot6 \pm 0\cdot7\%$ and $\delta^{18}O$ values of $-8\cdot9 \pm 1\cdot7\%_{ooV-PDB}$, although it was noted that the one sample below -12% $\delta^{18}O$ was altered (Fralick & Riding, 2015). Thus, the $\delta^{18}O$ values obtained from the Red Lake–Wallace Lake calcite samples have similarities with those from some other Archean carbonate platforms, but the O isotope values for the dolomite samples are lower and probably represent post-depositional alteration, possibly by meteoric fluids during dolomitization.

The presence of significant amounts of siliciclastic material in the carbonate samples can also affect the concentrations of elements of interest in the chemical sediments. Plotting relatively chemically immobile elements common in siliciclastics but not in carbonates against elements of interest will result in a positive correlation if there is a significant amount of siliciclastics in the samples. The Red Lake–Wallace Lake carbonate samples contain low amounts of Al and Ti, which show no correlation with Sr (Fig. 10A and B). In addition to affecting the concentration of elements such as Sr in the samples, contamination with siliciclastics can alter their Rare Earth Element (REE) pattern. Such contamination will limit primary sea water signatures, such as La, Ce, Eu, Gd and Y anomalies and cause flattening of the REE patterns (Kamber & Webb, 2001; Peter, 2003). Yttrium/holmium ratios can be used to test for siliciclastic

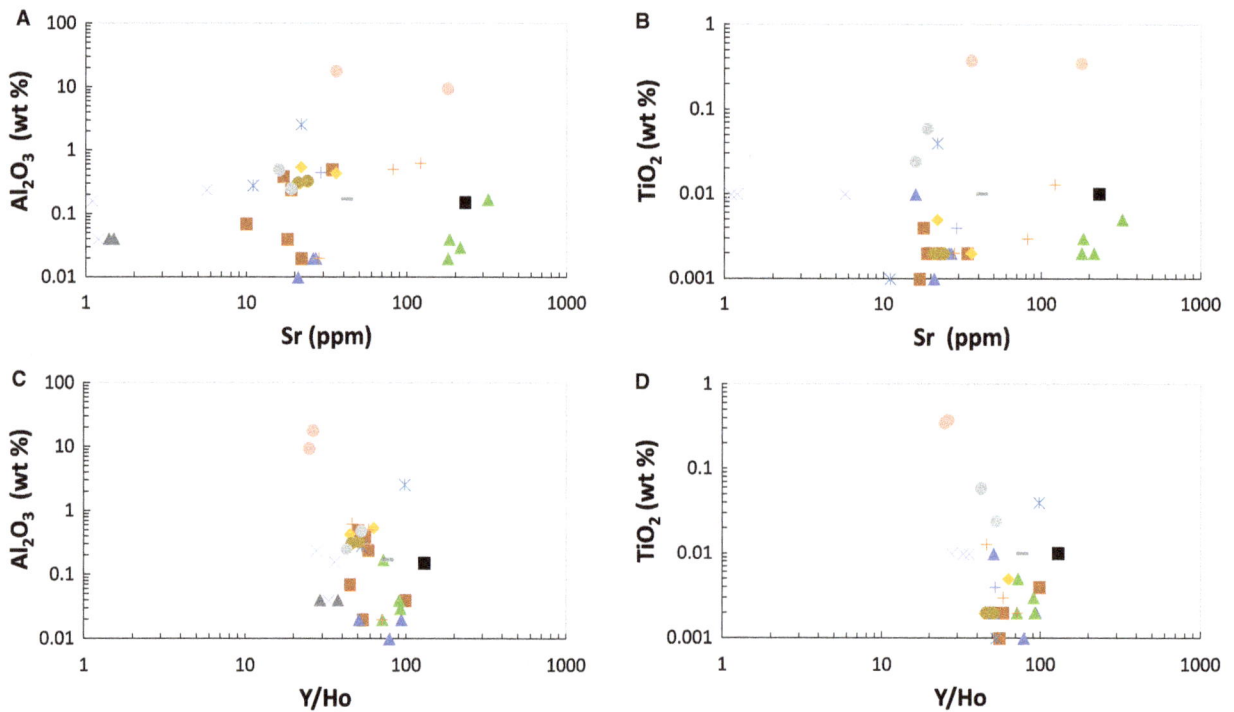

Fig. 10. (A and B) Sr does not vary systematically with elements contained in the siliciclastic component of the samples, such as Al and Ti. This indicates that siliciclastic levels in the majority of the samples are below the amount necessary to significantly impact the Sr content. However, a sample of herring-bone cement and the samples of magnetite from the carbonate-associated iron formation have relatively high amounts of Al and Ti, and Ti respectively, which indicate possible significant siliciclastic impact on their geochemistry. (C and D) Y/Ho weight ratios in most crustal rocks are *ca* 26, similar to the two siliciclastic samples. Most iron formation samples are between this value and 40, with the carbonate samples between 40 and 130. These are the ratios expected for those rock types. Significant amounts of siliciclastic contamination would lower the ratios for the chemical sediments, which, in general, do not appear to be the case. Symbols are the same as in Fig. 8.

contamination altering REE patterns. Crustal rocks typically have $Y_{p.p.m.}/Ho_{p.p.m.}$ ratios of *ca* 26 while ratios in carbonate precipitates are commonly greater than 44 and modern sea water has ratios between *ca* 44 and 120 (Nozaki *et al.*, 1997; Kamber & Webb, 2001; Bolhar *et al.*, 2004). Thus, contamination of carbonate by siliciclastics should lower their Y/Ho ratios. By plotting Y/Ho against insoluble elements that concentrate in siliciclastics, the development of a negative correlation will indicate significant contamination by siliciclastic material (Kamber & Webb, 2001; Bolhar *et al.*, 2004). The two slate samples have Y/Ho ratios of 25 and 26, i.e. chondritic ratios expected of crustal rocks. The iron formation samples have Y/Ho ratios of 33 ± 4, and the carbonates of 67 ± 22. There is no correlation between Y/Ho and either Al_2O_3 or TiO_2 indicating no significant contamination for most samples (Fig. 10C and D), as expected with such low levels of elements that concentrate in siliciclastics. However, three of the samples may contain sufficient siliciclastic material to significantly influence their geochemistry. One of the herring-bone cement samples has considerably more Al_2O_3 and TiO_2 than the other

chemical precipitates, and the two samples of magnetite from carbonate associated OF-IF have comparatively higher TiO_2 concentrations and low Y/Ho ratios indicating the possible influence of siliciclastics.

Studying the abundances of rare earth elements in chemical sediments can provide information on the fluids from which the precipitates formed (Kamber *et al.*, 2004; Nothdurft *et al.*, 2004; Webb *et al.*, 2009). Modern surface and groundwaters are commonly enriched in heavy REEs (Elderfield *et al.*, 1990), although some originating from organic-rich areas are middle REE enriched (Hannigan & Sholkovitz, 2001). Oxidation of Ce^{3+} to Ce^{4+} and its precipitation as CeO_2 or adherence to iron hydroxides in the weathering environment has also led to negative Ce anomalies in modern surface and groundwater. Archean chemical sediments derived from ocean water also commonly have heavy REE enrichment, but they only rarely exhibit a negative Ce anomaly. However, they do have a positive Eu anomaly caused by high temperature leaching from ocean crust basalts and gabbros by circulating fluids, similar to modern oceanic venting precipitates (Klinkhammer *et al.*, 1983, 1994; Derry & Jacobsen, 1990;

Danielson *et al.*, 1992). Some Archean chemical sediments also have positive La and Gd anomalies.

Rare earth elements were normalized using Post Archean Australian Shale (PAAS) and their patterns plotted. The REE patterns for magnetite and chert are similar, although with decreased abundance in the chert. They are also similar to other published REE patterns for Archean iron formations (Barrett *et al.*, 1988; Fralick *et al.*, 1989; Derry & Jacobsen, 1990; Danielson *et al.*, 1992; Alibert & McCulloch, 1993; Bau & Moller, 1993; Bau & Dulski, 1996). Both the magnetite and chert are heavy REE enriched, have a substantial positive Eu anomaly and no Ce anomaly (Fig. 11A and B). The OF-IF associated with carbonate has a somewhat different pattern with positive La and Gd anomalies, reduced Eu anomalies, and, aside from one sample, are not heavy REE enriched with a somewhat flat pattern (Fig. 11C). The samples of ribbon rock, both the wavy and more planar laminated carbonate, sheet crack cement, herring-bone cement, intraclastic carbonate lithoclasts and marl,

type-1 fans, stromatolites and atikokania have patterns very similar to the iron formation samples (Figs 12, 13 and 14). The notable differences being generally lower slopes, positive La anomalies in the carbonates and some carbonate samples have positive Gd anomalies. The one sample that shows a different pattern is the laminated carbonate directly below the atikokania, which has a negative Ce anomaly and extensive heavy REE enrichment (Fig. 14D). The positive slope that results from the enrichment forms the carbonate end member of a continuum when the overall slope of the patterns is plotted against slope of the medium to heavy REEs (Fig. 15). A herring-bone cement sample is the only other carbonate sample that has as high an overall slope, along with the iron formation magnetite and chert samples. A slate sample with slight light REE enrichment forms the other end of this continuum. The carbonate samples are spread out between these two end points with the magnetite from carbonate-associated OF-IF clustering at the slate end.

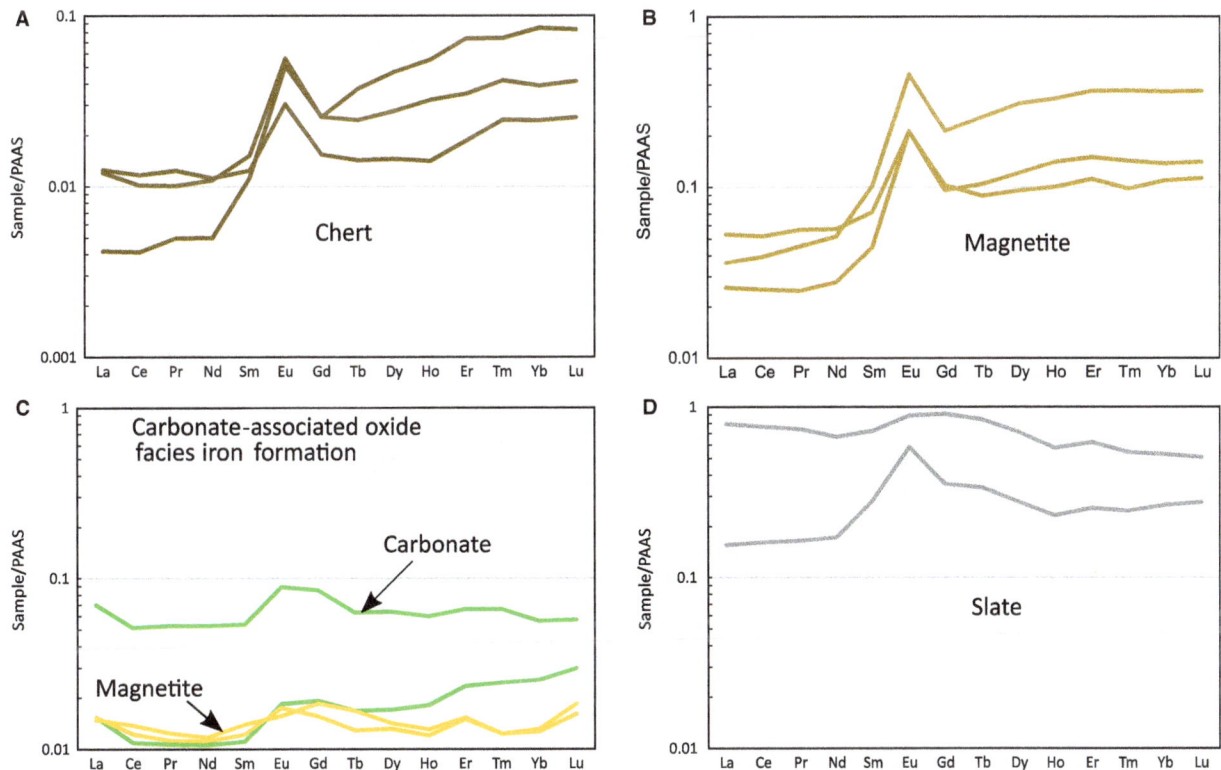

Fig. 11. Rare earth element patterns for the rock types associated with iron formation. (A and B) The magnetite and chert layers in samples from banded iron formation are depleted in light rare earth elements (REEs) and have substantial positive Eu anomalies. They do not have significant La, Ce or Gd anomalies. (C) The carbonate-associated and oxide facies iron formation (OF-IF) samples were from the iron formation layers and mostly consist of magnetite and chert with minor carbonate. They have flat to light depleted patterns with positive La, Eu and Gd anomalies and appear quite different from the iron formation not associated with carbonate layers. This may reflect their higher siliciclastic content (Fig. 10). (D) The carbonaceous slate samples are both light REE enriched and depleted, with moderate middle REE enrichment and one has a distinct positive Eu anomaly.

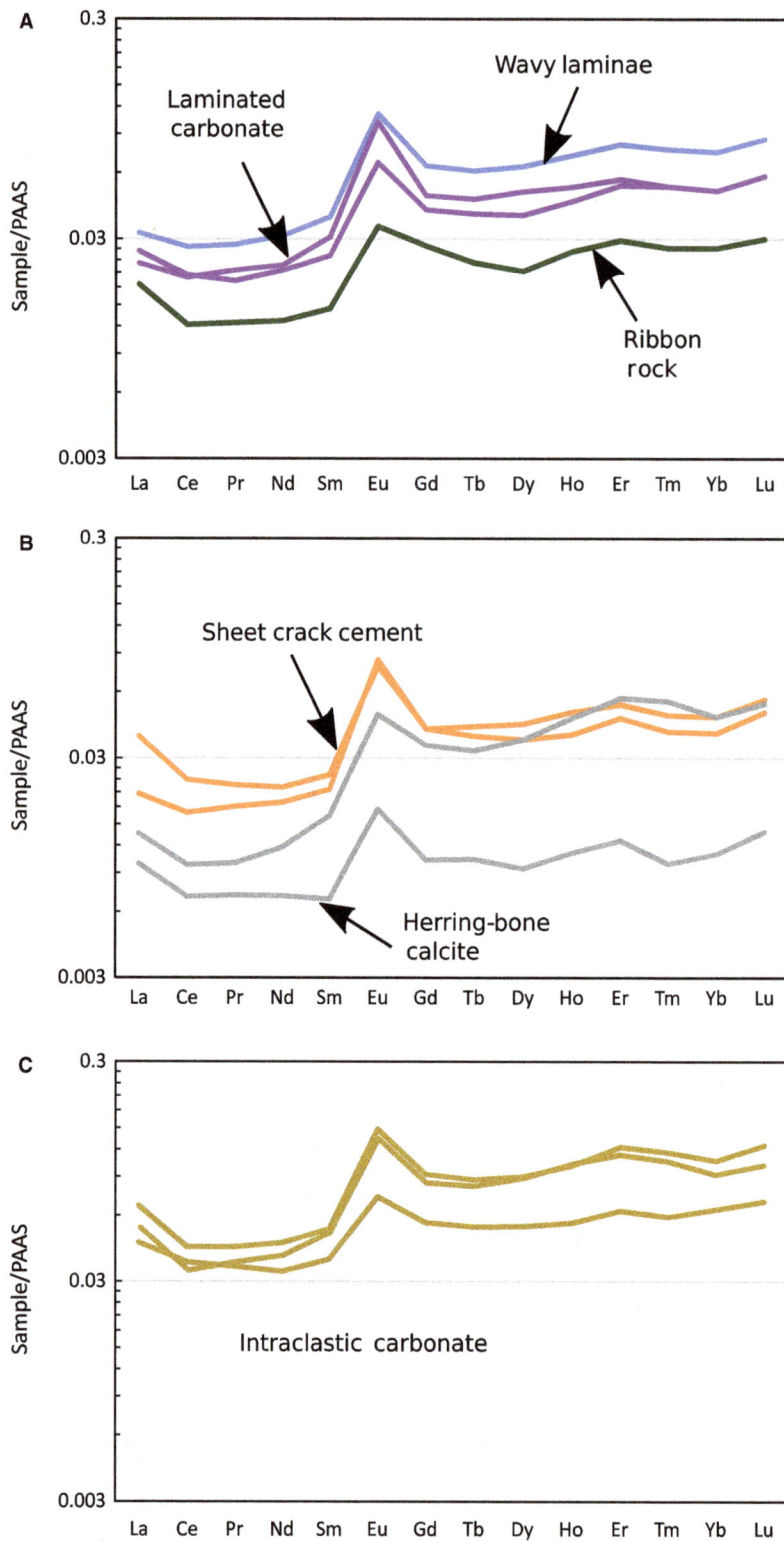

Fig. 12. The samples of various forms of laminated carbonate, wavy laminae, and ribbon rock (A), cement (B) and slumped material (C) are all similar with light rare earth element (REE) depletion, positive Eu anomalies and flat to slightly positive Gd anomalies. The sample of herring-bone calcite higher on the graph than the other has a much steeper slope and is also the only calcite sample in the figure, as the other herring-bone calcite has been dolomitized.

Comparing the Red Lake–Wallace Lake samples to those from the next oldest carbonate platform, the 2·8 Ga Steep Rock succession (Figs 13 and 14) highlights some similarities and differences. The Red Lake–Wallace Lake carbonates generally have larger positive Eu anomalies, fewer Gd anomalies, more subdued La anomalies and lower slopes. The most notable exception to this is the sample of laminated carbonate immediately below atikokania, which has a REE pattern very similar to that of the Steep Rock samples with negative Ce anomalies (compare Fig. 14B, C and D).

DISCUSSION

Stromatolite biogenicity

The stromatolites of the Red Lake–Wallace Lake carbonate platform likely formed by biomediated carbonate precipitation in a low energy intertidal environment. The laminae of the stromatolites are irregular and crinkly with

variable thicknesses characteristic of the fine-grained crusts of Riding (2008), and these crusts are a prerequisite for identifying Archean lithified microbial mats (Awramik & Grey, 2005; Fralick & Riding, 2015). In addition, laminoid fenestral fabrics and bird's eye vugs are ubiquitous in the stromatolites and can form from the desiccation and decomposition of microbial mats or sediment shrinkage during desiccation (Logan, 1974; Shinn, 1968; Demicco & Hardie, 1994; Flugel, 2010).

The internal features of the reticulate stromatactis-like structures and sheet cracks are similar to those found in the Doushantuo Cap Carbonates (DCC) (cf. Jiang et al., 2006). The stromatactis-like structures of the Red Lake–Wallace Lake carbonate platform differ from those of the DCC in that they are regularly confined by stromatolitic layers, whereas those in the DCC form extensive networks in plastically deformed detrital carbonate. Jiang et al. (2006) interpreted the stromatactis-like structures and sheet cracks to represent deformation caused by the injection of fluid into partially lithified sediment of low

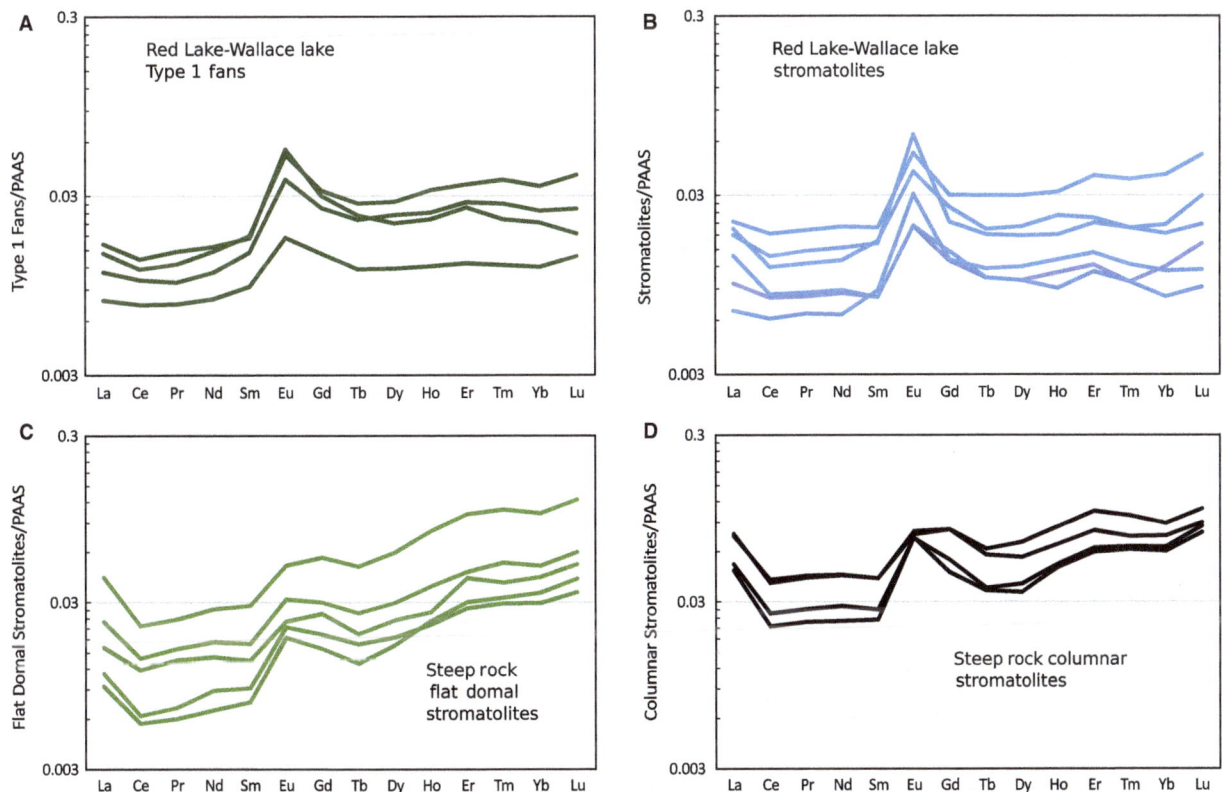

Fig. 13. Type 1 fans and stromatolites both have positive La, Eu and Gd anomalies and flat Ce. The type 1 fans exhibit a higher slope than the stromatolites. Compared to samples of Steep Rock flat, domal (C) and columnar (D) stromatolites the Red Lake–Wallace Lake samples have lower La and Gd anomalies but larger positive Eu anomalies. Also, their slope is not as high as the Steep Rock flat and domal stromatolites, which formed on a semi-restricted, 2·8 Ga carbonate platform. The Red Lake–Wallace Lake fans and stromatolites have rare earth element (REE) patterns that are more similar to Steep Rock columnar stromatolites probably developed during a flooding event that brought an influx of open ocean water onto the platform (Fralick & Riding, 2015).

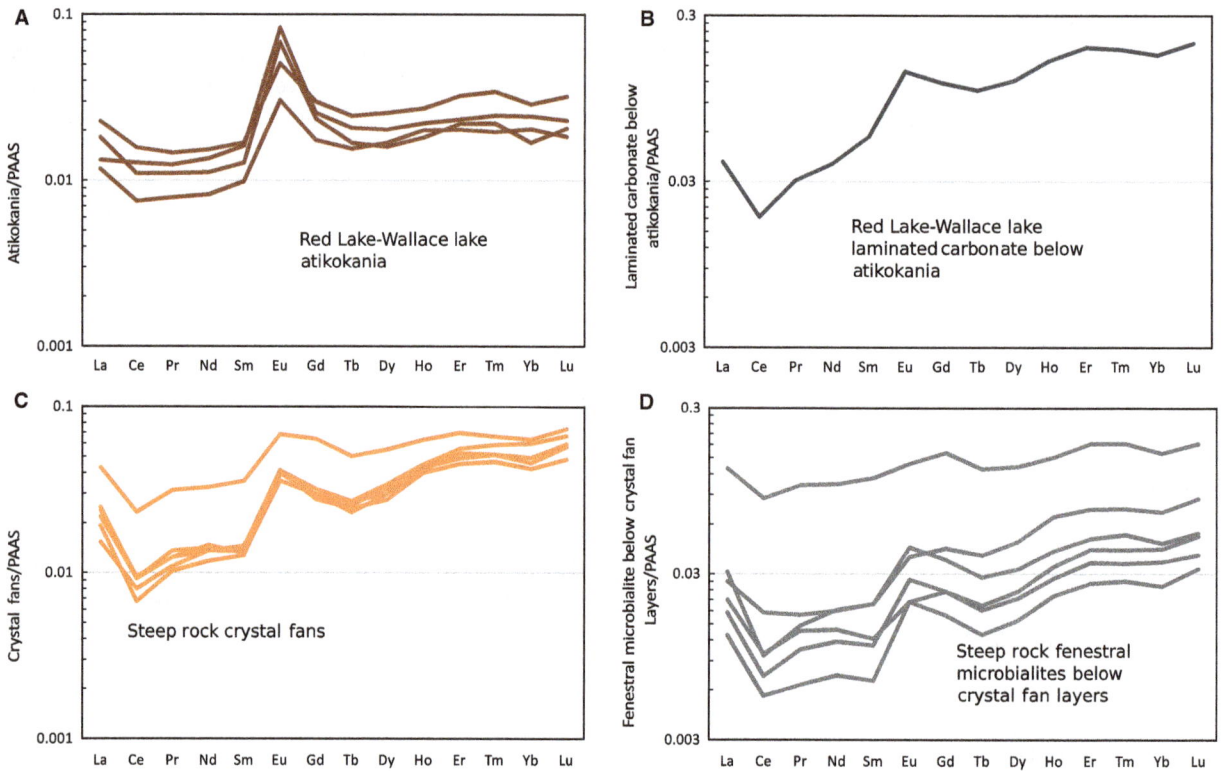

Fig. 14. Samples of atikokania (A) are light rare earth element (REE) depleted with positive La, Eu and lesser Gd anomalies, and flat Ce. Compared to samples of Steep Rock crystal fans (C) and cuspate fenestrae fabric (C.F.F.) (D) the Steep Rock samples have higher slopes, larger La and Gd anomalies, lower Eu, and most exhibit a negative Ce anomaly. The Steep Rock samples were probably deposited during periods of restricted circulation onto that carbonate platform (Riding et al., 2014; Fralick & Riding, 2015), whereas the atikokania and other REE patterns from the Red Lake–Wallace Lake platform are more similar to the Steep Rock columnar stromatolites (Fig. 13D) that were deposited during less restricted circulation. However, laminated carbonate from directly under an atikokania mound has a REE pattern (B) comparable to the Steep Rock platform samples deposited during a period of more restricted circulation.

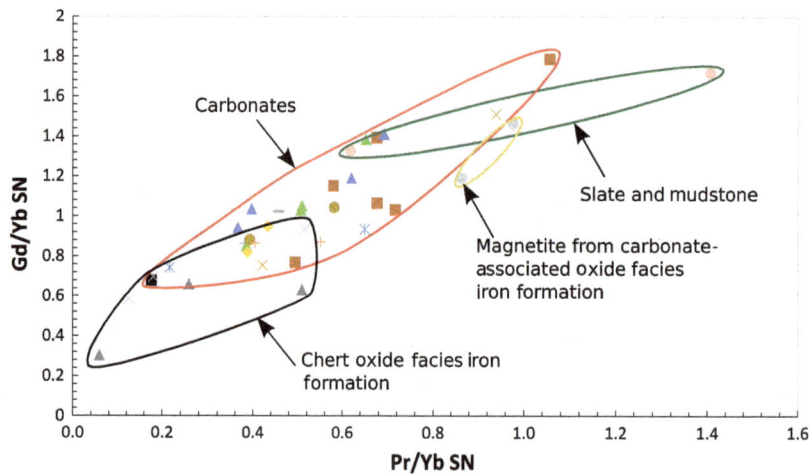

Fig. 15. Post Archean Australian Shale (PAAS) normalized (SN) ratios of Gd/Yb and Pr/Yb, representing slope of the heavy rare earth elements (REEs) and slope of the entire pattern, respectively. The chert and magnetite from the banded iron formation as a group have the highest slopes. The groups composed of slate, and magnetite from carbonate-associated iron formation have the lowest slopes. The carbonate samples form a continuum between these two end members. Symbols are the same as Fig. 8.

Fig. 16. Sedimentary structures and corresponding depositional environments. The interpretations are discussed in the text.

permeability. A similar process can be envisioned for the stromatactis-like structures in the Red Lake–Wallace Lake carbonate platform in which gases within the stromatolitic layering are confined by irregularly developed sparry crusts and the underlying leathery mats were deformed by pressurisation of gases within the stromatolites resulting in rounded and irregular voids with flat upper surfaces. Similarly, sheet cracks could have formed by fluid injected along layers of carbonate laminae. In addition, the limited extent of the structures implies localized fluid production in both stromatolites and wavy laminae lithofacies after deposition. However, the plastic deformation of stromatolitic laminae and wavy laminae indicates that fluid production occurred prior to or contemporaneous with lithification. Localized fluid production within the stromatolitic laminae would require the decomposition of organic material either by abiotic processes with CO_2 as a by-product (cf. Larralde *et al.*, 1995) or by several microbial pathways that are known to be operating by as early as the Neoarchean and have gaseous by-products (cf. Czaja *et al.*, 2010).

In summary, the development of thick microbial mats on the shallow sea floor was accompanied by periodic sparry crust precipitation, periodic *in situ* desiccation and partial early cementation of the microbial mat to produce laminoid fenestral fabrics and bird's eye vugs, and then burial and mat decomposition to produce stromatactis-like structures and sheet cracks through *in situ* gas production.

Evidence for low energy restricted environment

Stromatolite morphology

The energy of the environment in which the earliest stromatolites formed is important in understanding limiting factors affecting the geographic extent of the earliest life. Was life limited to environments that were shallow,

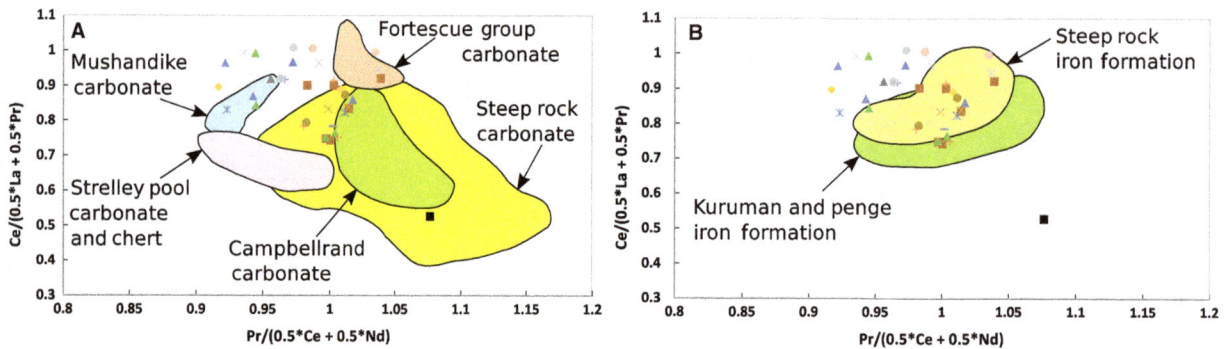

Fig. 17. (A) Post Archean Australian Shale (PAAS) normalized (*) La anomaly represented by Ce/(.5Pr+.5La) versus Ce anomaly represented by Pr/(.5Ce+.5Nd) compared to data from other Archean carbonate units. For an explanation of the use of these ratios, see Bau & Dulski (1996). The laminated carbonate sample from below an atikokania mound has the large negative Ce and positive La anomaly and overlying atikokania samples have positive Ce anomalies. This may indicate that the atikokania formed during periods of restricted circulation relative to the adjacent laminate allowing oxygen to build-up there and develop a redox boundary, resulting in these Ce anomalies, i.e. precipitation of oxidized Ce in the atikokania and delivery of the Ce deficient water to the sheltered area near it. Slight negative Pr anomalies cause two chert samples to plot with an apparent negative Ce anomaly. (B) Comparison of samples from the Red Lake–Wallace Lake Carbonate Platform with other Archean iron formations. Smaller to no La anomalies in some samples cause them to plot above the marked fields. Data from: Bau & Dulski (1996), Kamber & Webb (2001), Bolhar *et al.* (2004), Kamber *et al.* (2004) and Bolhar & van Kranendonk (2007). Symbols are the same as in Fig. 8.

restricted, low energy, and sheltered from destructive physical or chemical elements, or did life require current flux where high nutrient supply facilitated the development of microbial ecosystems? By examining the characteristics of stromatolite morphology related to the environment and applying this to the physical characteristics of the stromatolites in the present study and those of temporally related carbonate platforms provides insight into this question.

Dupraz *et al.* (2006) and Tice *et al.* (2011) show that stromatolite morphology is dependent on the physical characteristics of microbial mats (i.e. mat cohesion and response to current induced shear stress and the affinity of nutrients to the mats surface), sediment supply, and the limited diffusion and concentration of nutrients through and in the water that is in contact with the surface of microbial mats. Variations in these factors can produce four main morphologies given a range of peritidal environments (Dupraz *et al.*, 2006; Tice *et al.*, 2011). In the first scenario, in which current-induced shear stress is greater than mat cohesion on topographic highs, mats are limited to topographically low hydraulically protected areas, producing thin discontinuous mat lenses in a detrital matrix. If, on topographic highs, mat cohesion is greater than current-induced shear stress and mat cohesion between domes tends to be less than current induced shear stress, erosion of laminae and/or deposition of detritus in between domes will produce isolated domal stromatolites and branching stromatolites. The detritus supply is particularly important in the development of columnar and branching stromatolites (Dupraz *et al.*, 2006). If sediment supply is limited and the mat is

resistant to erosion on topographic highs and lows, the dominant limiting factor in stromatolitic growth is nutrient diffusion and water depth. Diffusion-limited processes (boundary Reynolds number, nutrient concentration and nutrient-mat affinity) may severely limit the growth of the mat between domes leading to massive and linked stromatolitic forms. Finally, if the nutrient-mat affinity is large, columnar stromatolites may form in the absence of sediment supply. As implied by the above discussion, given a consistent mat cohesiveness, the stromatolitic forms were presented in order of decreasing current energy.

In Archean carbonate formations, stromatolitic morphologies are consistent with the theoretical considerations above, such that, stromatolitic morphologies reflect the energy of the environment and higher energy environments tend to produce more diverse morphologies. The *ca* 3 Ga Chobeni Formation is a stromatolite bearing mixed siliciclastic-carbonate formation formed in a high energy tide-dominated shallow marine environment (Beukes & Lowe, 1989; Siahi *et al.*, 2016). Stromatolitic forms are diverse from supratidal stratiform, intertidal domal and subtidal/tidal channel linked to partially linked branching columnar, and laterally linked and isolated conical (conophyton) (Siahi *et al.*, 2016). Siahi *et al.* (2016) recognize the importance of physical processes in the morphology of the stromatolites in noting that strong tidal and storm currents provide an ideal environment for the formation of the conical and sporadic domal stromatolites. Columnar, domal and conical stromatolites occur as isolated forms or linked groups unevenly distributed in fine to coarse-grained clastic sediment between the domes and groups of stromatolites (Beukes & Lowe, 1989; Siahi *et al.*, 2016). This is consistent with mat survival rather than nutrient availability as a limiting factor in stromatolite development, and these stromatolitic forms are expected in a high energy environment.

Younger stromatolitic carbonates are morphologically diverse, yet conical, columnar, domal, and stratiform morphologies are still ubiquitous in carbonate platforms through the Neoarchean (e.g. Sumner & Grotzinger, 2004; Sakurai *et al.*, 2005; Awramik & Buchheim, 2009; Fralick & Riding, 2015). In the *ca* 2·8 Ga Mosher Carbonate columnar stromatolites, $\delta^{13}C$ values, Sr, Ba, Mn and Fe concentrations in the columnar and laterally linked domal stromatolites lie on a mixing line between evaporitic facies and facies that are interpreted as having formed when open marine water circulated through the platform (Fralick & Riding, 2015). The columnar stromatolites represent the open marine end-member component and laterally linked domal stromatolites trend more towards formation during periods of more restricted water circulation (Fralick & Riding, 2015), suggesting that laterally linked domal stromatolites were affected by diffusion limited nutrient availability whereas columnar stromatolites formed during periods of higher energy which probably resulted in a clastic sediment supply. The variability in the stromatolitic forms may have been related to varying current energy.

The *ca* 2·72 Ga Tumbiana Fromation is a lacustrine high energy, near-shore, shallow water, mixed carbonate-siliciclastic depositional environment, in which stromatolite morphology varies as a function of environmental energy (Awramik & Buchheim, 2009). In the highest energy, environment stromatolitic forms are rare consisting of isolated small domes and columns. With decreasing energy, stromatolite abundance increased and morphotypes included large columns, domes (linked clusters) and clusters of branching centimetre-scale columns. The increase in abundance of stromatolites within lower energy environments suggests that mat survival played an important role in the presence or absence of stromatolitic forms, and where microbial mats could propagate out of hydraulically protected areas, they form high energy morphotypes.

In the three stromatolitic carbonate occurrences discussed above, stromatolitic diversity and the survival of microbial mats is dependent on the energy of the environment, and all of these are indicative of mid to high energy environments. In contrast, the Red Lake–Wallace Lake, carbonate platform lacks many of the higher energy stromatolitic forms and stromatolitic diversity found in other Archean carbonate platforms suggesting a lower energy depositional environment. The main stromatolitic morphology, laterally linked domal stromatolites (*stratifera*), necessarily requires limited clastic sedimentation and an environment where diffusion of nutrients is the limiting process in stromatolite development. Whereas *Colleniella* are associated with herring-bone cross-stratification and formed within tidal channels. Because of the limited exposure of outcrops of *colleniella*, it is not possible to determine if these stromatolites began development within a tidal channel facies or were *stratifera* in which the development of stronger current activity destroyed low areas between the individual domes. Therefore, the extent to which *colleniella* can be used as an indicator of the depositional environment of the platform is limited. However, the paucity of higher energy stromatolitic morphologies, such as branching and non-branching columnar stromatolites, and the pervasive appearance of *stratifera* is indicative of sedimentation within a near-shore, low energy environment, possibly a very low slope coast with lagoons, similar in some ways to today's Trucial Coast.

In addition to an absence of high energy stromatolitic morphologies, the Red Lake–Wallace Lake carbonate platform lacks conical stromatolites. These stromatolitic forms have generally been interpreted as deep to shallow subtidal deposits formed on platform slopes (Grotzinger, 1989). Batchelor *et al.* (2004) have shown, through modelling, that the conical form occurs when phototactic microbial growth in stromatolites is much more rapid than mineral accretion. Consequently, conical stromatolites should only form within the photic zone. If the lack of conical stromatolites in the Red Lake–Wallace Lake carbonates is not the result of preferential preservation of other facies, using the reasoning of Batchelor *et al.* (2004), mineral accretion was too great to allow this stromatolitic form to occur. This supports a relatively restricted environment of carbonate precipitation. In addition, the ribbon rock and intraclastic carbonate lithoclasts and marl lithofacies are interpreted as platform proximal slope lithofacies. A lack of conical stromatolites in this environment indicates that these lithofacies formed below the photic zone and/or the platform slope was too unstable to allow the development of microbial mats. This would suggest a steep-sided flat-topped geometry for the platform.

Pseudomorphic crystal fans

Large Archean and Proterozoic sea floor precipitates occurred as crystal fan structures that are ubiquitous in platform carbonates and formed in restricted to open marine settings in upper intertidal to deep subtidal platform margin environments (e.g. Grotzinger, 1986; Sumner & Grotzinger, 1996, 2000, 2004; Grotzinger & James, 2000; Allwood *et al.*, 2007; Riding, 2008; Fralick &

Riding, 2015). Although variations in the types of Precambrian crystal fans can occur in the morphology, size and primary mineralogy, the earliest centimetre-scale crystal fans in Archean carbonate platforms formed in restricted low energy environments (Allwood et al., 2006; Fralick & Riding, 2015). In addition, neither the Chobeni Formation or Tumbiana Formation, which have both been interpreted as high energy environments, has crystal fans (cf. Beukes & Lowe, 1989; Sakurai et al., 2005; Awramik & Buchheim, 2009; Siahi et al., 2016). Although crystal fans in younger Archean carbonate platforms may have formed in open marine, high energy settings (cf. Grotzinger, 1986; Sumner & Grotzinger, 1996, 2000, 2004; Grotzinger & James, 2000), there is no indication that pseudomorph crystal fans formed in high energy and/or open marine environments prior to 2·8 Ga.

Although they are densely packed, the crystal fans of the Red Lake–Wallace Lake carbonate platform form delicate needle-like crystals with no evidence of current resistance. In addition, the similar morphology of the crystal fans found in the shallow to deep areas of the Red Lake–Wallace Lake carbonate platform indicates that the fans were not isolated to a single environment but span the entire peritidal range, and that the height of the fans was possibly controlled by water depth. Extrapolating from environments where similar fans formed at this time, it is likely that this platform wide fan development occurred only during periods where the platform was subject to restricted and evaporitic conditions.

Depositional environments

The lithofacies of the Red Lake–Wallace Lake carbonate platform have been grouped into six lithofacies assemblages in which the sedimentary structures have been related to specific depositional environments (Fig. 16). These depositional environments are supratidal, upper–mid-intertidal, lower intertidal, tidal channel and shallow subtidal, subtidal, and platform slope and distal environments.

Supratidal lithofacies assemblage

The wavy laminae lithofacies is characterized by millimetre-scale wavy and crinkly laminae containing teepee structures, tufa, cement crusts, pseudomorphs after gypsum, and desiccation cracks, features that are typical of supratidal environments in the Proterozoic and Archean (Grotzinger & Reed, 1983; Clough & Goldhammer, 2000; Bekker & Eriksson, 2003; Sumner & Grotzinger, 2004; Allwood et al., 2007). In addition, flat lying laminoid fenestral fabrics associated with wavy laminae were likely produced by desiccation of microbial mats in such an environment. The deformation features of these mats, roll-up structures and irregular folding, are from flooding of the supratidal flat. The presence of soft sediment deformation features indicates that some areas in this environment remained wet while at times relatively dry conditions elsewhere produced teepee structures, gypsum, desiccation cracks and colloform crusts. Wet conditions may have persisted in supratidal ponds as was suggest by Clough & Goldhammer (2000) in the Katakturuk Dolomite or the "wet" sedimentary structures may represent upper intertidal zone environments on a platform with limited topographic relief.

Upper–mid-intertidal lithofacies assemblage

Type 1 fans found in Red Lake–Wallace Lake carbonate platform are identical to crystal pseudomorphs occurring in the tidal flat environment of the Campbelrand-Malmani Platform (Sumner & Grotzinger, 2004) and the subtidal environment in the Cheshire Formation (Sumner & Grotzinger, 2000). Because the height of the fans appears to be controlled by water depth, the height and their association with the wavy laminae lithofacies places the fans in the upper to mid-intertidal flat environment. Type 1 fans are unique only in their association with thin carbonate laminae between the fan layers which may be a transition to wavy laminae lithofacies. In addition to crystal fans, large bioherms of statifera appear to be upwardly transitional to wavy laminae lithofacies. That is, the domal stromatolites lose relief up section to form wavy and irregular laminae characteristic of wavy laminae lithofacies. Although, statifera in the Red Lake–Wallace Lake carbonate platform are interpreted as lower intertidal facies, water depth may control the height of the domes. As the growth of the stromatolites reduces accommodation space, they lose topographic relief, begin to flatten, and form upper intertidal to supratidal lithofacies assemblages. Thus, low relief domes in these areas represent transitions from large statifera to wavy laminae lithofacies in the upper to mid-intertidal flat environment.

Lower intertidal lithofacies assemblage

Although not a definite environmental indicator (Hofmann et al., 2004), other Precambrian LLD stromatolites are interpreted as tidal flat to shallow subtidal depositional structures (Grotzinger, 1986; Clough & Goldhammer, 2000; Sumner & Grotzinger, 2004). Here, the statifera contain laminoid fenestral fabric and birds eye vugs indicative of some desiccation from periodic sub-aerial exposure or formation in an environment that was evaporitic (Logan, 1974; Shinn, 1968; Demicco & Hardie, 1994; Flugel, 2010). In addition, the statifera are associated with intertidal to supratidal lithofacies; type 1 fans and wavy laminae lithofacies. Furthermore, the domes are gradational to upper intertidal/supratidal environments. *Statifera* are also closely associated with tidal channel facies assemblages. Thus, *statifera* are interpreted as forming and were the dominant structure in the lower intertidal to shallow subtidal environment of the Red Lake–Wallace Lake carbonate platform.

Tidal channel and shallow subtidal lithofacies assemblage

Tidal channel lithofacies were identified by the presence of intraclastic mud-chip conglomerate with herring-bone cross-stratification. This lithofacies assemblage also contains some HBC and *colleniella*. As discussed above, the transition from laterally linked to isolated domal stromatolites is controlled by increasing turbulence in the water. Such an interpretation has also been applied to other Precambrian stromatolitic carbonates (e.g. Grotzinger, 1986; Hofmann *et al.*, 2004). Because most sedimentary structures in the Red Lake–Wallace Lake carbonates are indicative of an ambient low energy environment, the formation of *colleniella* is strongly suggestive of a unique microenvironment and most likely necessitates their formation in tidal channels. In addition, in the Red Lake–Wallace Lake carbonates, HBC is only found associated with the tidal channel lithofacies assemblage. If this association is not from preferential preservation in this area, then HBC may have formed in a zone of mixing between more shallow-supratidal/intertidal water and offshore water. This interpretation is supported by the variable trace element geochemistry in HBC (Fig. 12). It should also be noted that *statifera* are found in this assemblage, which may indicate their proliferation into the subtidal environment away from the influence of stronger currents in tidal channels.

Subtidal lithofacies assemblage

The subtidal environment is characterized by structures such as atikokania and *statifera* that require low turbulence to form. As discussed above, the size of crystal fans may be controlled by water depth, and the large size of atikokania that form metre-scale domes likely require that they form part of the subtidal assemblage. Type 2 crystal fans appear to be draped by sediment as they formed. These are not well preserved. However, these fans are like those found in the Campbellrand-Malmani Platform within the subtidal zone (cf. Sumner & Grotzinger, 2004) and are similarly interpreted here. In addition to the crystal fans and *statifera*, large low relief domes are also interpreted as having formed in the subtidal environment

likely stratigraphically below *statifera*. The laminae of the domes have a weak laminoid fenestral fabric with respect to *statifera* and wavy laminae facies. This likely indicates a relatively lower degree of desiccation in an evaporitic environment rather than being periodically sub-aerially exposed and these likely formed in an evaporitic lagoonal setting.

As is noted above, the proliferation of *stratifera* in the Red Lake–Wallace Lake carbonate platform has important implications for the depositional environment of the platform. *Colleniella* developed in microenvironments characterized by tidal channels and are isolated due to the higher turbulent water flow in this environment. In addition, in other Archean carbonates crystal fans are associated with low energy structures (Allwood *et al.*, 2007; Fralick & Riding, 2015). However, truncation surfaces in some stromatolites and roll-up structures in the supratidal zone indicate possibly infrequent storm events.

Platform slope and distal lithofacies assemblages

The platform slope lithofacies assemblage is defined by ribbon rock, disaggregated calcite beds and marl, carbonate associated iron formation, OF-IF, chert and slate. In the ribbon rock facies, the carbonate beds of the carbonate-shale couplets are likely thin bedded, platform derived detrital carbonate overlain by shale derived from sediment fallout, and the irregular bedding associated with these couplets is characteristic of syn-depositional folding by creep in an upper slope, peri-platform environment (Mcllearth & James, 1978). The intraclastic carbonate lithoclasts and marl contain fragments representing mixtures of the constituents of the ribbon rock lithofacies and these lithofacies are closely related. Eberli (1988) found the frequency of slope failure was increased and the angle of repose needed to produce slope failure decreased when peri-platform oozes hosted platform derived detrital carbonate as opposed to uniform sediments. Similarly, the density contrasts between different beds in the ribbon rock likely resulted in more prolific slope failure and slumping on the upper platform slope. Thus, intraclastic carbonate lithoclasts and marl lithofacies are likely slump-related structures on the upper platform slope.

The slump structures are found overlying slate-chert beds and carbonate-associated iron formation. These latter lithofacies are considered to have formed on the platform slope below ribbon rock facies. The carbonate beds are interpreted as being platform derived, chert and shale pelagic in origin, and iron formation produced from oxyhydroxide precipitation at a weak oxidative chemocline between the platform and more distal environments. The absence of slumping, carbonate laminae and turbidity current deposits in laminated chert and chert-oxide iron formation indicates that these lithofacies formed in deeper water than the carbonate-associated iron formation and considered a platform distal lithofacies.

Geochemistry

As the geochemistry of chemical sediments, and in particular their REE contents, can reflect the water from which they precipitated (Barrett et al., 1988; Derry & Jacobsen, 1990; Murray et al., 1992; Danielson et al., 1992; Alibert & McCulloch, 1993; Bau, 1993; Bau & Moller, 1993; Bau & Dulski, 1996; Kamber & Webb, 2001; Kato & Nakamura, 2003; Van Kranendonk et al., 2003; Bolhar et al., 2004; Klein, 2005; Planavsky et al., 2010; Bekker et al., 2010), they are key to understanding the ambient water masses associated with various sea floor precipitates and cements. The geochemistry also provides insight into the later alteration history of the sediment through the concentrations and isotopic ratios of elements more prone to exchange with the surrounding fluid, such as Sr, Ba and O.

Compared to limestones of the 2·8 Ga Steep Rock carbonate platform, which have not been significantly altered (Veizer et al., 1982; Fralick & Riding, 2015), the Sr and Ba concentrations in the Red Lake dolomites are low (Table 1). The $^{87}Sr/^{86}Sr$ ratios for the Red Lake dolomite samples are, on average, 0·0014‰ higher than those for the calcite samples and $\delta^{18}O$ values are significantly lower for the dolomite samples than the calcite samples (Fig. 9). This indicates that Sr, Ba and O were exchanged during alteration and the higher Sr isotopic ratios requires an alteration fluid that only affected the dolomitized samples and was derived, at least in part, from a cratonic source. The dolomitization itself is the prime candidate for this. However, there is no significant difference between the $\delta^{13}C$ values for the dolomite and calcite samples (Fig. 9), indicating that the dolomitization did not alter these.

Three of the samples are removed from the main grouping of $\delta^{13}C$ values (Fig. 9). These samples, the carbonate associated with iron formation and one of the slump deposit samples, formed in deeper water. Although data from previous studies have estimated the carbon isotopic composition of Archean sea water to have $\delta^{13}C$ values of +1·5 ± 1·5‰ (Veizer et al., 1989b and references therein), there is growing evidence that deeper water carbonates associated with other Archean carbonate platforms commonly have lower $\delta^{13}C$ values, such as those from the Hamersley (Becker & Clayton, 1972; Kaufman et al., 1990), Transvaal Supergroup (Beukes et al., 1990; Schneiderhan et al., 2006; Fischer et al., 2009) and Steep Rock (Fralick & Riding, 2015). Kaufman et al. (1990) believed that deep Archean sea water had a $\delta^{13}C$ value of ca −5‰ and diagenetic reactions between organic carbon

and iron hydroxide had further lowered it in their samples. A lower $\delta^{13}C$ value for deeper sea water was also suggested by Fralick & Riding (2015) who obtained −5·5‰ for a sharp sided siderite layer with no associated iron oxides. However, Fischer et al. (2009) and Heimann et al. (2010) have interpreted the lower values of the siderite in the Kuruman Formation overlying the Campbellrand carbonate platform as due to syn-depositional precipitation of siderite by microbial dissimilatory iron reduction of OF-IF. The carbonates of the Red Lake–Wallace Lake platform can offer insight into the nature of the cause of the variation in $\delta^{13}C$ values. The carbonate-associated oxide facies IF reflects the transitional facies between shallow carbonate deposits and distal chert-oxide facies IF. These contain the most negative $\delta^{13}C$ values of all the carbonate samples, −3·1‰ and −1·3‰. One of the slump samples, which was also deposited in deeper water, has the next lowest $\delta^{13}C$ value of −1·0‰. Because the carbonate in the carbonate-associated OF-IF is calcite and the carbonate in the slump is ferroan dolomite, the more negative values of these carbonates relative to the peritidal carbonates cannot be attributed to the microbial pathways described by Fischer et al. (2009) and Heimann et al. (2010) for siderite precipitation. It is more likely that the decreasing values for $\delta^{13}C$ going from the Red Lake–Wallace Lake carbonate shelf downslope to the slump deposits and the iron formation-associated carbonate indicates a gradient in the carbon isotopic ratio with depth.

There is a change in $\delta^{13}C$ values between the shallow water carbonates in the Mesoarchean and the Neoarchean; the ca 3·0 Ga Pongola carbonates have shallow water $\delta^{13}C$ values of 2 ± 1‰ (Veizer et al., 1990) and the 2·8 Ga Steep Rock platform has shallow water carbonates with $\delta^{13}C$ values of 1·6 ± 0·7‰ (Fralick & Riding, 2015), whereas the 2·55 Ga Campbellrand and Gamohaan have shallow water $\delta^{13}C$ values averaging −0·5 and −1·0‰ (Fischer et al., 2009; Heimann et al., 2010), respectively. The heavy carbon enrichment in the Mesoarchean shallow carbonate deposits is also reflected in $\delta^{13}C$ values for the shallow water carbonates of the Red Lake–Wallace Lake platform, which are 0·5 ± 0·4‰.

The REEs have been used extensively to infer attributes of the solutions from which chemical sediments have precipitated. Variations in Y/Ho ratios and PAAS normalized La, Ce, Eu and Gd anomalies, plus the slope of the pattern, all provide clues to fluid composition. The Y/Ho weight ratio in crustal rocks of about 26 (Kamber & Webb, 2001) is maintained as the two elements behave very similarly in this system (Nozaki et al., 1997; Bau, 1999). However, in aqueous solution Y and Ho behave differently. Dependent on the availability of particulate matter (Nozaki et al., 1997) and solution-surface complexation behaviour (Bau, 1999; Luo & Byrne, 2004; Quinn et al., 2004, 2006) the Y/Ho weight ratio of modern sea water varies between 44 and 120 (cf. Nozaki et al., 1997; Kamber & Webb, 2001). Similarly, the distribution of light REEs and heavy REEs depends on solution chemistry and the availability of particulate matter (Lee & Byrne, 1992; Bau, 1999; Luo & Byrne, 2004; Quinn et al., 2004, 2006). In solutions dominated by dissolved carbonate, REE carbonate complexation is dependent on pH and $[CO_3^{2-}]_t$ (total dissolved carbonate) (Luo & Byrne, 2004; Quinn et al., 2004, 2006). At lower pH, trivalent metal ions are more soluble and the formation of H_2CO_3 likely limits activation sites for complexation, resulting in REE carbonate complexation being relatively unselective at low pH and $[CO_3^{2-}]_t$ concentration (Quinn et al., 2004, 2006). With increasing pH and $[CO_3^{2-}]_t$ heavy REEs and Y preferentially form solution complexes with carbonate ions, but the degree of carbonate complexation of heavy REEs relative to Y is increased, making it more likely that Y will come out of solution rather than Ho possibly resulting in super-chonderitic Y/Ho ratios in precipitates. Also, solution complexation constants decrease from Nd to La in a curved path that is exponential with increasing pH and/or $[CO_2^{2-}]_t$ and the ability of Gd to form surface complexes, compared to its neighbours, decreases as well (Ohta & Kawabe, 2001; Quinn et al., 2004, 2006).

Surface complexation of REEs with precipitating iron hydroxides is also dependent on pH (Bau, 1999; Quinn et al., 2004, 2006) and $[CO_2^{2-}]_t$ (Sholkovitz et al., 1994; Quinn et al., 2004, 2006). Bau (1999) found that middle REEs were enriched in the hydroxides/oxides and Y was lower than Ho, also leading to higher Y/Ho ratios in the fluid.

Particulate organic matter is also important in REE scavenging. Lee & Byrne (1992) performed a study of REE complexation with organic ligands and found that heavy REEs formed surface complexes with organic surface ligands to a greater degree than light REEs. They also discovered that Gd complexed with organic surface ligands to a lesser degree than its neighbouring REEs. Similarly, Nozaki et al. (1997) found that Y behaves in a similar manner to light REEs in having low stability with organic surface complexes. However, Sholkovitz et al. (1994) discovered that precipitating oxides on particulate matter preferentially removed light REEs from solution.

Neoarchean and Palaeoproterozoic carbonates tend to be dominated by super-chondritic Y/Ho ratios, Pr/Yb_{PAAS} lower than or close to unity, positive Eu, La, and less commonly Gd anomalies, and no Ce anomalies (cf. Kamber & Webb, 2001; Bolhar et al., 2004, 2005; Kamber et al., 2004; Allwood et al., 2010; Planavsky et al., 2010). Neoarchean iron formations have similar trends, although commonly with lower Y/Ho ratios, Pr/Yb_{PAAS} <1, and Gd

and La anomalies are subdued to absent (cf. Bau & Dulski, 1996; Bolhar et al., 2004, 2005; Planavsky et al., 2010; Thurston et al., 2012). The Red Lake–Wallace Lake Carbonate Platform REE patterns for both iron formation and carbonates are similar to those in Neoarchean and Palaeoproterozoic deposits, with super-chondritic Y/Ho, Yb/Pr_{PAAS} close to unity to less than one, positive La (except the iron formation) and Eu anomalies, variable Gd anomalies and no Ce anomalies, except in one sample (Figs 11, 12, 13 and 14). The carbonate-associated iron formation samples are somewhat different with Gd anomalies of the same magnitude as their less pronounced Eu anomalies and flatter REE patterns.

From the previous discussion, it can be concluded that the enhanced ability of heavy REEs to form carbonate complexes would leave more free light REEs in solution and available for adsorption onto particulate matter (Bolhar et al., 2004; and references therein). This would have led to a depletion of the light REEs in Archean sea water, similar to today. The positive La and Gd anomalies in the carbonate samples may be due to their lower ability to complex with carbonate in solution and particulate matter, respectively. The super-chondritic Y/Ho ratios could be either the result of differences in the two element's ability to complex with carbonate in solution or differences in scavenging processes by particulate matter resulting in preferential removal of Ho upon delivery to the ocean (Zhang et al., 1994; Nozaki et al., 1997). Although the slope of the REE curves is less for the Red Lake–Wallace Lake chemical sediments, these trends are similar to average modern sea water.

The Red Lake–Wallace Lake samples have pronounced positive Eu anomalies and no negative Ce anomalies, except for one sample (Fig. 17). Positive Eu anomalies are common in Archean and early Palaeoproterozoic chemical sediments. Europium is the only REE that can exist in the $^{+2}$ state, and as such can substitute for Ca and Sr in minerals such as plagioclase. The REEs were leached from the ocean crust during hydrothermal circulation and these fluids developed a positive Eu anomaly (Klinkhammer et al., 1983, 1994; Derry & Jacobsen, 1990; Danielson et al., 1992). The REE contribution from this hydrothermal circulation dominated over REEs delivered in continental runoff during the Archean and, thus, the oceans at this time had excess Eu compared to other REEs. The lack of a negative Ce anomaly (Fig. 17) is the result of extremely low concentrations of oxygen present at this time. Cerium can exist in the Ce^{+4} state and when oxidized forms cerianite or is adsorbed onto iron hydroxides removing it from solution. On the modern Earth, oxidized soils have positive Ce anomalies and groundwater commonly develops negative Ce anomalies, leading to the ocean having a pronounced negative anomaly. In the

Archean, without common oxidizing agents Ce remained in the Ce $^{+3}$ state and behaved similarly to the other REEs. However, one sample has a negative Ce anomaly, an elevated Y/Ho ratio, and a higher positive slope than the others.

The sample of laminated calcite lying directly under an atikokania mound has a REE pattern very similar to patterns from crystal fan and cuspate fenestrate fabric samples from the 2·8 Ga Steep Rock carbonate platform (Fig. 14B, C and D). It is possible that the mounds provided a localized restricted environment where oxygen, produced by prolific stromatolite development, could build up without being overwhelmed by tidal flushing of relatively anoxic ocean water. This would have allowed a redox boundary to exist at the edge of this locally restricted area, Ce to preferentially be deposited there, and the more central area develops a negative Ce anomaly. One of the atikokania samples directly overlying the laminated carbonate has a positive Ce anomaly, further reinforcing this scenario. It is noteworthy that this sample and those deposited in areas of restricted circulation on the Steep Rock platform (Fralick & Riding, 2015) have steeper REE slopes than those deposited in less restricted settings. This may be caused by photosynthetic activity resulting in an increase in pH, which would then drive a draw-down of the HCO_3^- with delivery of HREEs complexed with it. The sample of laminated carbonate from directly under an atikokania mound is also enriched in iron oxides, which would have also delivered HREEs.

The sample discussed above is in contrast to the REE patterns for other carbonate samples from the Red Lake–Wallace Lake Carbonate Platform (Figs 12, 13A and B, and 14A), which have similar slopes to columnar stromatolite samples from the Steep Rock Platform (Fig. 13D). Of the carbonate samples, the stromatolite samples have the lowest slopes (Pr/Yb$_{(SN)}$ average 0·7), and the chert-OF-IF samples, which should reflect the basin water chemistry, have the greatest slopes (Pr/Yb$_{(SN)}$ average 0·2) (Fig. 15). The siliciclastic samples also have low slopes and it is possible that the low slopes of some of the chemical sediments are caused by siliciclastic contamination. This is probably applicable to the two samples of magnetite from the carbonate-associated iron formation, which in addition to flattened patterns, have low Y/Ho ratios and elevated concentrations of titanium. One of the samples of carbonate from the carbonate-associated iron formation and a stromatolite sample also have suspiciously flat slopes similar to the siliciclastics. However, the array of points, representing the carbonate samples, which stretch between the siliciclastic and iron formation end members on Fig. 15 likely is not caused by siliciclastic contamination as the samples have very low contents

of Al_2O_3 and TiO_2. This leaves the possibility that they are forming a continuum between oceanic sourced water and continentally sourced water. The majority of points are closer to the oceanic field, but even with this for continentally derived water to have an influence on at least a somewhat arid, tidally influenced coastline, current activity must have been sluggish. A setting similar to the present-day Trucial coast would be appropriate, although with somewhat increased runoff from the land.

CONCLUSIONS

The lithofacies assemblages of the Red Lake–Wallace Lake carbonate platform encompass structures that characterize supratidal flat to platform distal environments. Most of the preserved platform carbonates are peritidal in origin and formed in a low energy environment (possibly related to a low slope coastline with lagoon development) by a combination of the build-up of microbial mats within related biomediated sedimentary structures and *in situ* sea floor precipitation of sparry crusts. The low slope peritidal area would have had a variable topography produced by differential growth rates in biomediated structures which possibly allowed localized restricted conditions in which oxygen levels could build up. Dissolution features are present in some areas indicating periodic sub-aerial exposure of portions of the platform. The transition from subtidal to upper slope environment is not preserved. The upper platform slope is characterized by a combination of detrital platform derived carbonate and pelagic sedimentation that often resulted in slumping. The platform slope proper is similar to the upper slope but has an increase in pelagic sediment in combination with the introduction of OF-IF. The distal environment is limited to the deposition of chert-OF-IF. The pervasiveness of peritidal lithofacies assemblages and slumping in the slope environment indicates that the platform was flat-topped and steeply sided.

The sediments of the Red Lake–Wallace Lake carbonate platform precipitated from oceanic water that had relative REE concentrations similar to today's oceans (i.e. supra-condritic and variable Y/Ho ratios, PAAS normalized REE patterns that are positively sloping, and positive La and Gd anomalies). The notable exceptions include (1) the Mesoarchean ocean had a larger REE input from hydrothermal alteration of oceanic crust than today's, and (2) the lack of Ce anomalies in the chemical sediments indicates that the oceans were largely anoxic. In addition, the water from which the platform precipitated was chemically stratified; the deep water precipitates had a large chemical input from hydrothermal alteration of oceanic basalts while the shallow water precipitates formed from water with mixed deeper water signatures and chemical input from continental runoff. A significant finding, and in combination with the large areal extent of the platform, suggests that these carbonates precipitated on the margins of an extensive oceanic platform.

Dolomitization of many samples has reset O and Sr isotopic ratios but left C isotopic ratios largely unchanged from those in the precursor carbonate phase. In calcite samples, Sr and C isotopic ratios reflect this precursor phase and $^{87}Sr/^{86}Sr$ ratios are in agreement with estimates for Archean sea water at 2·9 Ga. The $\delta^{13}C$ values of the platform slope carbonates are light with respect to those of the peritidal areas, in agreement with Kaufman et al. (1990) and Fralick & Riding (2015) in that this suggests deep Archean ocean water was isotopically light relative to shallow water and further suggesting a chemically stratified water body.

The intermittent development and preservation of stromatolitic carbonates in the Archean from the earliest ca 3450 Myr old Strelley Pool Chert to the latest ca 2540 Myr old Campbellrand-Malmani carbonate platform spans almost 1 billion years of Earth's history. Despite extensive study of these carbonates, and those in the interim, we do not know details of important facts concerning Earth's surface processes during the Archean; the nature of the atmosphere, chemical and physical dynamics of the oceans, and how life fit into these systems are still largely mysteries. The sedimentary rock record is too sporadic and, in places, altered to allow high resolution of Earth's physical and chemical processes during this time interval, yet it is clear, and not unexpected, that such processes changed with frequencies on scales that were similar to those observed in the Phanerozoic rock record. Here, we document the physical and chemical character of a ca 2930 Myr old carbonate platform, the oldest yet discovered, and of which life formed an integral part, in hopes to fill some of the missing gaps.

ACKNOWLEDGEMENTS

We are grateful for NSERC's Discovery Grant Program for supporting this research and NSERC's Undergraduate Student Research Awards (USRA) Program. Comments by the associate editor Adrian Immenhauser and an anonymous reviewer helped improve the manuscript.

References

Abell, P.I., McClory, J., Martin, A. and Nisbet, E.G. (1985a) Archaean stromatolites from the Ngesi Group, Belingwe greenstone belt, Zimbabwe; preservation and stable isotopes – preliminary results. *Precambrian Res.*, 27, 357–383.

Abell, P.I., McClory, J., Martin, A., Nisbet, E.G. and Kyser, T.K. (1985b) Petrography and stable isotope ratios from Archaean stromatolites, Mushandike Formation, Zimbabwe. *Precambrian Res.*, 27, 385–398.

Alibert, C. and McCulloch, M.T. (1993) Rare earth element and neodymium isotopic compositions of the banded iron-formations and associated shales from Hamersley, Western Australia. *Geochim. Cosmochim. Acta*, 57, 187–204.

Allwood, A.C., Walter, M.R., Kamber, B.S., Craig, P.M.A. and Burch, I.W. (2006) Stromatolite reef from the Early Archaean era of Australia. *Nature*, 44, 714–718.

Allwood, A., Walter, M.R., Burch, I.W. and Kamber, B.S. (2007) 3.43 billion year old stromatolite reef from the Pilbara Craton of Western Australia; ecosystem scale insights to early life on Earth. *Precambrian Res.*, 158, 198–227.

Allwood, A.C., Kamber, B.S., Walter, M.R., Burch, I.W. and Kanik, I. (2010) Trace elements record depositional history of an Early Archean stromatolitic carbonate platform. *Chem. Geol.*, 270, 148–163.

Altermann, W. and Nelson, D.R. (1998) Sedimentation rates, basin analysis and regional correlations of three Neoarchaean and Palaeoproterozoic sub-basins of the Kaapvaal craton as inferred from precise U-Pb zircon ages from volcaniclastic sediments. *Sed. Geol.*, 120, 225–256.

Aubrecht, R., Schloegl, J., Krobicki, M., Wierzbowski, H., Matyja, B.A. and Wierzbowski, A. (2009) Middle Jurassic stromatactis mud-mounds in the Pieniny Klippen Belt (Carpathians); a possible clue to the origin of stromatactis. *Sed. Geol.*, 213, 97–112.

Awramik, S.M. and Buchheim, H.P. (2009) A giant, Late Archean lake system: the Meentheena Member (Tumbiana Formation; Fortescue Group), Western Australia. *Precambrian Res.*, 174, 215–240.

Awramik, S.M. and Grey, K. (2005) Stromatolites: biogenicity, biosignatures, and bioconfusion. *Proc. SPIE*, 5906, 59060P.

Barrett, T.J., Fralick, P.W. and Jarvis, I. (1988) Rare-earth-element geochemistry of some Archean iron formations north of Lake Superior, Ontario. *Can. J. Earth Sci.*, 25, 570–580.

Bartley, J., Knoll, A. H., Grotzinger, J. P. and Sergeev, V.N. (2000). Lithification and fabric genesis in precipitated stromatolites and associated peritidal carbonates, Mesoproterozoic Billyakah Group, Siberia. In: *Carbonate Sedimentation and Diagenesis in the Evolving Precambrian World* (Eds J.P. Grotzinger and N.P. James), *SEPM Spec. Publ.*, 33, 107–120.

Batchelor, M.T., Burne, R.V., Henry, B.I. and Jackson, M.J. (2004) A case for biotic morphogenesis of coniform stromatolites. *Phys. A*, 337, 319–326.

Bau, M. (1993) Effects of syn- and post-depositional processes on the rare-earth element distribution in Precambrian iron-formations. *Eur. J. Mineral.*, 5, 257–267.

Bau, M. (1999) Scavenging of dissolved yttrium and rare earths by precipitating iron oxyhydroxide: experimental evidence for Ce oxidation, Y-Ho fractionation, and lanthanide tetrad effect. *Geochim. Cosmochim. Acta*, **63**, 67–77.

Bau, M. and **Dulski, P.** (1996) Distribution of yttrium and rare-earth elements in the Penge and Kuruman iron-formations, Transvaal Supergroup, South Africa. *Precambrian Res.*, **79**, 37–55.

Bau, M. and **Moller, P.** (1993) Rare earth element systematics of the chemically precipitated component in Early Precambrian iron formations and the evolution of the terrestrial atmosphere-hydrosphere-lithosphere system. *Geochim. Cosmochim. Acta*, **57**, 2239–2249.

Becker, R.H. and **Clayton, R.N.** (1972) Carbon isotopic evidence for the origin of a banded iron-formation in Western Australia. *Geochim. Cosmochim. Acta*, **36**, 577–595.

Bekker, A. and **Erikkson, K.A.** (2003) A Paleoproterozoic drowned carbonate platform on the southeastern margin of the Wyoming Craton; a record of the Kenorland breakup. *Precambrian Res.*, **120**, 327–364.

Bekker, A., Slack, J.F., Planavsky, N., Krapez, B., Hofmann, A., Konhauser, K.O. and **Rouxel, O.J.** (2010) Iron formation: the sedimentary product of a complex interplay among mantle, tectonic, oceanic and biospheric processes. *Econ. Geol.*, **105**, 467–508.

Heimann, A., Johnson, C.M., Beard, B.L., Valley, J.W., Roden, E.E., Spicuzza, M.J. and **Beukes, N.J.** (2010) Fe, C, and O isotope compositions of banded iron formation carbonates demonstrate a major role for dissimilatory iron reduction in ~2.5 Ga marine environments. *Planet. Sci. Lett.*, **294**, 8–18.

Beukes, N.J. and **Lowe, D.R.** (1989) Environmental control on diverse stromatolite morphologies in the 3000 Ma Pongola Supergroup, South Africa. *Sedimentology*, **36**, 383–397.

Beukes, N.J., Klein, C., Kaufman, A.J. and **Hayes, M.** (1990) Carbonate petrology, kerogen distribution, and carbon and oxygen isotope variations in an early Proterozoic transition from limestone to iron-formation deposition, Transvaal Supergroup, South Africa. *Econ. Geol.*, **85**, 663–690.

Bolhar, R. and **van Kranendonk, M.J.** (2007) A non-marine depositional setting for the northern Fortescue Group, Pilbara Craton, inferred from trace element geochemistry of stromatolitic carbonates. *Precambrian Res.*, **155**, 229–250.

Bolhar, R., Kamber, B.S., Moorbath, S., Fedo, C.M. and **Whitehouse, M.J.** (2004) Characterisation of early Archaean chemical sediments by trace element signatures. *Earth Planet. Sci. Lett.*, **222**, 43–60.

Bolhar, R., van Kranendonk, M.J. and **Kamber, B.S.** (2005) A trace element study of siderite-jasper banded iron formation in the 3.45 Ga Warrawoona Group, Pilbara Craton – Formation from hydrothermal fluids and shallow seawater. *Precambrian Res.*, **137**, 93–114.

Clough, J. and **Goldhammer, R.K.** (2000) Evolution of the Neoproterozoic Katakturuk dolomite ramp complex, northeastern Brooks Range, Alaska. In: *Carbonate Sedimentation and Diagenesis in the Evolving Precambrian World* (Eds J.P. Grotzinger and N.P. James), *SEPM Spec. Publ.*, **67**, 209–241.

Corfu, F. and **Wallace, H.** (1986) U-Pb zircon ages for magmatism in the Red Lake greenstone belt, northwestern Ontario. *Can. J. Earth Sci.*, **23**, 27–42.

Czaja, A.D., Johnson, C.M., Beard, B.L., Eigenbrode, J.L., Freeman, K.H. and **Yamaguchi, K.E.** (2010) Iron and carbon isotope evidence for ecosystem and environmental diversity in the ~2.7 to 2.5 Ga Hamersley Province, Western Australia. *Earth Planet. Sci. Lett.*, **292**, 170–180.

Danielson, A., Moller, P. and **Dulski, P.** (1992) The europium anomalies in banded iron formation and the thermal history of the oceanic-crust. *Chem. Geol.*, **97**, 89–100.

Davis, D.W. (1994) *Report on the Geochronology of Rocks from the Rice Lake Belt, Manitoba: Toronto, Ontario, Canada.* Royal Ontario Museum Report, 11 pp.

Demicco, R.V. and **Hardie, L.A.** (1994) Sedimentary structures and early diagenetic features of shallow marine carbonate deposits. *SEPM Atlas Ser.*, **1**, 265.

Derry, L.A. and **Jacobsen, S.B.** (1990) The chemical evolution of Precambrian seawater – evidence from REEs in banded iron formations. *Geochim. Cosmochim. Acta*, **54**, 2965–2977.

Dupraz, C., Pattisina, R. and **Verrecchia, E.P.** (2006) Translation of energy into morphology: simulation of stromatolite morphospace using a stochastic model. *Sed. Geol.*, **185**, 185–203.

Eberli, G.P. (1988) Physical properties of carbonate turbidite sequences surrounding the Bahamas—implications for slope stability and fluid movements. In: *Proc. ODP Sci. Results Bahamas* (Eds J.A. Austin Jr. and W. Schlager), **11**, 305–314.

Eigenbrode, J.L. and **Freeman, K.H.** (2006) Late Archean rise of aerobic microbial ecosystems. *Proc. Natl Acad. Sci. USA*, **103**, 15759–15764.

Elderfield, H., Upstillgoddard, R. and **Sholkovitz, E.R.** (1990) The rare earth elements in rivers, estuaries and costal seas and their significance to the composition of ocean waters. *Geochim. Cosmochim. Acta*, **54**, 971–991.

Fischer, W.W., Schroeder, S., Lacassie, J.P., Beukes, N.J., Goldberg, T., Strauss, H., Horstmann, U.E., Schrag, D.P. and **Knoll, A.H.** (2009) Isotopic constraints on the Late Archean carbon cycle from the Transvaal Supergroup along the western margin of the Kaapvaal Craton, South Africa. *Precambrian Res.*, **169**, 15–27.

Flugel, E. (2010) *Microfacies of Carbonate Rocks.* Springer-Verlag, Berlin, 976 pp.

Fralick, P.W. and **Riding, R.** (2015) Steep Rock Lake: sedimentology and geochemistry of an Archean carbonate platform. *Earth Sci. Rev.*, **151**, 132–175.

Fralick, P.W., Barrett, T.J., Jarvis, K.E., Jarvis, I., Schnieders, B.R. and **van de Kemp, R.** (1989) Sulfide facies iron formation at the Archean Morley occurrence, northwestern Ontario: contrasts with ocean hydrothermal deposits. *Can. Mineral.*, **27**, 601–616.

Fralick, P., Hollings, P. and King, D. (2008) Stratigraphy, geochemistry, and depositional environments of Mesoarchean sedimentary units in western Superior Province; implications for generation of early crust. *Geol. Soc. Am. Spec. Pap.*, **440**, 77–96.

Godderis, Y. and Veizer, J. (2000) Tectonic control of chemical and isotopic composition of ancient oceans: the impact of continental growth. *Am. J. Sci.*, **300**, 434–461.

Grotzinger, J.P. (1986) Evolution of early Proterozoic passive-margin carbonate platform, Rocknest Formation, Wopmay Orogen, Northwest Territories, Canada. *J. Sed. Petrol.*, **56**, 831–847.

Grotzinger, J.P. (1989) Facies and evolution of Precambrian carbonate depositional systems: emergence of the modern platform archetype. In: *Controls on Carbonate Platform and Basin Development*, (Eds J.L. Wilson, F. Sarg and J.F. Read), *SEPM Spec. Publ.*, **44**, 79–106.

Grotzinger, J.P. and James, N.P. (2000) Carbonate sedimentation and diagenesis in the evolving precambrian world. *SEPM Spec. Publ.*, **67**, 123–144.

Grotzinger, J.P. and Reed, J.F. (1983) Evidence for primary aragonite precipitation, Lower Proterozoic (1.9 Ga) Rockport dolomite, Wopmay orogen, northwest Canada. *Geology*, **11**, 710–713.

Hannigan, R.E. and Sholkovitz, E.R. (2001) The development of middle rare earth element enrichments in freshwaters: weathering of phosphate minerals. *Chem. Geol.*, **175**, 495–508.

Hoffman, P. and McDonald, F.A. (2010) Sheet-crack cements and early regression in Marinoan (635 Ma) cap dolostones: regional benchmarks of vanishing ice-sheets?. *Earth Planet. Sci. Lett.*, **300**, 374–384.

Hofmann, H.J., Thurston, P.C. and Wallace, H. (1985) Archean stromatolites from Uchi greenstone belt, northwestern Ontario. In: *Evolution of Archean Supracrustal Sequences* (Eds L.D. Ayres, P.C. Thurston, K.D. Card and W. Weber), *Geol. Ass. Can. Spec. Pap.*, **28**, 125–132.

Hofmann, A., Dirks, P. and Jelsma, H.A. (2004) Shallowing-upward carbonate cycles in the Belingwe greenstone belt, Zimbabwe; a record of Archean sea-level oscillations. *J. Sed. Res.*, **74**, 64–81.

Hollings, P., Wyman, D. and Kerrich, R. (1999) Komatiite-basalt-rhyolite volcanic associations in northern Superior Province greenstone belts; significance of plume-arc interaction in the generation of the proto continental Superior Province. *Lithos*, **46**, 137–161.

Jacobsen, S.B. and Kaufman, A.J. (1994) Modeling the seawater Sr isotopic record. *EOS Trans. Am. Geophys. Union*, **75**, 140.

Jiang, G., Kennedy, M.J., Christie-Blick, N., Wu, H. and Zhang, S. (2006) Stratigraphy, sedimentary structures, and textures of the late Neoproterozoic Doushantuo cap carbonate in south China. *J. Sed. Res.*, **76**, 978–995.

Kamber, B.S. and Webb, G.E. (2001) The geochemistry of late Archaean microbial carbonate: implications for ocean chemistry and continental erosion history. *Geochim. Cosmochim. Acta*, **65**, 2509–2525.

Kamber, B.S., Bolhar, R. and Webb, G.E. (2004) Geochemistry of late Archaean stromatolites from Zimbabwe: evidence for microbial life in restricted epicontinental seas. *Precambrian Res.*, **132**, 379–399.

Kato, Y. and Nakamura, K. (2003) Origin and global tectonic significance of Early Archean cherts from the Marble Bar greenstone belt, Pilbara Craton, Western Australia. *Precambrian Res.*, **125**, 191–243.

Kaufman, A.J., Hayes, J.M. and Klein, C. (1990) Primary and diagenetic controls of isotopic compositions of iron-formation carbonates. *Geochim. Cosmochim. Acta*, **54**, 3461–3473.

Kendall, A.C. (1985) Radiaxial fibrous calcite: a reappraisal. In: *Carbonate Cements* (Eds N. Schneidermann and P.M. Harris), *SEPM Spec. Publ.*, **36**, 59–77.

Kendall, A.C. and Tucker, M.E. (1973) Radiaxial fibrous calcite; a replacement after acicular carbonate. *Sedimentology*, **20**, 365–389.

Kennedy, M., Christie-Blick, N. and Sohl, L.E. (2001) Are Proterozoic cap carbonates and isotopic excursions a record of gas hydrate destabilization following Earth's coldest intervals? *Geology*, **29**, 443–446.

Klein, C. (2005) Some Precambrian banded iron-formations (BIFs) from around the world: their age, geologic setting, mineralogy, metamorphism, geochemistry, and origins. *Am. Min.*, **90**, 1473–1499.

Klinkhammer, G., Elderfield, H. and Hudson, A. (1983) Rare earth elements in seawater near hydrothermal vents. *Nature*, **305**, 185–188.

Klinkhammer, G.P., Elderfield, H., Edmond, J.M. and Mitra, A. (1994) Geochemical implications of rare earth element patterns in hydrothermal fluids from mid-ocean ridges. *Geochim. Cosmochim. Acta*, **58**, 5105–5113.

Knoll, A.H. and Beukes, N.J. (2009) Introduction: initial investigations of a Neoarchean shelf margin-basin transition (Transvaal Supergroup, South Africa). *Precambrian Res.*, **169**, 1–14.

Kretz, R. (1982) A model for the distribution of trace elements between calcite and dolomite. *Geochim. Cosmochim. Acta*, **46**, 1979–1981.

Larralde, R., Robertson, M. and Miller, S. (1995) Rates of decomposition of ribose and other sugars: implications for chemical evolution. *Proc. Natl Acad. Sci. USA*, **92**, 8158–8160.

Lee, J.H. and Byrne, R.H. (1992) Examination of comparative rare earth element complexation behavior using linear free-energy relationships. *Geochim. Cosmochim. Acta*, **56**, 1127–1137.

Logan, B. (1974) Inventory of diagenesis in holocene-recent carbonate sediments, Shark Bay, Western Australia. *Am. Ass. Petrol. Geol. Mem.*, **22**, 195–248.

Luo, Y. and Byrne, R.H. (2004) Carbonate complexation of yttrium and the rare earth elements in natural waters. *Geochim. Cosmochim. Acta*, **69**, 691–699.

McIntyre, W.L. (1963) Trace element partition coefficients – a review of theory and applications to geology. *Geochim. Cosmochim. Acta*, **27**, 1209–1264.

Mcllearth, I.A. and James, N.P. (1978) Facies models, 13; Carbonate slopes. *Geosci. Canada*, **5**, 189–199.

Murray, R.W., Tenbrink, M.R.B., Gerlach, D.C., Russ, G.P. and Jones, D.L. (1992) Interoceanic variation in the rare earth, major, and trace-element depositional chemistry of chert – perspectives gained from the DSDP and ODP record. *Geochim. Cosmochim. Acta*, **56**, 1897–1913.

Nothdurft, L.D., Webb, G.E. and Kamber, B.S. (2004) Rare earth element geochemistry of Late Devonian reefal carbonates, Canning Basin, Western Australia: confirmation of a seawater REE proxy in ancient limestones. *Geochim. Cosmochim. Acta*, **68**, 263–283.

Nozaki, Y., Zhang, J. and Amakawa, H. (1997) The fractionation between Y and Ho in the marine environment. *Earth Planet. Sci. Lett.*, **148**, 329–340.

Ohta, A. and Kawabe, I. (2001) REE (III) adsorbtion onto Mn dioxide and Fe oxyoydroxide: Ce (III) oxidation by Mn dioxide. *Geochim. Cosmochim. Acta*, **65**, 695–703.

Ono, S., Kaufman, A.J., Farquhar, J., Sumner, D.Y. and Beukes, N. (2009) Lithofacies control on multiple-sulfur isotope records and Neoarchean sulfur cycles. *Precambrian Res.*, **169**, 58–67.

Peter, J. (2003) Ancient iron formations: their genesis and use in the exploration for strataform base metal sulfide deposits, with examples from the Bathurst Mining Camo. In, ed. D.R. Lentz, Geochemistry of Sediments and Sedimentary Rocks. *Geol. Ass. Canada Geotext*, **4**, 192–207.

Planavsky, N., Bekker, A., Rouxel, O.J., Kamber, B., Hofmann, A., Knudsen, A. and Lyons, T.W. (2010) Rare Earth Element and yttrium compositions of Archean and Paleoproterozoic Fe formations revisited: new perspectives on the significance and mechanisms of deposition. *Geochim. Cosmochim. Acta*, **74**, 6387–6405.

Quinn, K., Byrne, R.H. and Schijf, J. (2004) Comparative scavenging of yttrium and the rare earths in seawater: conpetitive influences of solution and surface chemistry. *Aquat. Geochem.*, **10**, 59–80.

Quinn, K., Byrne, R.H. and Schijf, J. (2006) Sorption of yttrium and rare earth elements by amorphous ferric hydroxide: influence of solution complexation with carbonate. *Geochim. Cosmochim. Acta*, **70**, 4151–4165.

Revez, K.M., Landwehr, J.M. and Keybl, J. (2001) *Measurement of $\delta^{13}C$ and $\delta^{18}O$ Isotopic Ratios of $CaCO_3$ Using a Thermoquest Finnigan GasBench II Delta Plus XL Continuous Flow Isotope Ratio Mass Spectrometer with Application to Devils Hole Core DH-11 Calcite.* USGS Open-File Report 01-257.

Ricketts, B.D. and Donaldson, J.A. (1981) Sedimentary history of the Belcher Group of Hudson Bay. In: *Proterozoic Basins of Canada* (Ed. F.H.A. Campbell), *Geol. Surv. Can. Pap.*, **81-10**, 235–254.

Riding, R. (2008) Abiogenic, microbial and hybrid authigenic carbonate crusts: components of Precambrian stromatolites. *Geol. Croat.*, **61**, 73–103.

Riding, R., Fralick, P. and Liang, L. (2014) Identification of an Archean marine oxygen oasis. *Precambrian Res.*, **251**, 232–237.

Rouxel, O.J., Bekker, A. and Edwards, K.J. (2005) Iron isotope constraints on the Archean and Paleoproterozoic ocean redox state. *Science*, **307**, 1088–1091.

Sakurai, R., Ito, M., Ueno, Y., Kitajima, K. and Maruyama, S. (2005) Facies architecture and sequence-stratigraphic features of the Tumbiana Formation in the Pilbara Craton, northwestern Australia: implications for depositional environments of oxygenic stromatolites during the Late Archean. *Precambrian Res.*, **138**, 255–273.

Sanborn-Barrie, M., Skulski, T. and Parker, J. (2001) Three hundred million years of tectonic history recorded by the Red Lake greenstone belt, Ontario. *Current Research 2001-C19-Geological Survey of Canada*.

Sasseville, C., Tomlinson, K.Y., Hynes, A. and McNicoll, V.J. (2006) Stratigraphy, structure, and geochronology of the 3.0-2.7 Ga Wallace Lake greenstone belt, western Superior Province, southeast Manitoba, Canada. *Can. J. Earth Sci.*, **43**, 929–945.

Satkoski, A.M., Fralick, P.W., Beard, B.L. and Johnson, C.M. (2017) Initiation of modern-style plate tectonics recorded in Mesoarchean marine chemical sediments. *Geochim. Cosmochim. Acta*, **209**, 216–232.

Schneiderhan, E.A., Gutzmer, J., Strauss, H., Mezger, K. and Beukes, N.J. (2006) The chemostratigraphy of a Paleoproterozoic MnF-BIF succession – the Voelwater Subgroup of the Transvaal Supergroup in Griqualand West, South Africa. *S. Afr. J. Geol.*, **109**, 63–80.

Schröder, S., Beukes, N.J. and Sumner, D.Y. (2009) Microbialite-sediment interactions on the slope of the Campbellrand carbonate platform (Neoarchean, South Africa). *Precambrian Res.*, **169**, 68–79.

Scott, C.T., Lyons, T.W., Bekker, A., Shen, S., Poulton, S.W., Chu, X. and Anbar, A. (2008) Tracing the stepwise oxygenation of the Proterozoic biosphere. *Nature*, **452**, 456–459.

Sherman, L.S., James, A., N.P., N. and Guy, M. (2000) Sedimentology of a late Mesoproterozoic muddy carbonate ramp, northern Baffin Island, Arctic Canada. In: *Carbonate Sedimentation and Diagenesis in the Evolving Precambrian World* (Eds J.P. Grotzinger and N.P. James), *SEPM Spec. Publ.*, **67**, 209–241.

Shinn, E.A. (1968) Practical significance of birdseye structures in carbonate rocks. *J Sed. Petrol.*, **38**, 215–223.

Shinn, E.A. (1983) Birdseyes, fenestrae, shrinkage pores, and loferites; a reevaluation. *J. Sed. Petrol.*, **53**, 619–628.

Sholkovitz, E.R., Landing, W.M. and Lewis, B.L. (1994) Ocean particle chemistry: The fractionation of rare earth elements between suspended particles and seawater. *Geochim. Cosmochim. Acta*, **58**, 1567–1579.

Siahi, M., Hofmann, A., Hegner, E. and Master, S. (2016) Sedimentology and facies analysis of Mesoarchaean stromatolitic carbonate rocks of the Pongola Supergroup, South Africa. *Precambrian Res.*, **278**, 244–264.

Simonson, B.M. and Carney, K.E. (1999) Roll-up structures: evidence of in situ microbial mats in late Archean deep shelf environments. *Palaios*, **14**, 13–24.

Simonson, B.M., Schubel, K.A. and Hassler, S.W. (1993) Carbonate sedimentology of the early Precambrian Hamersley Group of Western Australia. *Precambrian Res.*, **60**, 287–335.

Skotnicki, S. and Knauth, L.P. (2007) The middle Proterozoic Mescal Paleokarst, central Arizona, U.S.A.: Karst development, silicification, and cave deposits. *J. Sed. Res.*, **77**, 1046–1062.

Sumner, D.Y. (1996) Evidence for low late Archean atmospheric oxygen from oceanic depth gradients in iron concentration. *Geol. Soc. Am. Bull.*, **28**, 218.

Sumner, D.Y. and Grotzinger, J.P. (1996) Were the kinetics of calcium carbonate precipitation related to oxygen concentration? *Geol.*, **24**, 119–122.

Sumner, D. and Grotzinger, J.P. (2000) Late Archean aragonite precipitation; petrography, facies associations, and environmental significance. In: *Carbonate Sedimentation and Diagenesis in the Evolving Precambrian World* (Eds J.P. Grotzinger and N.P. James), *SEPM Spec. Publ.*, **33**, 107–120.

Sumner, D.Y. and Grotzinger, J.P. (2004) Implications for Neoarchaean ocean chemistry from primary carbonate mineralogy of the Campbellrand-Malmani platform, South Africa. *Sedimentology*, **51**, 1–27.

Swart, P.K. (2015) The geochemistry of carbonate diagenesis; the past, present and future. *Sedimentology*, **62**, 1233–1304.

Taylor, S.R. and McLennan, S. (1985) *The Continental Crust: Its Composition and Evolution; An Examination of the Geochemical Record Preserved in Sedimentary Rocks.* Blackwell Scientific, Oxford, 312 pp.

Thurston, P.C., Kamber, B.S. and Whitehouse, M. (2012) Archean cherts in banded iron formation: insights into Neoarchean ocean chemistry and depositional processes. *Precambrian Res.*, **214**, 227–257.

Tice, M. and Lowe, D.R. (2006) The origin of carbonaceous matter in pre-3.0Ga greenstone terrains: A review and new evidence from the 3.42Ga Buck Reef Chert. *Earth Sci. Rev.*, **76**, 259–300.

Tice, M., Thornton, D., Pope, M.C., Olszewski, T.D. and Gong, J. (2011) Archean microbial mat communities. *Annu. Rev. Earth Planet. Sci.*, **39**, 297–319.

Tomlinson, K.Y., Sasseville, C. and McNicoll, V. (2001). New U-Pb geochronology and structural interpretations from the Wallace Lake Greenstone belt (north Caribou terrane): implications for new regional correlations. In: *Western superior transect 7th annual workshop* (Eds R.M. Harrap and H. Helmstaedt), *Lithoprobe Rep.*, **80**, 8–9.

Van Kranendonk, M.J., Webb, G.E. and Kamber, B.S. (2003) New geological and trace element evidence from 3.45 Ga stromatolitic carbonates in the Pilbara Craton: support of a marine, biogenic origin and for a reducing Archean ocean. *Geobiology*, **1**, 91–108.

Veizer, J. (2003) Isotopic evolution of seawater on geological time scales: sedimentological perspectives. In: *Geochemistry of Sediments and Sedimentary Rocks* (Ed. D.R. Lentz), *Geol. Ass. Canada Geotext*, **4**, 53–68.

Veizer, J. and Jansen, S.L. (1979) Basement and sedimentary recycling and continental evolution. *J. Geol.*, **87**, 341–370.

Veizer, J., Compston, W., Hoefs, J. and Nielson, H. (1982) Mantle buffering of the early oceans. *Naturwissenschaften*, **69**, 173–180.

Veizer, J., Hoefs, J., Ridler, R.H., Jensen, L.S. and Lowe, D.R. (1989a) Geochemistry of Precambrian carbonates: I. Archean hydrothermal systems. *Geochim. Cosmochim. Acta*, **53**, 845–857.

Veizer, J., Hoefs, J., Lowe, D.R. and Thurston, P.C. (1989b) Geochemistry of Precambrian carbonates: II. Archean greenstone belts and Archean sea water. *Geochim. Cosmochim. Acta*, **53**, 859–871.

Veizer, J., Clayton, R.N., Hinton, R.W., von Brunn, V., Mason, T.R., Buck, S.G. and Hoefs, J. (1990) Geochemistry of Precambrian carbonates: 3-shelf seas and non-marine environments of the Archean. *Geochim. Cosmochim. Acta*, **54**, 2717–2729.

Voegelin, A.R., Nägler, T.F., Beukes, N.J. and Lacassie, J.P. (2010) Molybdenum isotopes in late Archean carbonate rocks: implications for early Earth oxygenation. *Precambrian Res.*, **182**, 70–82.

Walcott, C.D. (1912) Notes on fossils from limestone of Steeprock series, Ontario. *Can. Dept. Mines Geol. Surv. Branch Mem.*, **28**, 16–23.

Waldbauer, J.R., Sherman, L.S., Sumner, D.Y. and Summons, R.E. (2009) Late Archean molecular fossils from the Transvaal Supergroup record the antiquity of microbial diversity and aerobiosis. *Precambrian Res.*, **169**, 28–47.

Webb, G.E. and Kamber, B.S. (2000) Rare earth elements in Holocene reefal microbialite: a new shallow seawater proxy. *Geochim. Cosmochim. Acta*, **64**, 1557–1565.

Webb, G.E., Nothdurft, L.D., Kamber, B.S., Kloprogge, J.T. and Zhao, J. (2009) Rare earth element geochemistry of scleractinian coral skeleton during meteoric diagenesis: a sequence through neomorphism of aragonite to calcite. *Sedimentology*, **56**, 1433–1463.

de Wet, C.B., Frey, H.M., Gaswirth, S.B., Mora, C.I., Rahnis, M. and Bruno, C.R. (2004) Origin of meter-scale submarine cavities and herringbone calcite cement in a Cambrian microbial reef, Ledger Formation, USA. *J. Sed. Res.*, **74**, 914–923.

Wilks, M.E. (1986) *The Geology of the Steep Rock Group, N.W. Ontario: A Major Archaean Unconformity and Archaean Stromatolites.* Unpublished MS thesis, University of Saskatchewan, 206 pp.

Wilks, M.E. and **Nisbet, E.G.** (1985) Archaean stromatolites from the Steep Rock Group, northwestern Ontario, Canada. *Can. J. Earth Sci.*, **22**, 792–799.

Wilks, M.E. and **Nisbet, E.G.** (1988) Stratigraphy of the Steep Rock Group, northwest Ontario: a major Archaean unconformity and Archaean stromatolites. *Can. J. Earth Sci.*, **25**, 370–391.

Wright, D.T. and **Altermann, W.** (2000) Microfacies development in Late Archean stromatolites and oolites of the Ghaap Group of South Africa. *Geol. Soc. London. Spec. Publ.*, **178**, 51–70.

Zhang, J., Amakawa, H. and **Nozaki, Y.** (1994) The comparative behaviors of yttrium and lanthanides in the seawater of the north Pacific. *Geophys. Res. Lett.*, **21**, 2677–2680.

4

Discontinuity surfaces and microfacies in a storm-dominated shallow Epeiric Sea, Devonian Cedar Valley Group, Iowa

MARA BRADY (iD) and CHRISTOPHER BOWIE[1]

Department of Earth and Environmental Sciences, California State University, Fresno, 2576 East San Ramon Avenue, M/S ST-24, Fresno, CA 93740, USA (E-mail: mebrady@csufresno.edu)

Keywords
Carbonate, depositional environment, epeiric basin, non-depositional.

[1]Present address: Devon Energy Corporation, 333 West Sheridan Avenue, Oklahoma City, OK 73102, USA.

ABSTRACT

Discontinuity surfaces develop in carbonate successions in response to a range of environmental changes and represent an integral part of the stratigraphic record. In Palaeozoic shallow epeiric basins that are typified by extremely slow subsidence and intermittent sedimentation, discontinuity surfaces may represent the majority of the time-rock record. A depositional and sequence-stratigraphic model was developed through microfacies analysis and discontinuity surface characterization using three cores in a proximal to distal transect across the Middle to Upper Devonian Iowa Basin. Twelve microfacies are recognized, spanning supratidal to deep subtidal facies tracts. A total of 105 discontinuity surfaces were documented and classified as either submarine omission surfaces, subaerial exposure surfaces or submarine erosional surfaces. Omission surfaces increase in frequency basinward, indicating increased sediment starvation in the offshore direction. Exposure surfaces increase in frequency shoreward, indicating more frequent subaerial exposure in a shallower setting. Erosional surfaces are dominant in the inner and middle ramp and interpreted as the base of storm beds (tempestites); these surfaces are rare in the outer ramp due to its generally deeper setting below storm wave base. Moreover, discontinuity surfaces exhibit systematic groupings stratigraphically (vertically) across the three localities spanning the Devonian carbonate ramp. Zones of either exposure-dominated, erosion-dominated or omission-dominated surfaces were recognized and correlated with their landward or basinward equivalents (along with shifts in major facies belts) and interpreted in a sequence-stratigraphic context. This study highlights the importance of including a detailed characterization of both depositional facies and non-depositional discontinuity surfaces in order to better understand the stratigraphic history of a basin. The framework of analysis provided here is particularly useful for marine carbonate strata deposited in epeiric basins, which are especially common in the Palaeozoic and where non-deposition and erosion occur frequently, but can also be applied to other geological time periods and settings.

INTRODUCTION

Discontinuity surfaces can reveal information about environmental, eustatic and tectonic changes and are as important as the sedimentary units that they delimit (Hillgärtner, 1998). These stratal surfaces occur due to a hiatus in sedimentation and/or erosion of previously deposited material, and represent a break in the time-stratigraphic record of any duration (Clari *et al.*, 1995). Discontinuity surfaces (or discontinuities) are of interest due to their importance in resolving high-frequency sea-level fluctuations and basin evolution at scales finer than chronostratigraphic resolution typically available (Hillgärtner, 1998; Sattler *et al.*, 2005; Bishop *et al.*, 2010; Chow & Wendte, 2011; Christ *et al.*, 2012a,b; Brlek *et al.*, 2013a,b; Godet *et al.*, 2013). Furthermore, each surface can retain its own complex history revealing multiple geological events superimposed on a single stratigraphic break (Clari *et al.*, 1995; Hillgärtner, 1998; Sattler *et al.*, 2005; Rameil *et al.*, 2012). Documenting the geographical and stratigraphic occurrence of discontinuity surfaces can reveal important tectonic and eustatic processes, which are manifested by subtle regional and local environmental changes and can aid in sequence-stratigraphic analysis (Hillgärtner, 1998).

Palaeozoic inland seas (i.e. cratonic, epeiric or epicontinental seas) are perplexing to sedimentologists and stratigraphers because there are few to no adequate modern analogs (Irwin, 1965; Allison & Wells, 2006; Immenhauser, 2009). Epicontinental basins are characterised by slow subsidence (<10 m per Myr) and extremely poor stratigraphic completeness with, on average, only 10% of geological time represented by sedimentation (Sloss, 1996). This implies that the sedimentary record in such basins is dominated by non-deposition, and that the bulk of the time-rock record is represented by discontinuity surfaces. These conditions are not conducive to applying sequence-stratigraphic principles, a model that was developed by studying rapidly subsiding passive margins, and is largely based on assuming relatively constant and high rates of subsidence and sedimentation (Catuneanu, 2006).

The Devonian Iowa Basin strata are a remarkable example of shallow epeiric sea sedimentation. When compared with contemporaneous passive continental margin deposits in Western North America, the Cedar Valley Group strata are typified by both thinner and fewer depositional units, many of which are bounded by discontinuity surfaces (Brady, 2015). Previous studies have attributed this discontinuous and condensed section to a stressed benthic carbonate factory that was inhibited by restricted circulation due to its isolation from the open ocean, enhanced freshwater influx (and associated upwelling of nutrients), and/or increased terrigenous influx (and associated nutrient input) (Witzke, 1987; Witzke *et al.*, 1988, 1996; Witzke & Bunker, 1996, 1997; Brady, 2015). The Cedar Valley Group strata provide an excellent opportunity to apply discontinuity surface characterization to better constrain the basin's history in light of tectonic, eustatic and sedimentological drivers.

The purpose of this study was (1) to document and characterize microfacies and discontinuity surfaces in the Cedar Valley Group to better determine controls on sedimentation in the Middle Devonian Iowa Basin, and (2) to integrate these findings into a sequence-stratigraphic framework that takes into account discontinuity surfaces, depositional facies and their stacking patterns. This study finds that Devonian strata in the Iowa Basin show a systematic and cyclic pattern in the occurrence of discontinuity surfaces that is related to both depositional setting

and relative sea-level. The concepts presented here can be used to improve existing models of facies distribution and basin architecture in this and other epeiric basins, as well as other settings characterized by condensed strata. This study highlights the important information captured in discontinuity surfaces in addition to the depositional units they bound.

Geological and stratigraphic context

The Iowa Basin is centrally located within the North-American Craton (Laurentia) and is bound to the northwest by the Transcontinental Arch and by the Ozark uplift to the south (Witzke *et al.*, 1988; Fig. 1). The North-American Craton (Laurentia) has remained relatively tectonically stable since the development of the Precambrian Midcontinent rift system (Green, 1983; Van Schmus & Hinze, 1985; Davis & Paces, 1990; Paces & Miller, 1993). The rift ultimately failed at *ca* 1087 Ma after over 20 Myr of volcanism that filled the rift valley with up to 30 km of volcanic and clastic sedimentary rocks (Ojakangas *et al.*, 2001). The post-rift stability of the cratonic interior has preserved several large-scale 'supersequences' that deposited marine sediments across much of North America (Sloss, 1963). Johnson *et al.* (1985) divided the Devonian Kaskaskia supersequence of Sloss (1963) into six smaller scale transgressive-regressive (T-R) cycles with durations of 1 to 10 Myr. The strata targeted in this study were deposited during Johnson (1970) and Johnson *et al.*'s (1985) cycle IIa, also termed the Taghanic Onlap. The Taghanic Onlap is recognized as the single largest magnitude transgressive event of the Kaskaskia supersequence (Johnson, 1970). This transgression deposited shallow-marine carbonates across Iowa and adjacent states during the Middle Devonian and was the first transgressive stage to breach the Transcontinental Arch since the Early Devonian (Johnson, 1970; Johnson *et al.*, 1985). Additional higher order sequences have been recognized by previous workers in the Iowa Basin and are referenced herein as 'Iowa T-R cycles' (*sensu* Day *et al.*, 1996; Day, 2006; Witzke & Bunker, 1996; Figs 2 and 3).

The shallow Devonian strata onlap onto the Transcontinental Arch and deepen distally from the Sioux Ridge towards the southeast where the Sangamon Arch divides the Iowa Basin from the adjacent Illinois Basin (Witzke *et al.*, 1988). The Cedar Valley Group was deposited across a broad epeiric ramp with minimal tectonic influence (Witzke & Bunker, 1996), although some evidence suggests local variations in subsidence between Iowa and adjacent basins (Witzke *et al.*, 1988), and minor tectonic upwarping along the Sangamon Arch (Whiting & Stevenson, 1965).

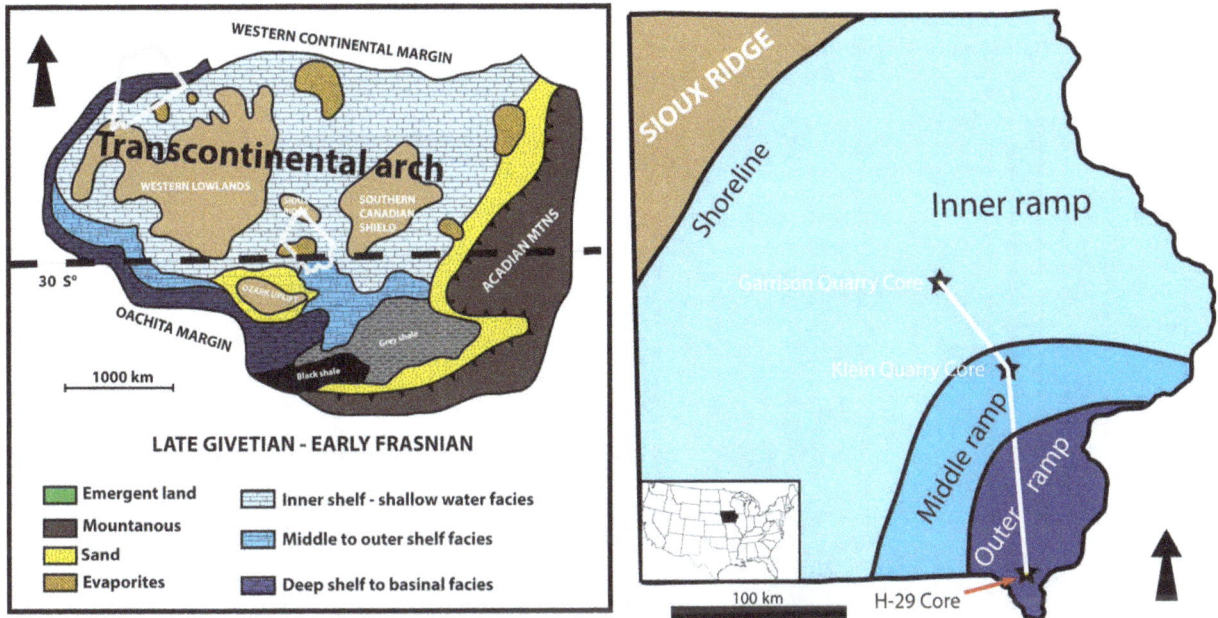

Fig. 1. Palaeogeographic map of North America and present day study area (left). North America (then Laurasia) during the Middle-Late Devonian showing major depositional regimes. Palaeonorth (arrow) and palaeolatitude are shown (modified from Bunker & Witzke, 1992) (right). Eastern half of Iowa showing generalized facies belts and locations of cores used in this study.

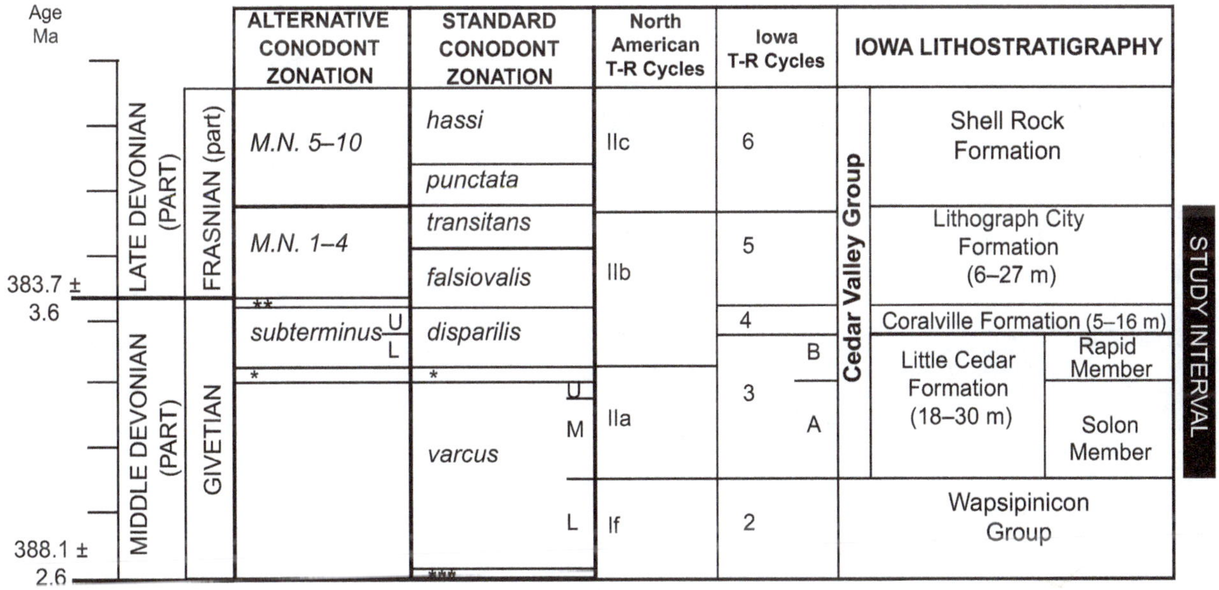

Fig. 2. Chronostratigraphic and lithostratigraphic chart of Middle-Late Devonian time periods covered in this study [modified from Brady, 2015). Based on Day (2006) and Morrow and Sandberg (2008). Ages (mya) from Kaufmann (2006). Chronostratigraphic placement of lithostratigraphic units after Witzke *et al.* (1988), Day (2006), Morrow and Sandberg (2008), Warme *et al.* (2008), Sandberg (2009)]. Thicknesses of each lithostratigraphic unit measured in this study are indicated in parentheses. Unlabelled conodont zones: * = *hermanni*, ** = *norrisi*, *** = *hermiansatus*.

The Cedar Valley Group strata is characterized by shallow restricted subtidal carbonate and evaporite (sabkha) facies that transition into open-marine shallow to deep subtidal facies across an expansive epeiric ramp (Witzke & Bunker, 1996, 1997). Diverse coral-stromatoporoid biostromes characterize the 'middle shelf margin' which

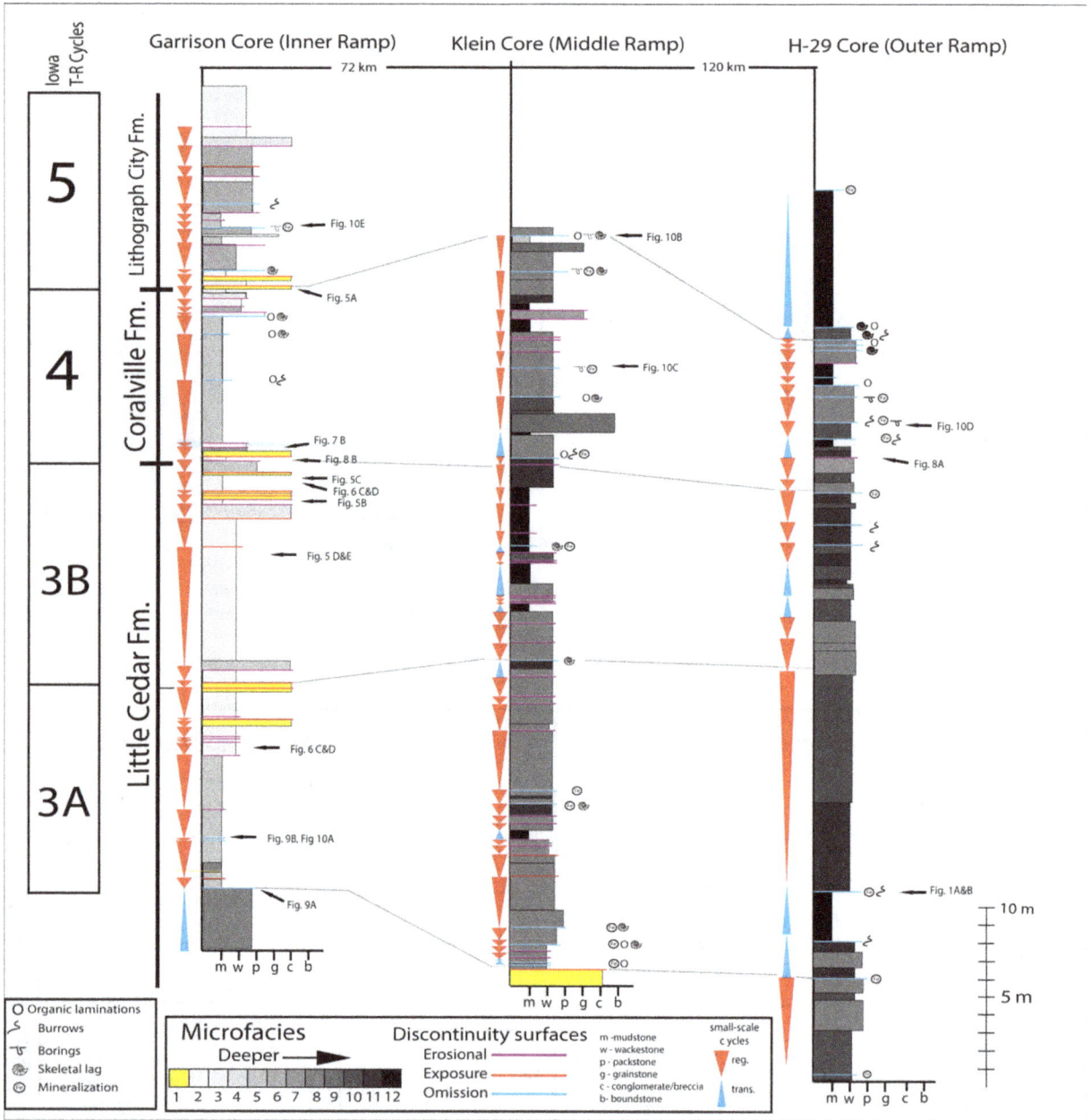

Fig. 3. Stratigraphic logs of microfacies variations, discontinuity surfaces and small-scale cycles described across three cores in this study. Iowa T-R Cycles are from Witzke & Bunker (1996). Solid grey lines between cores show correlation of T-R Cycles across the carbonate ramp. See Tables 1 and 2 for microfacies and discontinuity surface categories. Small-scale depositional cycles are interpreted based on facies tract excursions and delineated with transgressive and regressive energy polygyons (blue and red triangles). Core recovery was good for all cores (all ca >90%).

transition distally into argillaceous and low-diversity wackestones and mudstones of the 'outer shelf' (Witzke et al., 1988; Witzke & Bunker, 1996, 1997). For this study, the Cedar Valley strata are divided into inner, middle and outer ramp facies belts (Fig. 1), which correspond to the inner, middle and outer-middle 'shelf' belts of Witzke & Bunker (1996, 1997), respectively. The formations of the Cedar Valley Group are defined by

transgressive – regressive cycles (Iowa T-R Cycles) and are considered third-order (1 to 3 Myr) depositional sequences based on their depositional time scale as determined by biostratigraphic controls (Witzke & Bunker, 1996; Fig. 2).

Previous workers have documented that Iowa T-R Cycles are generally capped by subaerial exposure surfaces in the inner ramp that transition laterally into submarine hardgrounds in the distal areas (Witzke & Bunker, 1996). Low long-term rock accumulation rates (6 to 12 m per Myr), condensed facies and depositional cycles (relative to continental margin coeval records) and regionally widespread hardgrounds suggest a suppressed carbonate factory with intermittent sedimentation due to environmental stresses associated with epeiric seas. Poor circulation, hypersalinity and episodic anoxia (associated with enhanced nutrient flux) have been suggested as mechanisms that suppressed the shallow subtidal carbonate factory (Witzke et al., 1988; Witzke & Bunker, 1996; Brady, 2015).

MATERIALS AND METHODS

Three cores (totalling over 55 m, cumulatively) were described in detail with coverage through most of the Little Cedar and Coralville Formations, and part of the Lithograph City Formation (Iowa T-R Cycles 3A, B, 4, and 5; as defined in Witzke & Bunker, 1996) in a proximal to distal transect across the Iowa Basin. Core descriptions and interpretations were compared with previously described sections for the same study interval (Witzke et al., 1988; Witzke & Bunker, 1996; Brady, 2015). Microfacies and discontinuity surface descriptions were supplemented with field checks of local outcrops and with supplemental cores in the middle ramp area (Johnson County, Iowa) (Fig. 3).

Thin sections were impregnated with a blue dyed epoxy to observe porosity and stained with a mixed Alizarin Red S and potassium ferricyanide solution (after Dickson, 1966) to distinguish among carbonate minerals. Microfacies were characterized by a combination of polished faces of split core and thin section ($n = 71$) observations following Flügel (2004) and described according to Dunham's (1962) classification with an emphasis on identifying the primary sediment components. Bioturbation was classified in terms of ichnofabric index (Droser & Bottjer, 1986). When appropriate, dolomite microfabric was described using the nomenclature of Sibley & Gregg (1987).

Discontinuity surfaces in carbonate strata are generally interpreted as either the result of erosion, subaerial exposure, a hiatus in sedimentation leading to lithification of the sea floor, or some combination thereof (Clari et al., 1995; Hillgärtner, 1998; Sattler et al., 2005). In this study, discontinuities are assigned to one of three general categories using the surface morphology, biological activity, mineralization, facies contrasts and diagenetic contrast between underling and overlying units: (i) exposure surfaces, showing evidence of subaerial exposure and alteration; (ii) erosional surfaces, associated with evidence of removal of previously deposited material or (iii) omission

surfaces, showing evidence of a submarine sedimentary hiatus (following Hillgärtner, 1998; Sattler et al., 2005; Brady, 2015). A total of 105 discontinuity surfaces were documented in the study interval; 43 in the inner ramp, 43 in the middle ramp and 19 in the outer ramp.

FACIES AND DISCONTINUITY SURFACE DESCRIPTIONS AND INTERPRETATIONS

Carbonate ramp depositional system

Microfacies and depositional facies tracts

Based on thin section analysis and core description, the inner, middle and outer ramp areas are divided into six depositional facies tracts, and a total of 12 individual microfacies are recognized (see Table 1 for detailed descriptions). Within the study area, this portion of the epeiric ramp is characteristic of a carbonate ramp (Fig. 4). The inner ramp is comprised of a supratidal sabkha and tidal flat facies dissected by tidal channel deposits and low-diversity, restricted lagoonal muddy environments with brachiopods, ostracods, stromatoporoids and rare phylloid algae. The restricted shallow marine waters of the inner ramp transition into more open-marine and high-energy grainstone shoal and biostromal facies of the middle ramp consisting of crinoids, corals, bryozoans and brachiopods. The outer ramp is characterized by intermediate to deep subtidal marine wackestone and mudstones with crinoid, byrozoan and brachiopod debris, primarily concentrated in thin beds (3 to 10 cm) with scoured bases. Deep-marine facies are primarily mudstones with scattered fine-grained skeletal debris derived from the middle ramp (commonly fine crinoid debris), are more argillaceous and contain organic-rich stringers with rare articulated brachiopods.

The observed distribution of sedimentary components indicates that the inner and middle ramps were the primary sources of sediment generation and accumulation, whereas the distal outer ramp was sediment starved. Sediment was sourced from the inner and middle ramp, redistributed across the ramp by storms, and delivered to the outer ramp via fine-grained (muddy) sediment gravity flows. The slope was likely similar to that of the present-day lower Coralville contact, which is 0·0007 (0·045°) between the middle and outer ramp cores (based on depth to this contact in both cores and present-day horizontal distance between cores). The facies associations, low-angle slope and lack of a reef rim indicate that the morphology is consistent with a homoclinal carbonate ramp (Burchette & Wright, 1992). The low-angle slope precludes slumping or earthquake triggered mass wasting events such as reef apron debris flows and grain flows due to overfilling. The most likely mechanism for

Table 1. Microfacies characterized in this study, their common features described from thin sections (n = 71) and polished faces of split core, and interpreted depositional environments

Facies tract (interpretation)	Facies ID	Microfacies	Sedimentary structures	Sedimentary components and fossils (in order of abundance)	Dominant mineralogy	Microfabric	Depositional environment
1 (Supratidal)	MF1	(1) Collapse breccia	Vugs, solution seams, crystal silt geopetal fill, chert nodules and breccia	Recrystallized, rare quartz grains	Dolomite	Medium crystalline, Planar-e and Planar-s dolomite	Salt flat/sabkha
	MF2	(2) Laminated dolomudstone	Planar, curviplanar, and fenestral laminations, chert nodules and breccia	Recrystallized, rare quartz grains	Dolomite	Medium to fine crystalline, Planar-e, Planar-s, and nonplanar dolomite	Salt flat/sabkha
2 (Intertidal)	MF3	(3) Laminated peloidal mudstone and wackestone	Planar and planar-cross-laminated, microlenses	Peloids, ostracods, oncoids, gastropods, phylloid algae, rare quartz grains	Dolomite or calcite	Micrite	Tidal flat or low-energy beach
	MF4	(4) Intraclast pebble conglomerate	Imbricated intraclasts	Intraclasts, peloids, ostracods, oncoids, gastropods, rare quartz grains	Dolomite or calcite	Micrite	Tidal channel/surge channel
3 (Restricted shallow subtidal)	MF5	(5) Bioturbated mudstone	Bioturbated (ichnofabric 3 to 5)	Branching stromatoporoids common. Sparse brachiopods, crinoids, corals, gastropods, peloids, rare quartz grains	Dolomite or calcite	Micrite, fine to coarse crystalline planar-e, planar-s and anhedral dolomite	Coastal lagoon or embayment
	MF6	(6) Stromatoporoid-gastropod wackestone and packstone	Massive, bioturbated, *skolithos*	Gastropods, peloids, ostracods, branching and domal stromatoporoids, red algae, rare quartz grains	Calcite	Micritic peloids, dolomite-replaced fossils	Patch reefs/biostromes in coastal lagoon or embayment
4 (Open-marine shallow subtidal)	MF7	(7) Crinoidal Grainstone	Thinly-bedded, cross-bedded, burrowed	Crinoids, bryozoans, tentaculites, tabulate and rugose corals, tentaculites, stromatoporoids, rare quartz grains	Calcite	Calcite spar cement, syntaxial overgrowths, partially micritized in outer-middle ramp	Shoal/patch reef in high-energy open-marine environment
	MF8	(8) Laminar Stromatoporoid boundstone	Boundstone	Laminar stromatoporoids, brachiopods, bryozoans, mixed skeletal debris, serpulids, rare quartz grains	Calcite	Micrite	Quiet water open marine, below Fair weather wave base

(Continued)

Table 1. Continued

Facies tract (interpretation)	Facies ID	Microfacies	Sedimentary structures	Sedimentary components and fossils (in order of abundance)	Dominant mineralogy	Microfabric	Depositional environment
5 (Intermediate subtidal)	MF9	(9) Bryozoan-crinoid wackestone and packstone	Laminated to burrow-mottled	Bryozoans, crinoids, tabulate and rugose corals, stromatoporoids, brachiopods, rare quartz grains	Calcite	Micrite with scattered dolomite rhombs, rare dedolomite	Open ramp near ramp margin, shallow to intermediate, below fairweather wave base
	MF10	(10) Mixed skeletal packstone and wackestone	Laminated to burrow-mottled	Crinoids, bryozoans, brachiopods, skeletal hash corals, stromatoporoids, tentaculites, trilobites, rare quartz grains	Calcite	Micrite with scattered dolomite rhombs, rare dedolomite	Open ramp, fore-reef/biostrome debris
	MF11	(11) Brachiopod packstone and wackestone	Laminated to burrow-mottled or massive	Brachiopods, rarely articulated, skeletal hash, trilobites, tentaculites, rare quartz grains	Calcite	Micrite with scattered dolomite rhombs, rare dedolomite	Open ramp, intermediate depth
6 (Deep subtidal)	MF12	(12) Fine-grained mixed skeletal mudstone	Massive	Skeletal hash, articulated brachiopods, brachiopod valves, crinoids, bryozoans, surpulid worm tubes, trilobites, tentaculites, rare quartz grains	Calcite or dolomite	Micrite with scattered dolomite rhombs, rare dedolomite	Distal open ramp, deep, below storm wave base.

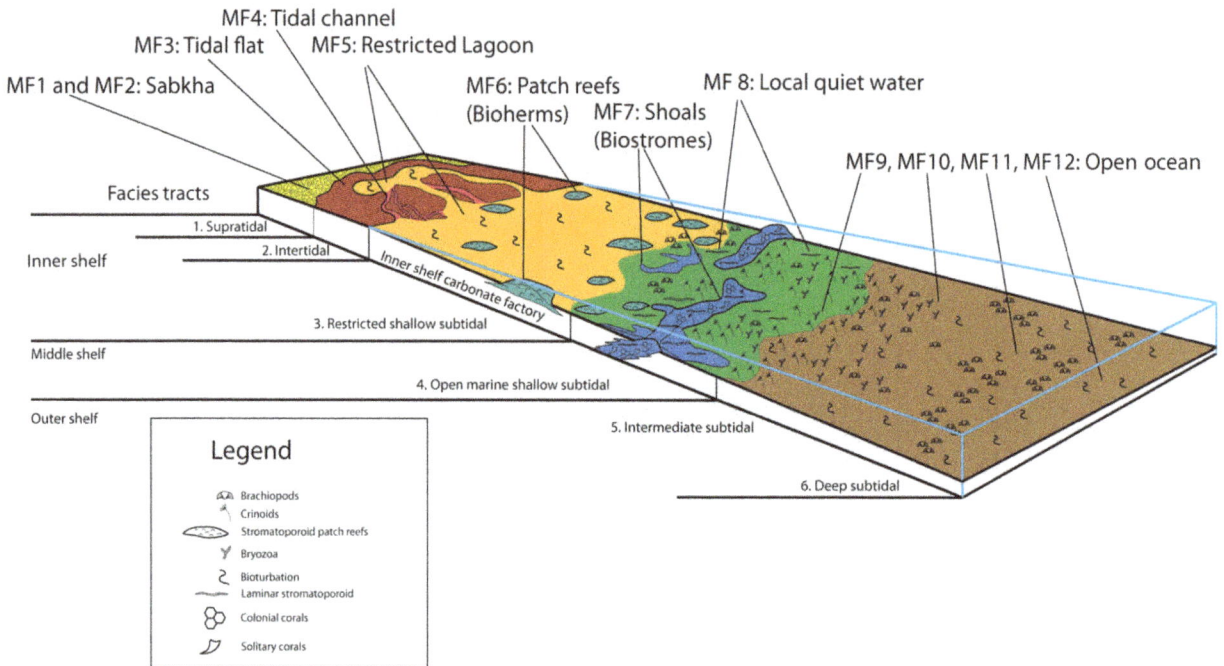

Fig. 4. Schematic block diagram showing the interpreted distribution of microfacies within facies tracts for the study interval (modified after Plocher, 1990; Witzke & Bunker, 1997; Da Silva *et al.*, 2011).

sediment transport to the outer ramp is through storm-generated turbidity currents (Aigner,1982; Osleger, 1991; Bádenas & Aurell, 2001). Storms have been suggested as a sediment transport mechanism in previous studies (Witzke *et al.*, 1988; Witzke & Bunker, 1997) and are consistent with observations in this study (hummocky cross-stratification, microlenses, toppled coral heads and stromatoporoids within grainstone units). In addition, fully articulated crinoids as well as other well preserved delicate fossils have been attributed to rapid mud burial in the middle ramp area, with storm-generated density currents as a likely mechanism (Witzke & Bunker, 2010).

Small-scale depositional cycles

Here, we recognize 'small-scale depositional cycles' defined by stacked beds and facies with gradational internal trends that are interrupted by discontinuity surfaces, which reflect an abrupt abrupt vertical juxtaposition of non-contiguous facies tracts, observations that indicate non-deposition, erosion, subaerial exposure or some combination of these processes. These metre-scale cycles (*ca* 0·5 to 10 m thick) were delineated based core descriptions and thin section interpretation of discontinuity surfaces. As described and defined here, these depositional cycles need not, and often do not, exhibit a cyclical or strictly shallowing-upward pattern. These small-scale cycles can exhibit different kinds of internal facies

stacking patterns (shallowing, deepening, no change in interpreted water depth or some combination of trends) and can be bounded by different types of discontinuity surfaces (Fig. 3 and discussed below). The scale and definition of cycles used here most closely follow Strasser *et al.* (1999)'s 'small-scale depositional sequence' and is consistent with those described in carbonate-dominated records by several other authors (e.g. Schlager, 1993; Spence & Tucker, 2007; Bishop *et al.*, 2010).

Characterization of discontinuity surfaces

All discontinuity surfaces were classified into one of three general categories: exposure surfaces, erosional surfaces or omission (hiatal) surfaces. Within these three categories, nine discontinuity surface types were recognized and ranked according to prominent physical characteristics, which can be interpreted as reflecting the temporal significance of the sedimentary hiatus or degree of erosional removal (Table 2; see Table S1 for details on each discontinuity surface encountered the study).

Subaerial exposure surfaces

The primary diagnostic features of these surfaces are solution seams, vugs, zones of collapse breccia or stylobreccia, indicative of karsted horizons. Subaerial-exposure surfaces commonly cap intertidal or supratidal small-scale cycles,

Table 2. Discontinuity surface types and subclassifications, their descriptions and interpreted processed of formation. Total number of surfaces observed listed under each broad facies belt

Discontinuity Type	Subclass	Interpreted Temporal Significance of Hiatus	Features	Interpretation	Total Inner Ramp	Total Middle Ramp	Total Outer Ramp
Subaerial exposure surfaces							
Ex1	Sabkha surface	Minor	Pervasive dolomitization, solution collapse breccia, vadose-zone diagenesis	Progradation and periodic flooding of supratidal zones	5	0	0
Ex2	Epikarst	Minor or Major	Solution collapse breccia, vugs, solution seams, meteoric diagenesis	Drop in relative sea-level, often subaerially exposing subtidal sediment	5	3	0
Ex3	Palaeosol	Major	Meteoric diagenesis, pedogenisis, solution seams	Prolonged exposure leading to soil development	1	0	0
Erosional surfaces							
Er1	Tempestites	Minor	Rip-up clasts, skeletal pavements, flute casts, graded beds and ripple cross laminations above surface	High-energy event, scouring of soft sediment, no significant removal and return to normal sedimentation	13	20	0
Er2	Simple change	Minor or Major	Sharp contact between different lithologic units, may be stylolitic, rip-up clasts, argillaceous partings	Change in sedimentation patterns or shift in facies belts, transgressive surfaces or flooding surfaces	8	5	0
Er3	Composite disconformity	Major	Sharp contact between different lithologic units, may be stylolitic, strong diagenetic contrast	Major erosional truncation, often associated with subaerial exposure in underlying unit	2	0	2
Omission surfaces							
O1	Diastem	Minor	Phosphatic and glauconitic lag, pyritized horizon organic laminations	Very slow sedimentation to minor temporal hiatus	4	1	5
O2	Firmground to incipient hardground	Minor or major	Burrows, phosphatic and glauconitic lag, weak mineralization, organic laminations	Submarine hiatus with partial sea floor lithification	2	2	2
O3	Hardground	Major	Borings, burrows, phosphatic and glauconitic lag, strong mineralization, biological encrustation	Submarine hiatus with complete lithification and prolonged exposure on the sea floor	3	12	10

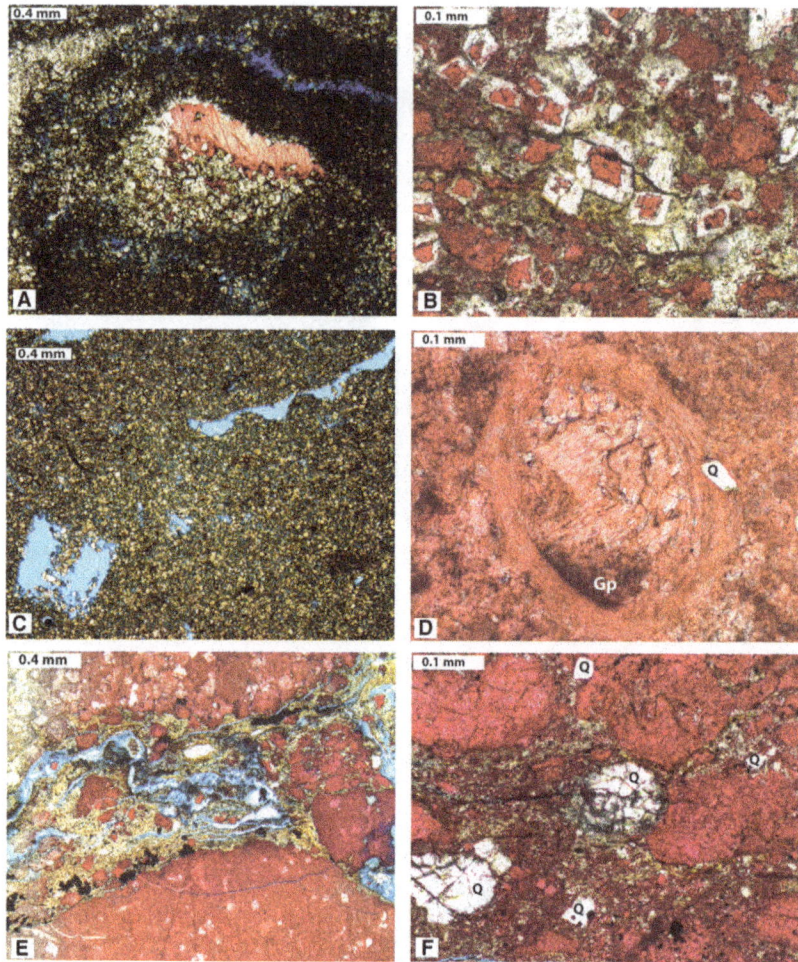

Fig. 5. Photomicrographs of subaerial exposure surface features (A) Dolomite crystal silt forming a geopetal mound in a vug that was later filled by calcite spar. (B) Dedolomite cores (red = calcite) with dolomite rims. (C) Mouldic porosity in subhedral dolomite matrix below Coralville-Little Cedar sequence boundary showing crinoid segment (lower left) and brachiopod valve (upper right) moulds. (D) Geopetal fill (Gp) in a vug that was once open and later healed by calcite spar. This may have originally been an ostracod fossil, or possibly a root cast. Residual insoluble material collected on the floor of the vug after original internal material dissolved or decomposed. The concentric micritic coating enveloping a quartz grain (Q) is likely of microbial origin. (E) Solution collapse breccia in solution seam/vug with void filling argillaceous material, note insoluble pyrite and Fe-oxides collecting on the floor of the vug. Brecciation may have resulted from wetting and drying of expansive clays. (F) Detrital quartz sand grains (Q) in an argillaceous matrix, some grains appear to be fractured *in situ*, with the cracks infilled by clay. This is interpreted as the result of expansive clays shrinking and swelling under alternating dry and wet conditions in the vadose zone. See Fig. 3 for sample locations corresponding to A-F.

and more rarely occur within subtidal facies. The rocks underlying most subaerial exposure surfaces are commonly dolomitized and may include chert pseudomorphs after evaporite. Petrographic evidence of subaerial exposure is indicated by vadose-zone diagenesis and includes geopetal mounds of dolomite crystal silt and Fe-oxides lining the base of vugs and solution seams, dedolomite and meniscus cement (Fig. 5A-F). Most secondary porosity associated with vugs and seams is occluded by calcite spar; however, open mouldic and vuggy porosity is associated with some surfaces that define formation

boundaries (Fig. 3C). Additional evidence of vadose-zone diagenesis and possibly pedogenesis includes brecciated quartz grains with clay fillings, carbonate breccia in clay-filled solution seams, calcrete with rhizocretions and pisoids with concentric micrite laminations (Fig. 5D-F; Wright, 1994).

Three distinct types of exposure surfaces are recognized in the Cedar Valley Group strata in this study and ranked in terms of the interpreted duration of subaerial exposure: sabkha surfaces (type Ex1), epikarst surfaces (type Ex2) and palaeosols (type Ex3).

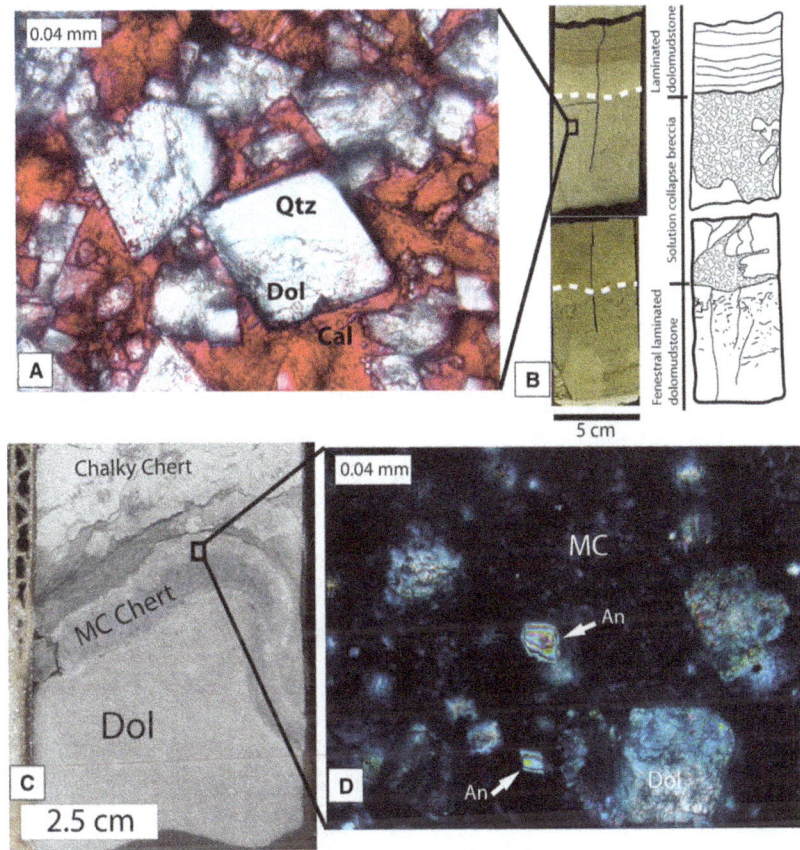

Fig. 6. Examples of Sabkha (Ex1) surfaces in core and thin section. (A) 630× magnification of centre dolomite rhomb in B, showing internal geopetal mound (Dol). Poorly ordered dolomite, or protodolomite, commonly precipitates in evaporitic environments and is prone to dissolution and replacement (Scholle & Ulmer-Scholle, 2003). Geopetal mounds form from the residual dolomite material collecting on the floor of the rhombehedral void (Dol), which was later occluded by silica cement (Qtz). The remaining intercrystalline porosity was later occluded by calcite spar (Cal). (B) MFI: collapse breccia. (C) Enterolithic anhydrite that has been replaced by microcrystalline (MC) chert. (D) Anhydrite (An) and dolomite (Dol) rhombs floating in microcrystalline chert matrix, visible under XPL and 630× magnification. See Fig. 3 for sample locations corresponding to A-D.

Sabkha surfaces (Ex1)

Ex1 surfaces associated with sabkha facies (Facies Tract 1, Table 1) typically consist of thin collapse breccia beds (3 to 10 cm) (MF1, Table 1) interbedded with laminated dolostones. Dedolomite with calcite and silica replacement is common and often associated with geotedal mounds and chert pseudomorphs after evaporite (Fig. 6A-D). Supratidal facies are, by definition, subaerially exposed most of the time, and virtually all MF1 units are delineated by a minor exposure surface. To be more precise, the sabkha surface (Ex1 surface) is placed at the top of each MF1 unit and delimits individual sabkha depositional packages (*ca* 10 to 200 cm thick sequences of MF2 and MF1 units) (Fig. 6B). These surfaces do not show evidence of prolonged subaerial exposure, but they are important for recognizing and interpreting the sabkha facies and delimiting small-scale depositional cycles.

Epikarst (Ex2)

Ex2 surfaces occur within subtidal units and are characterized by evidence of solution collapse breccia within karst cavities and evidence of vadose-zone diagenesis. Compaction results in dissolution and suturing between brecciated clasts in some epikarst surfaces (i.e. stylobreccia). Ex2 surfaces are interpreted as reflecting a relative sea-level fall that resulted in subaerial exposure of the sea floor (Hillgärtner, 1998). The result is typically a shallow subtidal small-scale cycle truncated by an epikarst surface, i.e. a 'catch-down' cycle (Soreghan & Dickinson, 1994). The interpreted formation of these surfaces differs from Ex1 surfaces in that the formation of the exposure surface forms independently of sedimentation rate, such that subtidal facies (as opposed to supratidal or intertidal facies) are truncated by subaerial exposure features.

Palaeosols (Ex3)

Both Ex1 and Ex2 surfaces have the potential to develop into Ex3 surfaces (palaeosols) under prolonged subaerial exposure. Ex3 surfaces are underlain by calcrete that may exhibit rhizocretions, root traces and pisoids. Brecciated grains and cavities filled with green clay are also indicative of subaerial exposure and may suggest shrinking and swelling of expanding clays due to wetting and drying in the vadose zone (Fig. 5E and F) (Deconinck & Strasser, 1987). Concentric micrite laminations surrounding rhizocretions or other grains can displace or envelop the surrounding grains (Fig. 5D), and may be a microbial precipitate or the result of evaporation in the soil zone (Wright, 1994). The presence of palaeosols, in addition to the apparent activity of expanding clays, suggests a semi-arid climate (Wright, 1994).

Erosional surfaces

Erosional surfaces show evidence of a change in hydrodynamic energy level and removal of previously deposited material at any scale. Evidence of traction deposition, flame structures, scours, flute casts and rip-up clasts indicate minor erosional events that displace or remove soft to semi-lithified sediment. Such discontinuities are often associated with a change in grain-size or texture; bioclasts and/or intraclasts are common above many erosional surfaces and may show crude upward grading. Imbricated brachiopod valves and shelter porosity are also associated with increased hydrodynamic energy and traction deposition.

Sharp contacts juxtaposing two different facies tracts (non-Waltherian contacts) typically show diagenetic contrasts between underlying and overlying units and indicate regional erosional events. Petrographic observations of these major erosional surfaces show that the discontinuity is defined by an Fe-oxide rich parting that may be stylolitic in part with resistant grains protruding from the underlying unit.

Three erosional surface subdivisions are proposed for this study interval, based primarily on the interpreted magnitude of the erosional event. These range from minor and short-lived, high-energy events (Er1) or changes in hydrodynamic regime (Er2) to major migrations of facies belts punctuated by regional erosional events (Er3).

Tempestite surfaces (Er1)

Er1 surfaces are characterized by scouring of the underlying bed by flute cast or flame structures marking the base of decimetre-scale sequences. The surface is typically overlain by rip-up clasts or shell lags, depending on the depositional environment and are sometimes capped by planar and/or ripple cross laminations (Fig. 7A). There is generally no change in depositional environment and the complete interval over which they are expressed is typically no

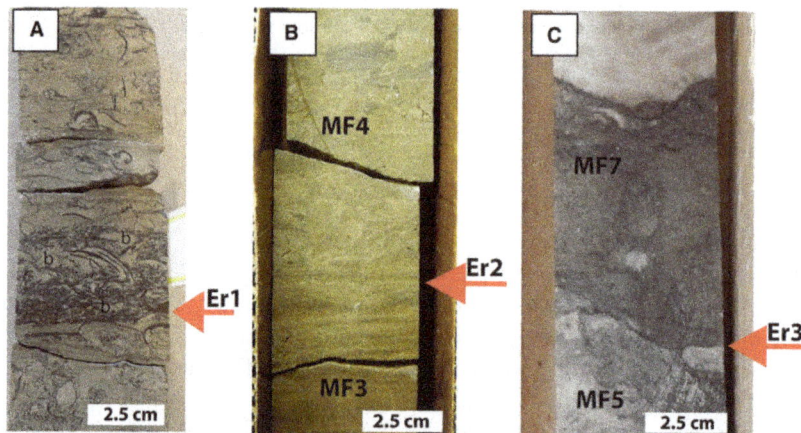

Fig. 7. Examples of erosional surfaces in core. (A) High-energy event deposit in an outer-middle ramp wackestone/packstone (MF10) interpreted as a tempestite (Er1). Note basal scour surface with imbricated brachiopod shells, escape burrows (b) in overlying skeletal hash, and an upper ripple cross-laminated muddy wackestone unit. (B) Simple change surface (Er2) in intertidal laminite (MF3), overlying pebble conglomerate (MF4) consists of rip-up clasts from underlying unit. Interpreted as tidal channel deposit incised into a tidal flat mudstone. Sample is from the Little Cedar – Coralville Formation contact in the middle ramp found in a supplemental core near the Klein Core highlighted in this study (Mid-America Pipeline, Iowa City terminal, core 4; NE NW SE SE sec. 27, T79N, R5W, Johnson Co., reposited in the Department of Natural Resources Geological Survey Bureau, Iowa City, Iowa; Witzke & Bunker, 1997). (C) Composite disconformity (Er3) in the inner ramp juxtaposing crinoidal grainstone (MF7) with coral head on top of lagoonal mudstone (MF5). This surface is interpreted as the contact between the Little Cedar (underlying) and Coralville (overlying) Formations. See Fig. 3 for sample locations corresponding to A and C.

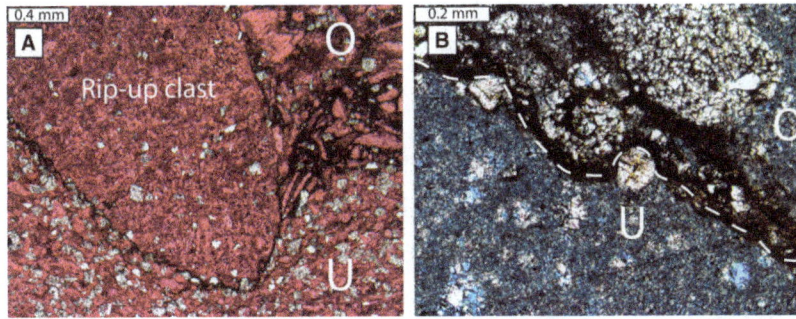

Fig. 8. Photomicrographs of Er3 surfaces showing diagenetic contrasts between underlying (U) and overlying (O) units. (A) Er3 surface in the outer ramp. Underlying unit shows pervasive recrystallization (dolomite and calcite) that is also observed in the intraclast rip-up. Overlying unit is an argillaceous packstone. Note that compaction has sutured the rip-up clast into the top of the underlying unit along a stylolitic contact. (B) Contact between Coralville and Little Cedar Formations in the inner ramp. Resistant grains protruding into overlying unit define pre-existing microtopography and suggest underling unit was lithified before erosional event. Note vuggy porosity (blue) in underlying unit. See Fig. 3 for sample locations corresponding to A and B.

more than 10 cm thick. These intervals are compositionally different at the inner and middle ramp localities. The inner ramp skeletal lags typically consist of restricted marine fauna and rip-up clasts in a peloidal matrix, whereas the middle ramp lags are commonly coral fragments, bryozoans, brachiopod and crinoid debris in a micritic matrix.

The sedimentary structures and bed forms overlying most Er1 surfaces are consistent with storm-event deposits (tempestites) described in the literature (Aigner, 1982; Dott & Bourgeois, 1982; Seilacher & Aigner, 1991). Therefore, Er1 surfaces are interpreted as the base of tempestites and are not thought to represent significant removal of material or prolonged break in sedimentation. These surfaces are interpreted as reflecting abrupt and short-lived, high-energy events followed by a return to normal sedimentation. Their abundance in the study interval supports storm currents as an important mechanism for sediment redistribution across the ramp (Witzke & Bunker, 2010).

Simple change (Er2)

Er2 surfaces are typified by a sharp contact juxtaposing two different microfacies (Fig. 7B). These are dominated by microfacies transitions within the same facies tract and are generally interpreted as autogenic processes operating within a given depositional environment, e.g. migration of tidal channels through the tidal flat, or a prograding shoreline. Er2 surfaces are defined by scoured surfaces, flame structures, and/or rip-up clasts and may be accentuated by stylolites (Hillgärtner, 1998). Er2 surfaces are similar to Er1 surfaces, but are distinguished by a shift in facies reflecting a change in depositional environment, rather than a single sedimentary event. It is possible, however, that these changes can be triggered by a single storm

event, as hurricanes and typhoons are capable of rearranging the shoreline and obliterating reef habitats (Flügel, 2004). Some Er2 surfaces may mark basin-wide deepening events and are interpreted as transgressive or flooding surfaces that delimit depositional sequences.

Composite disconformity (Er3)

Er3 surfaces are characterized by a sharp juxtaposition of two different microfacies commonly showing a diagenetic contrast between underlying and overlying units (Fig. 7C). These surfaces are usually defined by a sharp non-Waltherian contact that juxtaposes different facies tracts (e.g. outer ramp on top of inner ramp). Er3 surfaces are exemplified by erosional surfaces in the inner ramp that are underlain by rocks showing evidence of meteoric diagenesis (mouldic and vuggy porosity) suggesting that the inner ramp was subaerially exposed prior to erosion. The lithofacies contrast across these surfaces reflects a relative deepening. Resistant grains protruding off of these surfaces reflect pre-existing microtopography and indicate that the underlying unit was lithified prior to the erosional event (Fig. 8). Er 3 surfaces in the outer ramp are similarly associated with diagenetic contrasts, but display no evidence of subaerial exposure. Er3 surfaces are interpreted as regionally widespread erosional surfaces associated with a relative sea-level fall and subsequent rise (where associated with deepening across the surfaces). Diagenetic contrasts between underlying and overlying units suggest Er3 surfaces represent major temporal hiatuses and a lithified substrate prior to erosional truncation.

Omission surfaces

Omission surfaces show evidence of a submarine hiatus in sedimentation. Depositional breaks are indicated by

Fig. 9. Photomicrographs of glauconite and pyrite concentrated at omission surfaces. (A) Inner ramp: Glauconite peloids and fossil casts; glauconitized crinoid debris (Cr) is also partially pyritized. (B) Inner Ramp: Lenticular glauconite (microlense) in coarse crystalline planar-e dolomite matrix. See Fig. 3 for sample locations corresponding to A and B. (C) Middle ramp: High magnification of HCl-etched thin section showing a crinoid segment that has been completely replaced by glauconite preserving internal pore structure. (D) Middle ramp: Example of skeletal fabric preserving pyrite; euhedral crystals mimic the original brachiopod shell structure. Photomicrographs in C and D are from supplemental outcrop samples in Coralville Formation: Mehaffey Bridge location, Johnson County, Iowa (Brady, 2015).

clay and/or organic stringers, mineralized horizons and hardgrounds that exhibit evidence of biological colonization of the sea floor. Glauconite is common at most omission surfaces in the form of microlenses, glaebules, peloids, grain coatings, fossil casts, disseminated in the rock matrix or draped over the discontinuity surface (Fig. 9). Glauconitized grains concentrated at omission surfaces are typically intermixed with phosphatic skeletal debris and pyrite in the form of euhedral crystals, framboids, partial to complete fossil casts, and/or dispersed in the matrix. Mineralized crusts are diagnostic of major omission surfaces and can penetrate a few centimetres into the underlying unit. Some omission surfaces, particularly in the outer ramp, also consist of stylolite swarms or anastomosing sets of stylolites and are often associated with organic partings.

Evidence of biological activity includes borings and burrows (distinguishable by sharp vs. fuzzy burrow walls, respectively), and encrusting fossils of sessile benthic fauna (Fig. 10). In many cases, burrows and borings are filled by sediment of the overlying unit and in rare cases are filled with a completely different lithofacies than the overlying unit (Fig. 10D. Taxonomic composition of skeletal lags resting on omission surfaces vary across

depositional environments following these general trends: branching stromatoporoids at the inner ramp; corals, crinoids and bryozoans at the middle ramp and articulated brachiopods in the outer ramp. Several omission surfaces preserve concentrations of organic matter at or around the discontinuity. Black organic material occurs either as single partings less than 1 cm in extent or as zones of fine wavy laminations that can extend up to 30 cm above or below the surface.

Based on the evidence described above, three omission surface types are differentiated and ranked in terms of increasing magnitude of the hiatus represented (*adapted from* Christ *et al.*, 2012a). Type O1 surfaces (diastems) are minor hiatuses in comparison to Type O2 (firmgrounds to incipient hardgrounds). Type O3 surfaces (hardgrounds) are the most prominent omission surfaces in the study interval and are interpreted as representing the longest duration depositional hiatuses.

Diastems (O1)

In this study, these surfaces are marked by phosphatic lags and concentrated glauconite grains. Pyrite and glauconite-draped O1 surfaces indicate dysoxic to anoxic and

Fig. 10. Core photos of omission surfaces. (A) Inner ramp minor condensation surface (O1) defined by a pyritic horizon (dashed white line) and thin chert parting after evaporite. Dark grains above discontinuity are glauconitized fossils and phosphatic debris. (B) Organic laminations at omission surface (middle ramp), coral skeleton lag resting on the surface (c), note burrows penetrating organic laminations (dashed white line). (C) Mineralized omission surface cut by borings, which are filled with sediment from the overlying unit. (D) Mineralized irregular surface (white dashed line). Burrows (Bu) in lower mudstone unit are filled with an anomalous packstone (missing facies); unit overlying discontinuity is a wackestone. (E) Hardground on an encrusting domal stromatoporoid (white dashed line); this organism has been bored (Bo) and mineralized (min), with the borings filled with sediment of the overlying unit. Burrows (Bu) stratigraphically higher are also consistent with condensed sedimentation associated with omission surfaces. See Fig. 3 for sample locations corresponding to A-E.

reducing conditions, particularly when organic laminations are also present. These surfaces often mark a break in bioturbation, with burrows generally re-appearing several centimetres above the break. These surfaces are common in the inner and outer ramp cores, and less common in the middle ramp. Chert pseudomorphs after evaporites are generally limited to the inner ramp, suggesting a shift to hypersalinity caused by isolation from open ocean currents (Fig. 10A). The key characteristic of O1 surfaces is that the sea floor sediment remained unlithified although brittle pyrite and evaporite crusts may have developed at the sediment-water interface. Based on these observations, these surfaces indicate a minor submarine depositional hiatus due to environmental change that inhibited the production and/or deposition of carbonate sediment.

Firmgrounds to incipient hardgrounds (O2)

O2 surfaces can exhibit all of the same features as diastems with the addition of bioturbation cutting across the surface. Burrows or boreholes penetrating into the underlying unit are commonly filled with sediment of the

overlying unit (Fig. 10C). Skeletal lags commonly rest on the discontinuity surface and weak mineralization may penetrate into the underlying unit. Mineral impregnation coupled with burrows and boreholes penetrating underling units indicate a submarine depositional hiatus where the sea floor sediment was partially to fully lithified, implying a longer duration hiatus relative to O1 surfaces (Hillgärtner, 1998).

Hardgrounds (O3)

O3 surfaces exhibit strong mineralization, which penetrates several centimetres into the underlying unit, as well as borings and biological encrustation indicative of a fully lithified sea floor. In addition to the features described for O1 and O2 surfaces, borings in O3 surfaces are generally filled with sediment from the overlying unit and may show multiple generations of biological activity (Fig. 11). These observations indicate that these surfaces represent major depositional hiatuses with a greater magnitude of depositional change and a longer non-depositional duration than O1 and O2 surfaces.

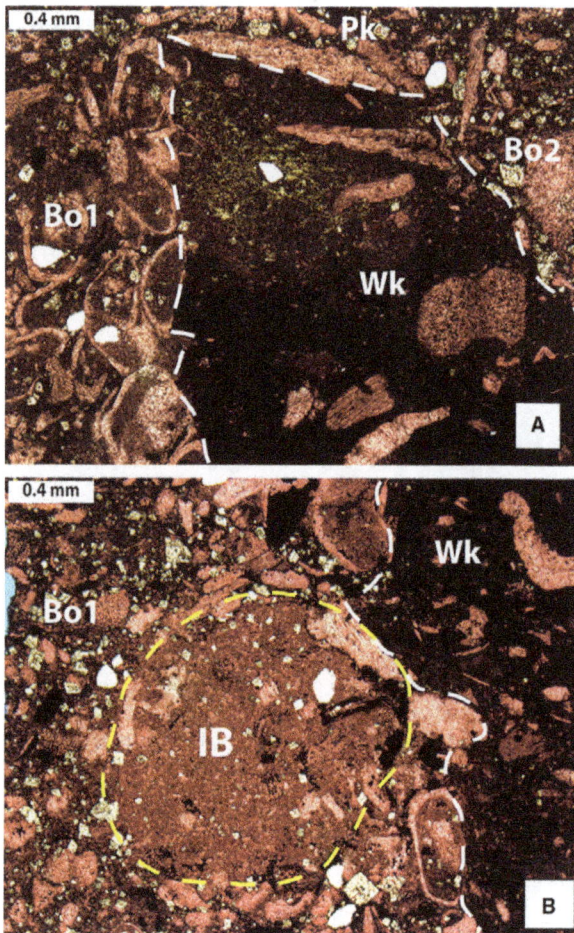

Fig. 11. Photomicrographs of mineralization and biological activity at omission surfaces. (A) Mineralized and bored wackestone (Wk) hardground, overlying packstone (Pk) fills borings, the walls of boring Bo1 is encrusted by serpulid worm tubes. (B) Photo extension of boring Bo1 showing internal burrow (IB) within the boring fill. See Fig. 3 for sample locations corresponding to A-C.

DISCUSSION

Geographical distribution of discontinuity surfaces

The occurrence and frequency of discontinuity surfaces differs between the inner, middle and outer ramp (Fig. 12). These variations give insight into the balance of controlling factors operating across the study area, particularly with regards to sedimentation, erosion, transportation and creation or loss of accommodation space. Each facies belt has a unique signature with respect to the types and frequency of discontinuity surfaces described in this study.

Exposure surfaces are most frequent in the inner ramp, rare in the middle ramp and absent in the outer ramp

(Fig. 12A). These observations are consistent with expectations that subaerial exposure should decrease distally. The peritidal environments that characterize the inner ramp would presumably be more prone to subaerial exposure from minor relative sea-level oscillations. The middle ramp is dominated by shallow to intermediate subtidal facies and is generally interpreted as a higher energy environment. Consequently, vertical aggradation would have been limited by wave sweeping as accumulating sediment approached wave base (Osleger, 1991). Subaerial exposure surfaces in the middle ramp were likely the result of a significant drop in relative sea-level. The three exposure surfaces that occur in the middle ramp are epikarst (Ex2 surfaces) and may actually represent subaerially exposed hardgrounds (Hillgärtner, 1998). The absence of exposure surfaces in the outer ramp suggests that it remained below sea-level for the duration of the study interval.

Erosional surfaces are the dominant surface type at the inner and middle ramp, and rare in the outer ramp (Fig. 12A). Most erosional surfaces in the inner and middle ramp are interpreted as the basal surface of tempestites (Er1 surfaces) (Fig. 12B). These observations support previous studies that suggested storms as mechanism for sediment transport (Witzke et al., 1988; Witzke & Bunker, 1997; Groves, 2004; Preslicka et al., 2010; Witzke & Bunker, 2010; Day et al., 2013). The two erosional surfaces recognized in the outer ramp are major erosional disconformities (Er3 surfaces) that likely developed during maximum sea-level lowstands as they are the only evidence that suggests wave base got low enough to scour the sea floor. There are no tempestite (Er1) surfaces in the outer ramp, presumably because deposition was infrequent enough that the sea floor was lithified between depositional events or bioturbation may have obscured original sedimentary structures.

The frequency of omission surfaces increases from the inner ramp towards the outer ramp (Fig. 12A), indicating increased sediment starvation and more frequent submarine hiatuses in the offshore direction. These observations are consistent with the inner and middle ramps acting as the primary sources of sediment supply and sedimentary hiatuses increasing away from the locus of sedimentation. The abundance of hardgrounds (O2 and O3 surfaces) in the outer ramp suggest that these sedimentary hiatuses were longer in duration relative to O1 surfaces, allowing for sea floor cementation, biological colonization and mineralization.

Taken together, the distribution of discontinuities is characterized by storm-dominated deposition in shallow water and sediment starvation in deeper water. The abundance of omission surfaces in all studied sections supports previous interpretations of a suppressed or stressed benthic carbonate factory (Witzke *et al.*, 1988; Witzke & Bunker, 1996; Brady, 2015). Lower carbonate production rates could be caused by several environmental and ecological factors. The frequent occurrence of omission surfaces, associated minerals (glauconite, pyrite, phosphate), abundance of organic-rich and argillaceous rich beds documented here are consistent with lower carbonate production rates resulting from anoxic or dysoxic conditions (Figs 9 and 10). Excess nutrients due to upwelling and/or increased terrigenous sediment input in this epeiric setting could lead to occasional benthic oxygen stress and attenuated light levels factory (Hallock & Schlager, 1986; Witzke, 1987; Hallock, 1988; Heckel, 1991; Pope & Steffen, 2003; Cramer & Saltzman, 2007; Baird *et al.*, 2012; Brady, 2015). Based on these factors, sedimentation would have been intermittent; and the sediment that was generated in shallow water was frequently redistributed by storm currents.

Stratigraphic distribution of discontinuity surfaces and sequence-stratigraphic interpretations

The stratigraphic distribution of discontinuities appears to exhibit a systematic grouping of related surfaces. That is, rather than randomly distributed through the section, clusters of similar surfaces (erosional, omission or exposure) occur together and are also related to the depositional units they delimit. Therefore, zones of either exposure-dominated, erosion-dominated or omission-

dominated surfaces can be recognized and correlated with their landward or basinward equivalents (along with shifts in major facies belts) and interpreted in a sequence-stratigraphic context (Fig. 13). The sequence-stratigraphic interpretation of systems tracts and associated surfaces is presented here in the context of previously defined T-R cycles to show how this study relates to previous work (Day *et al.*, 1996; Witzke & Bunker, 1996). Note that T-R cycles are defined based on identifying transgressive (deepening) and regressive (shallowing) trends, but those same trends can also be interpreted in the context of sequence-stratigraphic terminology as discussed below, especially when considering the significance and distribution of stratigraphic surfaces.

Inner ramp

The inner ramp strata exhibit cycle stacking patterns and zones of discontinuity surfaces that define three depositional sequences, corresponding to the third-order Iowa T-R cycles 3A through 5 (Day *et al.*, 1996; Witzke & Bunker, 1996). Each sequence is separated by an interval of closely spaced exposure surfaces that define decimetre-scale supratidal cycles. These 'exposure-dominated zones' can be interpreted as sequence boundary zones (SBZ) (*sensu* Montañez & Osleger, 1993; Hillgärtner, 1998; Bover-Arnal & Strasser, 2013). Each SBZ marks a longer term trend of falling sea-level, where the sabkha and intertidal facies prograded across the inner ramp. Palaeosols (Ex3 surfaces) and regional erosional (Er3) surfaces

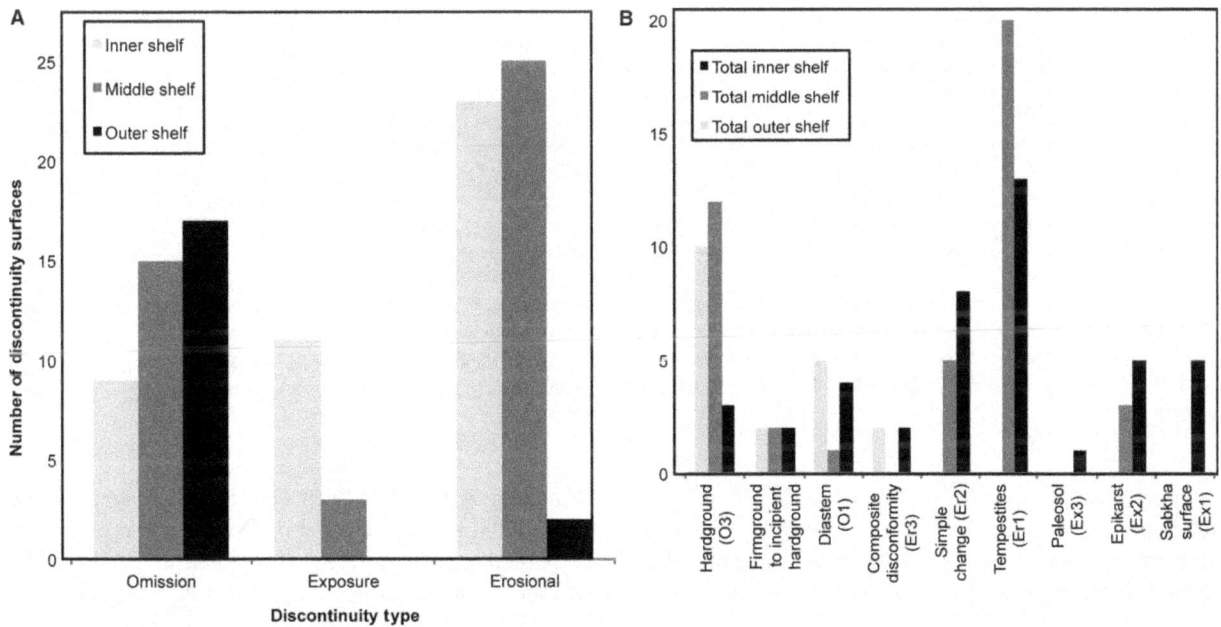

Fig. 12. Frequency of discontinuity surfaces across major depositional settings: by (A) major categories and (B) subtypes (see Table 2).

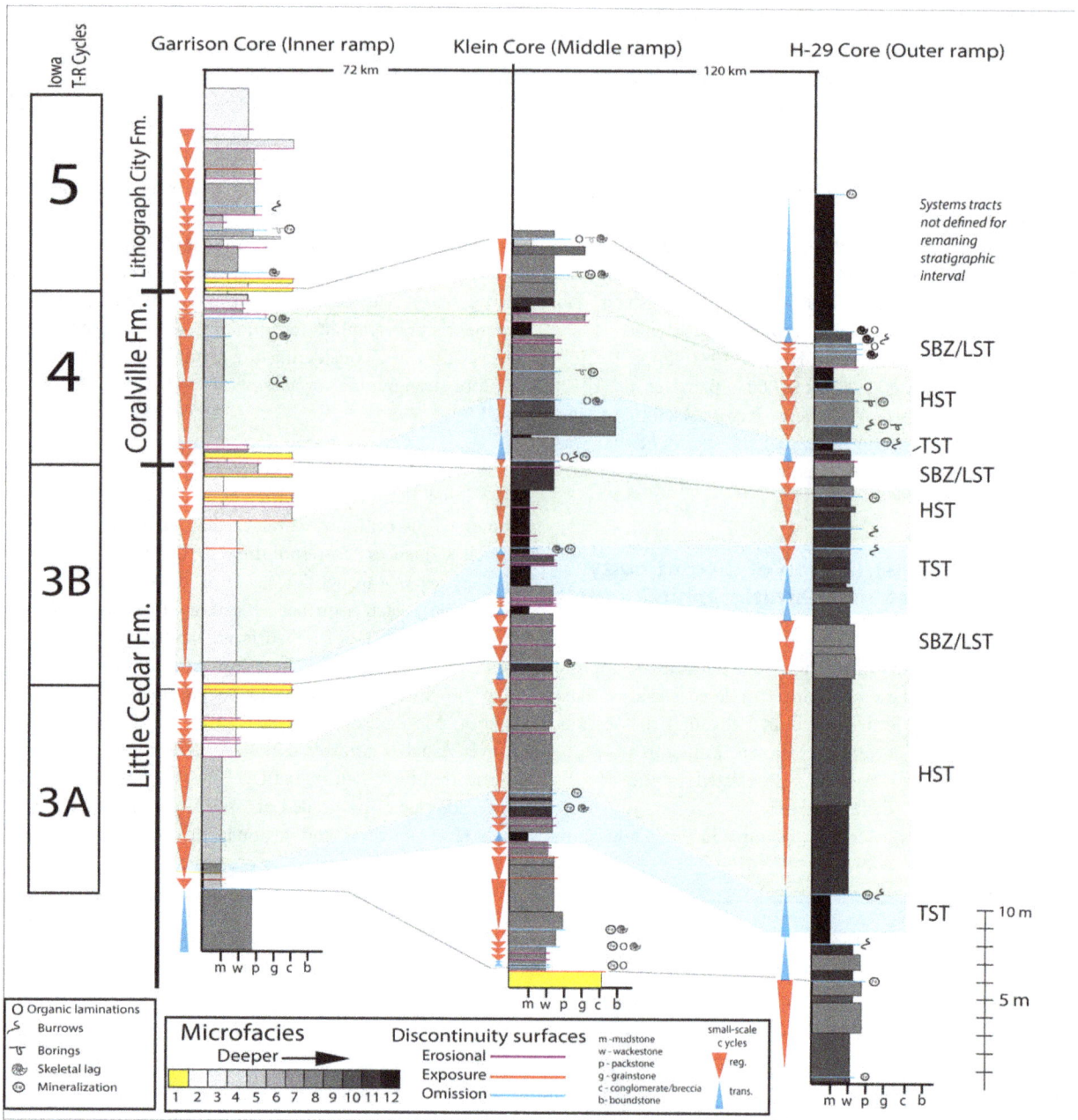

Fig. 13. Stratigraphic logs of microfacies variations, discontinuity surfaces and small-scale cycles from Fig. 3 overlain with sequence-stratigraphic interpretations. Sequence-stratigraphic interpretations of systems tracts and associated surfaces/zones are based on correlating discontinuity surfaces/zones and patterns within and across small-scale depositional cycles. Coloured backgrounds correspond to systems tracts: TST = Transgressive Systems Tract (blue), HST = Highstand Systems Tract (green), SBZ/LST = Sequence Boundary Zone + Lowstand Systems Tract (white).

developed during episodes of prolonged exposure. Each SBZ separates a relatively thick succession of metre-scale restricted shallow subtidal cycles that generally thin upward. Upward-thinning cycles in shallow-water facies are generally considered an indicator of decreasing accommodation space (Read & Goldhammer, 1988;

Montañez & Osleger, 1993; Sadler *et al.*, 1993; Husinec *et al.*, 2008) and suggest that the overall (third-order) rate of sea-level rise was decelerating while these cycles were deposited. These subtidal intervals between SBZs are considered most representative of the highstand systems tract (HST), which was deposited as the rate of sea-level rise

Flat topped platform model
(after Hillgärtner, 1998)

Carbonate ramp model
(This study)

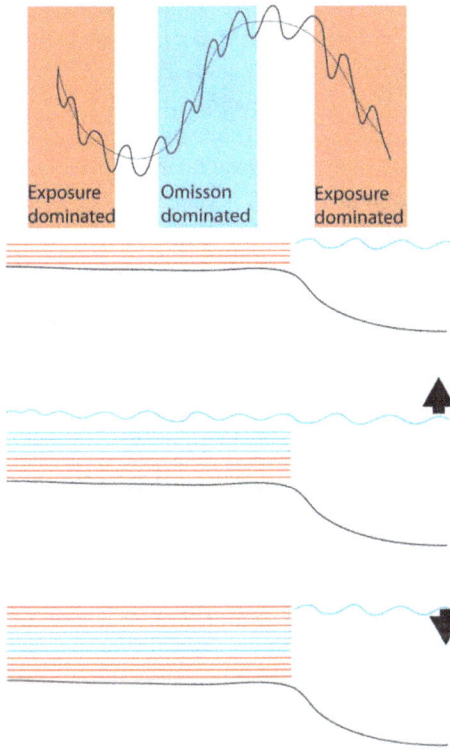
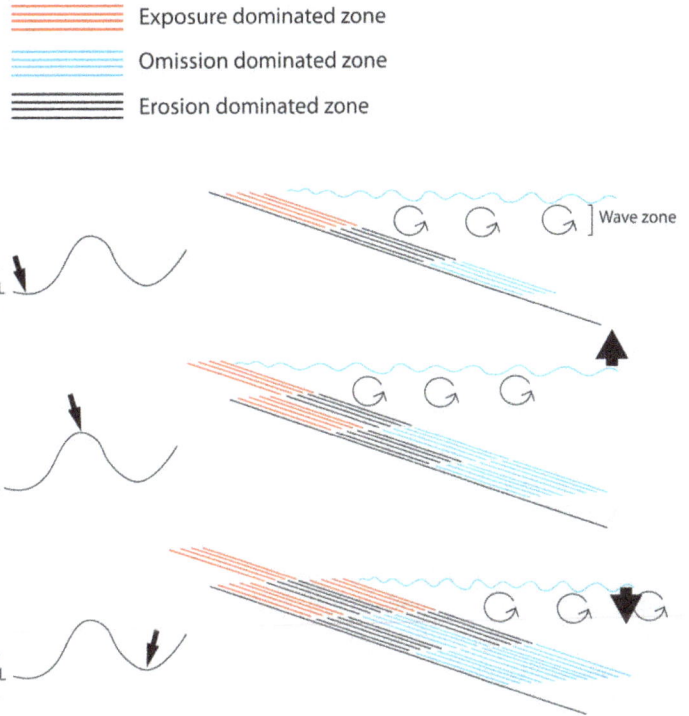

Fig. 14. Discontinuity surface zones models for a flat-topped carbonate platform (after Hillgärtner, 1998) and the for carbonate ramps (this study). Discontinuity zones in carbonate ramps are controlled by bathymetry and shift landward and basinward as sea-level rises and falls. Note that this model is a schematic (not to scale) and that relative sea-level amplitudes are vertically exaggerated. Arrows above and below relative sea-level in B and C (left and right) indicate sea-level rise and fall, respectively.

decelerated, and restricted lagoonal sediments began accumulating in the inner ramp (Fig. 13).

Together, each inner ramp sequence exhibits an overall progradational stacking pattern consisting of thinning upward shallow subtidal cycles (HST) that transition into supratidal cycles (SBZ), which is consistent with the highstand prograding wedge, typical of homoclinal carbonate ramps (Burchette & Wright, 1992). The absence of retrogradational stacking in the inner ramp suggests the transgressive systems tract (TST) is not represented here, though a discontinuity capping each SBZ marks a slight deepening and correlates to the TST down depositional dip (Fig. 13). The presence of Er3 surfaces within two of the SBZ showing evidence of vadose-zone meteoric diagenesis in the underling units suggest the inner ramp was emergent for much of the time when the lowstand system tract (LST) and TST were deposited in deeper settings, and these Er3 surfaces correspond to regionally extensive erosional surfaces that mark the sequence boundaries between formations (Witzke *et al.*, 1988; Witzke &

Bunker, 1997). The postulated ramp morphology and observations presented here suggest that the higher elevation inner ramp was only submerged during maximum sea-level highstands.

Middle ramp

The middle ramp is dominated by subtidal cycles delimited by hardgrounds (O3 surfaces) or tempestites (Er1 surfaces) (Figs 3 and 13). These observations are consistent with expectations for a storm-dominated carbonate ramp, where currents limit sediment accumulation above wave base (Osleger, 1991; Bádenas & Aurell, 2001; Christ *et al.*, 2012a). These limits to sediment accumulation would be compounded in the Devonian Iowa Basin where a stressed benthic carbonate factory (due to enhanced nutrient flux and/or siliciclastic input in the epeiric setting) could not always keep pace with rates of sea-level rise (Pomar, 2001; Allison & Wright, 2005; Brady, 2015). Moreover, facies and discontinuity surface interpretations, and the depositional cycles they delimit, become increasingly important for these sections because systems tracts in homoclinal carbonate ramps are generally only represented by landward and basinward shifts in facies tracts (Tucker *et al.*, 1993; Schlager, 2005).

Sequence boundaries are represented in the middle ramp by prominent hardgrounds (O3 surfaces) that correlate to regional erosional (Er3) surfaces and subaerial exposure surfaces in the inner ramp (Witzke *et al.*, 1988; Witzke & Bunker, 1996). Although there are no apparent SBZs in the middle ramp section due to a lack of subaerial exposure surfaces, there appears to be zones dominated by erosional surfaces that are interpreted here as the basinward equivalent of the SBZs found in the inner ramp. These surfaces would form more readily during relative sea-leavel fall and associated lowering of storm wave base. Furthermore, these erosional-surface-dominated zones alternate stratigraphically with zones dominated by omission surfaces in the middle ramp section. The omission-dominated intervals can be interpreted as reflecting maximum flooding zones (MFZs) and associated condensed intervals during the TST. Indeed, similar discontinuity surface types (exposure, erosional or omission) seem to occur together in stratigraphic intervals (zones) that reflect a dominant non-depositional process, and are characterized by either subaerial exposure, erosion or a submarine hiatus in sedimentation (Fig. 13). The TST, HST and LST are therefore identified in the middle ramp by deepening and shallowing trends and formational boundaries that represent sequence boundaries (Fig. 13).

Outer ramp

The outer ramp is characterized by intermediate to deep subtidal deposition (mostly facies tract 5 and 6), with relatively thin shallow-subtidal intervals that occur in the appropriate stratigraphic position to be the basinward equivalent of the SBZ and LST (Fig. 13). The entire section is dominated by omission surfaces with only two erosional surfaces present. TST's are defined in this section based on the presence of the deepest facies tracts and associated omission surfaces. These relatively thin, deep intervals, and surfaces with glauconite, pyrite and organic-rich laminae, are consistent with condensed intervals that form during maximum transgression. These observations are in contrast to the inner and middle ramp section where alternating discontinuity surface zones are present. The dominance of omission surfaces throughout the outer ramp indicates that subaerial exposure and erosion are rare or absent in the deepest water settings. This makes interpretation of the distinct erosional surfaces all that more essential as they are likely valuable for sequence-stratigraphic interpretation.

The two erosional surfaces are interpreted as Er3 (composite disconformities) and actually mark the top and base of the Coralville Formation (Iowa T-R cycle 4). The diagenetic contrast across these surfaces indicates that the underlying units were affected by meteoric phreatic fluids (e.g. Fig. 8B). The freshwater meteoric lens can extend for a considerable distance past the shoreline during sea-level lowstand, and discharge at the sea floor (Chafetz *et al.*, 1988; Burchette & Wright, 1992). There is no evidence, such as vadose-zone diagenesis, to suggest the outer ramp was exposed subaerially in this study interval.

Summary

Based on the observations of vertical and lateral trends in lithofacies and discontinuity surfaces, a sequence-stratigraphic model has been developed (Fig. 13) that builds on previously published interpretations in the study interval (Witzke & Bunker, 1997; Brady, 2015). In summary, the TST thins or pinches out landward due to the morphology of the epeiric ramp and may be represented in the inner ramp by a single erosional flooding surface. In the middle and outer ramp, basinward shifts in facies and groups of omission surfaces typically define the TST and comprise condensed intervals that form during maximum flooding. Most sediment was accumulated during the HST, as the rate of sea-level rise slowed and sedimentation could keep pace (Witzke & Bunker, 1997; Brady, 2015). Towards the end of the HST, the peritidal zone prograded towards the basin. Frequent subaerial exposure and erosion typify the SBZ in the inner ramp, whose basinward equivalent (the LST) is marked by a shoreward shift in facies tracts and erosional surfaces in the middle ramp. During sea-level fall and lowstands, the inner ramp was left emergent, whereas the freshwater meteoric lens altered submarine sediments in the middle and outer ramp, which were subject to submarine erosion during the lowstand.

Implications for epeiric depositional settings and sequence stratigraphy

Inhibited sedimentation due to environmental stresses is one of the defining characteristics of the Cedar Valley Group carbonates (Witzke *et al.*, 1988, 1996; Witzke & Bunker, 1997; Brady, 2015), thus making it an ideal study interval to show how the stratigraphic distribution of non-depositional surfaces can be used to illustrate variations in sedimentation and accommodation across the carbonate ramp, and how these variables change through time.

The systematic variation in discontinuity surfaces across the study interval provides insight into the carbonate ramp morphology and sedimentary processes operating at various locations in a proximal to distal transect. The recognition of genetically similar groupings of discontinuity surfaces has been documented in other carbonate strata and has proven useful in sequence-stratigraphic correlations and interpretations (Hillgärtner, 1998; Sattler *et al.*, 2005; Christ *et al.*, 2012a,b; Godet *et al.*, 2013; Strasser, 2015), especially in epeiric platforms that are particularly sensitive to environmental changes (Morales *et al.*, 2016). This study contrasts with Hillgärtner's (1998) observations of alternating zones dominated by either omission or exposure surfaces that could be correlated across a carbonate platform (see Fig. 14). Across a flat-topped carbonate platform influenced primarily by eustatic sea-level change, one would expect to observe a consistent and predictable stratigraphic distribution of discontinuities. Instead, the Cedar Valley Group ramp can be divided laterally into three distinct zones across a relatively small geographic area: an exposure-dominated inner ramp zone, the erosion-dominated middle ramp zone and the outer ramp zone which is dominated by omission surfaces (Fig. 14). The geographic and stratigraphic variation in the frequency of observed discontinuity surfaces is consistent with a complex interaction between relative sea-level change, bathymetric zone and depositional energy across carbonate ramp settings in epeiric seas (Christ *et al.*, 2012a; Strasser, 2015).

These zones are analogous to facies belts in that they should migrate landward and basinward as relative sea-level rises and falls. Moreover, the different discontinuity zones are intimately linked to specific depositional facies and should follow Walther's Law with respect to their vertical and lateral predictability. Recognition of these zones of genetically similar discontinuity surface serves as an additional tool, which can be used for sequence-stratigraphic analysis (Figs 13 and 14). This is particularly useful in cases such as this study's Cedar Valley Group, where slow and intermittent sedimentation often results in an absence of systematic facies stacking patterns. In summary, documenting and characterizing discontinuity surfaces can provide additional valuable information that may not be apparent by focusing on lithologic units alone. Discontinuity surfaces are as important to the sedimentary record as the sedimentary units they delimit, and should not be regarded as gaps in the sedimentary record, but considered as supplemental non-depositional facies that are essential for accurate sedimentological and stratigraphic interpretations (Hillgärtner, 1998; Sattler *et al.*, 2005; Christ *et al.*, 2012a,b; Godet *et al.*, 2013; Brady, 2015; Strasser, 2015; Morales *et al.*, 2016).

CONCLUSIONS

The Iowa Basin is an extraordinary example of epeiric sea sedimentation, which has no modern analog. The Cedar Valley Group strata described here are characterized as a sediment starved, storm-dominated epeiric carbonate ramp based on the distribution of microfacies and discontinuity surfaces: (i) Peritidal facies dominate the inner ramp, whereas the middle to outer ramp zones are characterized by intermediate to deep subtidal facies; (ii) Submarine omission surfaces increase in frequency basinward, indicating increased sediment starvation in the offshore direction; (iii) Exposure surfaces increase in frequency landward, indicating more frequent subaerial exposure associated with a shallower depositional setting; and (iv) The predominant discontinuity surface type in the inner and middle ramps are erosional surfaces; and most of these are interpreted as the erosional base of tempestites.

This study demonstrates a distinct distribution of surfaces for the Cedar Valley Group carbonate ramp, which differs markedly from the expected distribution on flat-topped carbonate platforms (Hillgärtner, 1998). Discontinuity surfaces in the Cedar Valley Group ramp occur in bathymetrically controlled zones: the inner ramp zone is prone to subaerial exposure, the middle ramp zone has more frequent submarine erosion and the outer ramp zone is characterized by more frequent submarine sedimentary omissions. Due to the ramp morphology, these zones shift back and forth like facies tracts as sea-level rises and falls (Fig. 14). Importantly, recognition of these discontinuity surface zones and their lateral equivalents is essential for sequence-stratigraphic analysis. This work provides an important framework for conducting sequence-stratigraphic analysis with particular attention to the characterization and distribution of discontinuity surface in addition to the depositional units they delimit. This approach proves especially useful in epeiric strata, which may be exemplified by slow and intermittent sedimentation, and exhibit distinct facies and surface stacking patterns compared to continental margin strata for which sequence stratigraphy was first developed.

ACKNOWLEDGEMENTS

The authors thank Brian Witzke and Bill Bunker for valuable discussions and feedback on this work as well as assistance with accessing field localities and core repository samples housed at the Iowa Geological Survey. Thank you to Robert Dundas and Mathieu Richaud, who provided constructive feedback on earlier drafts of this manuscript. Associate Editor James Klaus, André Strasser and an anonymous reviewer provided constructive feedback that greatly improved the manuscript. This study was generously supported by American Association of Petroleum Geologists (AAPG) Grants-in-Aid Program and California State University, Fresno.

References

Aigner, T. (1982) Calcareous tempestites: storm-dominated stratification in Upper Muschelkalk limestones (Middle Trias, SW-Germany). In: *Cyclic and Event Stratification* (Eds G. Einsele and A. Seilacher), pp. 180–198. Springer, Berlin.

Allison, P.A. and Wells, M.R. (2006) Circulation in large ancient epicontinental seas: what was different and why? *Palaios*, 21, 513–515.

Allison, P.A. and Wright, V.P. (2005) Switching off the carbonate factory: a-tidality, stratification and brackish wedges in epeiric seas. *Sed. Geol.*, 179, 175–184.

Bádenas, B. and Aurell, M. (2001) Proximal–distal facies relationships and sedimentary processes in a storm dominated carbonate ramp (Kimmeridgian, northwest of the Iberian Ranges, Spain). *Sed. Geol.*, 139, 319–340.

Baird, G.C., Zambito, J.J. and Brett, C.E. (2012) Genesis of unusual lithologies associated with the Late Middle Devonian Taghanic biocrisis in the type Taghanic succession of New York State and Pennsylvania. *Palaeogeogr. Palaeoclimatol. Palaeoecol.*, 367, 121–136.

Bishop, J.W., Montañez, I.P. and Osleger, D.A. (2010) Dynamic Carboniferous climate change, Arrow Canyon, Nevada. *Geosphere*, 6, 1–34.

Bover-Arnal, T. and Strasser, A. (2013) Relative sea-level change, climate, and sequence boundaries: insights from the Kimmeridgian to Berriasian platform carbonates of Mount Salève (E France). *Int. J. Earth Sci.*, 102, 493–515.

Brady, M.E. (2015) Stratigraphic completeness of carbonate-dominated records from continental interiors versus continental margins: stratigraphic thinning occurs via condensation and omission at multiple scales. *J. Sed. Res.*, 85, 337–360.

Brlek, M., Korbar, T., Tešović, B.C., Glumac, B. and Fuček, L. (2013a) Stratigraphic framework, discontinuity surfaces, and regional significance of Campanian slope to ramp carbonates from central Dalmatia, Croatia. *Facies*, 59, 779–801.

Brlek, M., Korbar, T., Košir, A., Glumac, B., Grizelj, A. and Otoničar, B. (2013b) Discontinuity surfaces in Upper Cretaceous to Paleogene carbonates of central Dalmatia (Croatia): glossifungites ichnofacies, biogenic calcretes, and stratigraphic implications. *Facies*, 1–21.

Bunker, B.J. and Witzke, B.J. (1992) An upper Middle through lower Upper Devonian lithostratigraphic and conodont biostratigraphic framework of the Midcontinent Carbonate Shelf area, Iowa. In: *The stratigraphy, paleontology, depositional and diagenetic history of the Middle-Upper Devonian Cedar Valley Group of central and eastern Iowa*, Vol. 16, pp. 3–26 (Eds J. Day and B.J.

Bunker). Iowa Department of Natural Resources, Guidebook Series, Iowa City, IA.

Burchette, T. and Wright, V. (1992) Carbonate ramp depositional systems. *Sed. Geol.*, 79, 3–57.

Catuneanu, O. (2006) *Principles of Sequence Stratigraphy*. Elsevier, Amsterdam, 376 pp.

Chafetz, H.S., McIntosh, A.G. and Rush, P.F. (1988) Freshwater phreatic diagenesis in the marine realm of Recent Arabian Gulf carbonates. *J. Sed. Res.*, 58, 433–440.

Chow, N. and Wendte, J. (2011) Palaeosols and palaeokarst beneath subaerial unconformities in an Upper Devonian isolated reef complex (Judy Creek), Swan Hills Formation, west-central Alberta, Canada. *Sedimentology*, 58, 960–993. https://doi.org/10.1111/j.1365-3091.2010.01191.x.

Christ, N., Immenhauser, A., Amour, F., Mutti, M., Tomás, S., Agar, S.M., Alway, R. and Kabiri, L. (2012a) Characterization and interpretation of discontinuity surfaces in a Jurassic ramp setting (High Atlas, Morocco). *Sedimentology*, 59, 249–290. https://doi.org/10.1111/j.1365-3091.2011.01251.x.

Christ, N., Immenhauser, A., Amour, F., Mutti, M., Preston, R., Whitaker, F.F., Peterhänsel, A., Egenhoff, S.O., Dunn, P.A. and Agar, S.M. (2012b) Triassic Latemar cycle tops – subaerial exposure of platform carbonates under tropical arid climate. *Sed. Geol.*, 265, 1–29.

Clari, P.A., Dela Pierre, F. and Martire, L. (1995) Discontinuities in carbonate successions: identification, interpretation and classification of some Italian examples. *Sed. Geol.*, 100, 97–121. https://doi.org/10.1016/0037-0738 (95)00113-1.

Cramer, B.D. and Saltzman, M.R. (2007) Fluctuations in Epeiric Sea carbonate production during Silurian positive carbon isotope excursions: a review of proposed paleoceanographic models. *Palaeogeogr. Palaeoclimatol. Palaeoecol.*, 245, 37–45.

Da Silva, A.C., Kershaw, S. and Boulvain, F. (2011) Sedimentology and stromatoporoid palaeoecology of Frasnian (Upper Devonian) carbonate mounds in southern Belgium. *Lethaia*, 44, 255–274.

Davis, D. and Paces, J. (1990) Time resolution of geologic events on the Keweenaw Peninsula and implications for development of the Midcontinent Rift system. *Earth Planet. Sci. Lett.*, 97, 54–64.

Day, J. (2006) Overview of the Middle-Upper Devonian Sea Level History of the Wapsipinicon and Cedar Valley Groups, with Discussion of New Conodont Data from the Subsurface Cedar Valley Group of Southeastern Iowa. In: *New Perspectives and Advances in the Understanding of the Lower and Middle Paleozoic Epeiric Carbonate Depositional Systems of the Iowa and Illinois Basins* (Eds J. Day, J. Luczaj and R. Anderson), pp. 1–22. Iowa Geological Survey Guidebook, Iowa City, IA.

Day, J., Uyeno, T., Norris, W., Witzke, B.J. and Bunker, B.J. (1996) Middle-Upper Devonian relative sea-level histories of central and western North American interior basins. *Geol. Soc. Am. Spec. Pap.*, **306**, 259–275.

Day, J., Witzke, B.J. and Bunker, B.J. (2013) Overview of Middle and Upper Devonian Northern Shelf Facies in the Iowa Basin. In: *Aspects of the Paleozoic History of Epeiric Seas of the Iowa Basin* (Eds J. Day and S. Lundy), *Iowa Geol. Water Surv. Guidebook*, **29**, 1–40.

Deconinck, J.F. and Strasser, A. (1987) Sedimentology, clay mineralogy and depositional environment of Purbeckian green marls (Swiss and French Jura). *Eclogae Geol. Helv.*, **80**, 753–772.

Dickson, J. (1966) Carbonate identification and genesis as revealed by staining. *J. Sed. Res.*, **36**, 491–505.

Dott, R. and Bourgeois, J. (1982) Hummocky stratification: significance of its variable bedding sequences. *Geol. Soc. Am. Bull.*, **93**, 663–680.

Droser, M.L. and Bottjer, D.J. (1986) A semiquantitative field classification of ichnofabric: research method paper. *J. Sed. Res.*, **56**, 558–559.

Dunham, R.J. (1962) Classification of carbonate rocks according to depositional textures. In: *Classification of Carbonate Rocks* (Ed. W.E. Ham), *AAPG Mem.*, **1**, 108–121.

Flügel, E. (2004) *Microfacies of Carbonate Rocks: Analysis, Interpretation and Application.* Springer Verlag, Berlin.

Godet, A., Föllmi, K.B., Spangenberg, J.E., Bodin, S., Vermeulen, J., Adatte, T., Bonvallet, L. and Arnaud, H. (2013) Deciphering the message of Early Cretaceous drowning surfaces from the Helvetic Alps: what can be learnt from platform to basin correlations? *Sedimentology*, **60**, 152–173. https://doi.org/10.1111/sed.12008.

Green, J.C. (1983) Geologic and geochemical evidence for the nature and development of the Middle Proterozoic (Keweenawan) Midcontinent Rift of North America. *Tectonophysics*, **94**, 413–437.

Groves, J.R. (2004) Bedrock geology of the Cedar Falls/Waterloo Area. In: *From Ocean to Ice: An examination of the Devonian Bedrock and Overlying Pleistocene Sediments at Messerly & Morgan Quarries, BlackHawk County, Iowa* (Eds J. Walters, J. Groves and S. Lundy), *Geol. Soc. Iowa Guidebook*, **75**, 29–36.

Hallock, P. (1988) The role of nutrient availability in bioerosion – consequences to carbonate buildups. *Palaeogeogr. Palaeoclimatol. Palaeoecol.*, **63**, 275–291.

Hallock, P. and Schlager, W. (1986) Nutrient excess and the demise of coral reefs and carbonate platforms. *Palaios*, **1**, 389–398.

Heckel, P.H. (1991) Thin widespread Pennsylvanian black shales of Mid-Continent North America: a record of cyclic succession of widespread pycnoclines in a fluctuating epeiric sea. In: *Modern and Ancient Continental Shelf Anoxia* (Eds R.V. Tyson and T.H. Pearson), *Geol. Soc. Spec. Publ.*, **58**, 259–273.

Hillgärtner, H. (1998) Discontinuity surfaces on a shallow-marine carbonate platform (Berriasian, Valanginian, France and Switzerland). *J. Sed. Res.*, **68**, 1093–1108.

Husinec, A., Basch, D., Rose, B. and Read, J.F. (2008) FISCHERPLOTS: An Excel spreadsheet for computing Fischer plots of accommodation change in cyclic carbonate successions in both the time and depth domains. *Comput. Geosci.*, **34**, 269–277.

Immenhauser, A. (2009) Estimating palaeo-water depth from the physical rock record. *Earth Sci. Rev.*, **96**, 107–139.

Irwin, M.L. (1965) General theory of epeiric clear water sedimentation. *AAPG Bull.*, **49**, 445–459.

Johnson, J.G. (1970) Taghanic onlap and the end of North American Devonian provinciality. *Geol. Soc. Am. Bull.*, **81**, 2077–2106.

Johnson, J.G., Klapper, G. and Sandberg, C.A. (1985) Devonian eustatic fluctuations in Euramerica. *Geol. Soc. Am. Bull.*, **96**, 567–587.

Kaufmann, B. (2006) Calibrating the Devonian time scale: a synthesis of U-Pb ID–TIMS ages and conodont stratigraphy. *Earth-Sci. Rev.*, **76**, 175–190.

Montañez, I.P. and Osleger, D.A. (1993) Parasequence stacking patterns, third-order accommodation events, and sequence stratigraphy of middle to upper Cambrian Platform Carbonates, Bonanza King Formation, Southern Great Basin. In: *Carbonate Sequence Stratigraphy: Recent Developments and Applications* (Eds R. Loucks and J. Sarg), *AAPG Mem.*, **57**, 305–326.

Morales, C., Spangenberg, J.E., Arnaud-Vanneau, A., Adatte, T. and Föllmi, K.B. (2016) Evolution of the northern Tethyan Helvetic Platform during the late Berriasian and early Valanginian. *Depositional Rec.*, **2**, 47–73.

Morrow, J.R. and Sandberg, C.A. (2008) Evolution of Devonian carbonate-shelf margin, Nevada. *Geosphere*, **4**, 445–458.

Ojakangas, R.W., Morey, G.B. and Green, J.C. (2001) The Mesoproterozoic midcontinent rift system, Lake Superior region, USA. *Sed. Geol.*, **141**, 421–442.

Osleger, D. (1991) Subtidal carbonate cycles: implications for allocyclic vs. autocyclic controls. *Geology*, **19**, 917–920.

Paces, J.B. and Miller, J.D. (1993) Precise U-Pb ages of Duluth complex and related mafic intrusions, northeastern Minnesota: geochronological insights to physical, petrogenetic, paleomagnetic, and tectonomagmatic processes associated with the 1.1 Ga midcontinent rift system. *J. Geophys. Res. Solid Earth*, **98**, 13997–14013.

Plocher, O.W. (1990) *Biotic, Petrographic and Diagenetic Analysis of Strata that Bound the Contact Between the Little Cedar and Coralville Formations (Middle Devonian) East-central Iowa.* Unpublished Masters Thesis. University of Iowa, Iowa City, IA.

Pomar, L. (2001) Ecological control of sedimentary accommodation: evolution from a carbonate ramp to rimmed shelf, Upper Miocene, Balearic Islands. *Palaeogeogr. Palaeoclimatol. Palaeoecol.*, **175**, 249–272.

Pope, M.C. and Steffen, J.B. (2003) Widespread, prolonged late Middle to Late Ordovician upwelling in North America: a proxy record of glaciation? *Geology*, **31**, 63–66.

Preslicka, J.E., Newsom, C.R., Blume, T.E. and Rocca, G.A. (2010) Cephalopods of the Lower Cedar Valley Group: a general overview. In: *The Geology of Klein and Conklin Quarries, Johnson County, Iowa* (Eds J. Marshall and C. Fields), *Geol. Soc. Iowa Guidebook*, **87**, 99–116.

Rameil, N., Immenhauser, A., Csoma, A.E. and Warrlich, G. (2012) Surfaces with a long history: the Aptian top Shu'aiba Formation unconformity, Sultanate of Oman. *Sedimentology*, **59**, 212–248.

Read, J.F. and Goldhammer, R.K. (1988) Use of Fischer plots to define third-order sea-level curves in Ordovician peritidal cyclic carbonates, Appalachians. *Geology*, **16**, 895–899.

Sadler, P.M., Osleger, D.A. and Montañez, I.P. (1993) On the labeling, length, and objective basis of Fischer plots. *J. Sediment. Res.*, **63**, 360–368.

Sandberg, C. (2009) *Recognition of Fox Mountain Formation solves enigmatic Devonian stratigraphy in central and northern Utah*, Vol. 41, p. 15. Geological Society of America Abstracts with Programs, Boulder, CO, 15 p.

Sattler, U., Immenhauser, A., Hillgärtner, H. and Esteban, M. (2005) Characterization, lateral variability and lateral extent of discontinuity surfaces on a carbonate platform (Barremian to Lower Aptian, Oman). *Sedimentology*, **52**, 339–361.

Schlager, W. (1993) Accommodation and supply – a dual control on stratigraphic sequences. *Sed. Geol.*, **86**, 111–136. https://doi.org/10.1016/0037-0738(93)90136-s.

Schlager, W. (2005) *Carbonate sedimentology and sequence stratigraphy, SEPM Concepts in Sedimentology and Paleontology*, Vol. **8**. SEPM Society for Sedimentary Geology, Tulsa, OK, 200 p.

Scholle, P.A. and Ulmer-Scholle, D.S. (2003) *A color guide to the petrography of carbonate rocks: grains, textures, porosity, diagenesis*. American Association of Petroleum Geologists, Tulsa, OK, 474 pp.

Seilacher, A. and Aigner, T. (1991) Storm deposition at the bed, facies, and basin scale: the geologic perspective. In: *Cycles and Events in Stratigraphy* (Eds G. Einsele, W. Ricken and A. Seilacher), pp. 249–267. Springer Verlag, Berlin.

Sibley, D.F. and Gregg, J.M. (1987) Classification of dolomite rock textures. *J. Sed. Res.*, **57**, 908–931.

Sloss, L.L. (1963) Sequences in the cratonic interior of North America. *Geol. Soc. Am. Bull.*, **74**, 93–114.

Sloss, L.L. (1996) Sequence stratigraphy on the craton: caveat emptor. *Geol. Soc. Am. Spec. Pap.*, **306**, 425–434.

Soreghan, G.S. and Dickinson, W.R. (1994) Generic types of stratigraphic cycles controlled by eustasy. *Geology*, **22**, 759–761.

Spence, G.H. and Tucker, M.E. (2007) A proposed integrated multi-signature model for peritidal cycles in carbonates. *J. Sed. Res.*, **77**, 797–808.

Strasser, A. (2015) Hiatuses and condensation: an estimation of time lost on a shallow carbonate platform. *Depositional Rec.*, **1**, 91–117.

Strasser, A., Pittet, B., Hillgärtner, H. and Pasquier, J.B. (1999) Depositional sequences in shallow carbonate-dominated sedimentary systems: concepts for a high-resolution analysis. *Sed. Geol.*, **128**, 201–221.

Tucker, M., Calvet, F. and Hunt, D. (1993) Sequence stratigraphy of carbonate ramps: systems tracts, models and application to the Muschelkalk carbonate platforms of eastern Spain. In: *Sequence Stratigraphy and Facies Association, Sequence Stratigraphy and Facies Association* (Eds H.W. Posamentier, C. P Summerhayes, B.U. Haq and G.P. Allen), *Int. Assoc. Sediment. Spec. Publ.*, **18**, 397–415.

Van Schmus, W.R. and Hinze, W.J. (1985) The midcontinent rift system. *Annu. Rev. Earth Planet. Sci.*, **13**, 345–383.

Warme, J.E., Morrow, J.R. and Sandberg, C.A. (2008) Devonian carbonate platform of eastern Nevada: Facies, surfaces, cycles, sequences, reefs, and cataclysmic Alamo Impact Breccia. In: *Field Guide to Plutons, Volcanoes, Faults, Reefs, Dinosaurs, and Possible Glaciation in Selected Areas of Arizona, California, and Nevada* (Eds E.M. Duebendorfer and E.I. Smith), *Geol. Soc. Am. Field Guide*, **11**, 215–247.

Whiting, L.L. and Stevenson, D.L. (1965) *The Sangamon Arch, State of Illinois*. Department of Registration and Education, Illinois State Geological Survey, Urbana, IL.

Witzke, B.J. (1987) Models for circulation patterns in epicontinental seas applied to Paleozoic facies of North America craton. *Paleoceanography*, **2**, 229–248.

Witzke, B.J. and Bunker, B.J. (1996) Relative sea-level changes during Middle Ordovician through Mississippian deposition in the Iowa area, North American Craton. *Geol. Soc. Am. Spec. Pap.*, **306**, 307–330.

Witzke, B.J. and Bunker, B.J. (1997) Sedimentation and stratigraphic architecture of a Middle Devonian (late Givetian) transgressive-regressive carbonate-evaporite cycle, Coralville Formation, Iowa Area. *Geol. Soc. Am. Spec. Pap.*, **321**, 67–88.

Witzke, B.J. and Bunker, B.J. (2010) Devonian stratigraphy of Johnson County, Iowa; Conklin-Klein Quarries and Surrounding Area. In: *The Geology of Klein and Conklin Quarries, Johnson County, Iowa* (Eds J. Marshall and C. Fields), *Geol. Soc. Iowa Guidebook*, **87**, 33–72.

Witzke, B., Bunker, B. and Rogers, F. (1988) Eifelian through lower Frasnian stratigraphy and deposition in the Iowa area, central Midcontinent, USA. In: *Devonian of the World, Volume I: Regional Syntheses* (Eds N.J. McMillan, A.F. Embry and D.J. Glass), *Can. Soc. Petrol. Geol. Memoir*, **14**, 221–250.

Witzke, B.J., Ludvigson, G.A. and Day, J. (1996) Introduction: paleozoic applications of sequence stratigraphy. *Geol. Soc. Am. Spec. Pap.*, **306**, 1–6.

5

Methane seepage in a Cretaceous greenhouse world recorded by an unusual carbonate deposit from the Tarfaya Basin, Morocco

DANIEL SMRZKA*, JENNIFER ZWICKER*, SADAT KOLONIC†, DANIEL BIRGEL‡, CRISPIN T.S. LITTLE§, AKMAL M. MARZOUK¶, EL HASSANE CHELLAI**, THOMAS WAGNER†† and JÖRN PECKMANN*,‡ (iD)

*Department für Geodynamik und Sedimentologie, Universität Wien, 1090 Wien, Austria
†Shell Petroleum Development Company of Nigeria, Port Harcourt, Nigeria
‡Institut für Geologie, Universität Hamburg, Bundesstraße 55, 20146 Hamburg, Germany (E-mail: joern.peckmann@uni-hamburg.de)
§School of Earth and Environment, University of Leeds, Leeds, LS2 9JT, UK
¶Geology Department, Faculty of Science, Tanta University, 31527 Tanta, Egypt
**Department of Geology, Faculty of Sciences Semlalia, Cadi Ayyad University, Marrakech, Morocco
††Lyell Centre, Heriot-Watt University, Edinburgh, EH14 4AS, UK

Keywords

Black shales, carbonate authigenesis, cold seeps, oceanic anoxic events, Tarfaya Basin.

ABSTRACT

During the Cretaceous major episodes of oceanic anoxic conditions triggered large scale deposition of marine black shales rich in organic carbon. Several oceanic anoxic events (OAEs) have been documented including the Cenomanian to Turonian OAE 2, which is among the best studied examples to date. This study reports on a large limestone body that occurs within a black shale succession exposed in a coastal section of the Tarfaya Basin, Morocco. The black shales were deposited in the aftermath of OAE 2 in a shallow continental sea. To decipher the mode and causes of carbonate formation in black shales, a combination of element geochemistry, palaeontology, thin section petrography, carbon and oxygen stable isotope geochemistry and lipid biomarkers are used. The ^{13}C-depleted biphytanic diacids reveal that the carbonate deposit resulted, at least in part, from microbially mediated anaerobic oxidation of methane in the shallow subseafloor at a hydrocarbon seep. The lowest obtained $\delta^{13}C_{carbonate}$ values of $-23.5‰$ are not low enough to exclude other carbon sources than methane apart from admixed marine carbonate, indicating a potential contribution from *in situ* remineralization of organic matter contained in the black shales. Nannofossil and trace metal inventories of the black shales and the macrofaunal assemblage of the carbonate body reveal that environmental conditions became less reducing during the deposition of the background shales that enclose the carbonate body, but the palaeoenvironment was overall mostly characterized by high productivity and episodically euxinic bottom waters. This study reconstructs the evolution of a hydrocarbon seep that was situated within a shallow continental sea in the aftermath of OAE 2, and sheds light on how these environmental factors influenced carbonate formation and the ecology at the seep site.

INTRODUCTION

The Cretaceous was a period in Earth history when major episodes of oceanic anoxic conditions led to large scale accumulation of organic matter in marine sediments (Arthur & Sageman, 1994; Larson & Erba, 1999; Jones & Jenkyns, 2001). These organic-rich deposits have been termed black shales, and phases of their coeval and repeated occurrence in the Phanerozoic rock record have been referred to as oceanic anoxic events (OAEs, Schlanger & Jenkyns, 1976; Weissert et al., 1979; Schlanger et al., 1987; Kerr, 1998; Jenkyns, 2003; Meyer & Kump, 2008). Sea-level was markedly elevated during some periods of the Cretaceous compared to the present due to the warm, equable climate (Sames *et al.*, 2016 and references therein). The expansion of shallow epicontinental seas, where upwelling and enhanced nutrient flux from the continents increased primary production, enabled the accumulation of organic carbon-rich sediments in water depths of several hundreds of metres (Hallam & Bradshaw, 1977; Weissert, 1981; Kuhnt *et al.*, 1990; Arthur & Sageman, 1994). Two processes held primarily responsible for the deposition of black shales are (1) pulses of enhanced primary productivity accompanied by elevated atmospheric carbon dioxide concentrations and sea surface temperatures (Barron, 1983; Ingall *et al.*, 1993; Forster *et al.*, 2007), and (2) the increased preservation of organic matter within restricted, stratified, oxygen-deficient basins (Mascle *et al.*, 1997; Meyer & Kump, 2008). With supporting evidence for both scenarios, it is likely that both production and preservation in stratified basins increased sequestration rates of organic carbon in Cretaceous marine sediments (Jenkyns *et al.*, 2007; Owens *et al.*, 2013).

Several OAEs have been documented during the Cretaceous including the Aptian to Albian OAE 1, the Cenomanian to Turonian OAE 2 and the Coniacian to Santonian OAE 3 (Jenkyns, 2010 and references therein). The OAE 2, or Bonarelli event (Bonarelli, 1891; Schlanger *et al.*, 1987), is among the best studied examples to date. High rates of sea floor spreading and volcanic activity – both linked to the breakup of Gondwana with the formation of early ocean basins as carbon sinks and the evolution of ocean gateways – led to increasing atmospheric carbon dioxide concentrations followed by a significant turnover of ocean chemistry and extinctions of marine biota (Brumsack, 1980; Arthur *et al.*, 1985; Larson, 1991; Kaiho, 1994; Wagner & Pletsch, 1999; Gale *et al.*, 2000; McAnena *et al.*, 2013). Black shales related to OAE 2 have been reported from many parts of the world including the proto southern North Atlantic and eastern Tethys Oceans (Lüning *et al.*, 2004). During OAE 2 the Tethys and Atlantic Oceans were affected by sluggish circulation of bottom waters caused by the shallow and partly closed connections to neighbouring oceans (Kuypers *et al.*, 2002). Replenishment of oxygen was impeded within these partially isolated basins, favouring the development of anoxic and, possibly, euxinic conditions even in the photic zone (Herbin *et al.*, 1986; Baudin, 1995; Sinninghe Damsté & Köster, 1998; Handoh *et al.*, 1999). Large areas of the continental shelf and slope along the northern parts of Algeria, Libya, Tunisia and Morocco were deprived of oxygen, and enabled the rapid accumulation of shales and limestones rich in organic carbon linked to cyclic perturbations of redox-nutrient cycles and orbital forcing (Leine, 1986; El Albani *et al.*, 1999a; Nederbragt & Fiorentino, 1999; Kolonic *et al.*, 2005; Poulton *et al.*,

2015; for Atlantic-wide teleconnections see Wagner *et al.*, 2013 and references therein).

The Cretaceous has yielded several unusual carbonate deposits occurring before, during and after the OAEs. Some of these deposits have been ascribed to carbonate precipitation related to chemosynthesis-based microbial activity, and include examples from the Western Interior Seaway (Kauffman *et al.*, 1996; Metz, 2010; Kiel *et al.*, 2012) and the northern Tethyan and North African continental margins (Layeb *et al.*, 2012, 2014). Limestones from the Western Interior Seaway feature shallow-water methane-seep deposits that formed during OAE 2 in warm bottom waters (Kiel *et al.*, 2012), and are characterized by ^{13}C-depleted carbonate mineral phases and a high-abundance but low-diversity faunal assemblage. The Campanian organic-rich Pierre Shales comprise numerous peculiar carbonate bodies, collectively known as the Tepee Buttes (Shapiro & Fricke, 2002; Metz, 2010). These deposits are rich in chemosymbiotic macrofaunal communities and exhibit diagnostic ^{13}C-depleted mineral fabrics and ^{13}C-depleted molecular fossils of the anaerobic oxidation of methane (AOM) consortium, confirming that carbonate precipitation was driven by microbial oxidation of methane (Kauffman *et al.*, 1996; Birgel *et al.*, 2006). The fossiliferous Tepee Butte limestones possibly formed during periods of enhanced bottom water oxygenation, supported by the presence of molecular fossils of aerobic methanotrophic bacteria (Birgel *et al.*, 2006). Layeb *et al.* (2012) documented carbonate build-ups from the Lower Cretaceous Fahdene Basin in Tunisia. These limestones reveal a thrombolitic fabric and formed within organic-rich black shales deposited during the Late Albian OAE 1d. The fabric and texture of these limestones suggest the involvement of microbes during their formation below a water body with eutrophic, anoxic waters. Layeb *et al.* (2014) further reported on mound-shaped carbonate structures associated with OAE 2 within Late Cenomanian black shales from the Bahloul Formation in Northern Tunisia. The genesis of the Tunisian limestones possibly involved microorganisms mineralizing large amounts of organic matter, as suggested by ^{13}C-depleted carbonate cements and thrombolitic fabrics (Layeb *et al.*, 2014).

This study reports on a large limestone body associated with black shales deposited in the aftermath of OAE 2 exposed along a coastal section of the Tarfaya Basin, Morocco. A combination of petrography, macro- and micropalaeontology, carbon and oxygen stable isotope and trace element geochemistry, as well as lipid biomarkers has been used to decipher the mode of carbonate accretion that led to the unusual morphology of this limestone body. The potential triggers for carbonate precipitation within the mid-Cretaceous organic-rich deposits, and the impact of the prevailing oceanographic and geochemical conditions on the benthonic environment and faunal communities are assessed. Although the Amma Fatma limestone body reveals only a limited number of lithological features typical of hydrocarbon seep carbonates and apparently no endemic seep fauna, it is shown to have resulted from methane seepage. The lack of obligate, endemic seep fauna is interpreted to have been caused by (1) shallow water depth – a factor known to discriminate against seep taxa (Kiel, 2010) – and (2) the eutrophic palaeoenvironment in the aftermath of OAE 2 that favoured photosynthetic primary production and subsequent consumption of organic matter by heterotrophs. This work reveals that seep deposits are difficult to identify in high productivity settings, where chemosynthesis-based seep fauna is likely to be outcompeted by heterotrophic background fauna.

GEOLOGICAL SETTING

Deposits from the Cenomanian–Turonian (C/T) boundary in Northern Africa have been intensively studied in Tunisia and Morocco. This is partly due to the fact that these regions feature up to 800 m thick successions of organic-rich shales and limestones (Kuhnt et al., 2001; Kolonic

et al., 2002), making them prolific petroleum source rocks (Leine, 1986) and high resolution climate archives (Kolonic et al., 2005; Kuhnt et al., 2005). The southern Moroccan passive continental margin includes the Tarfaya Basin, one of several palaeo-shelf basins located on the coasts of Senegal, Morocco and Tunisia (Figs 1 and 2A; Lüning et al., 2004; Kolonic et al., 2005). It covers an area of around 170 000 km^2 and features the highest accumulation rates of organic matter from the C/T boundary known to date, as well as one of the largest oil shale deposits in the world (Amblés et al., 1994; Kuhnt et al., 2001). The Tarfaya Basin is situated to the south of the Anti-Atlas Mountains and extends into the Tindouf Basin to the East, the Senegal Basin to the South, and is bordered by the East Canary Ridge to the West (Kolonic et al., 2002). It has been suggested that the basin was a distal section of the proto southern North Atlantic with outer shelf to upper bathyal water depths of 200 to 300 m (Figs 1 and 2B; El Albani et al., 1999b; Kuhnt et al., 2001). It experienced intense upwelling, extended oxygen minimum zones and recurrent pulses of high primary production since at least the Late Cenomanian (Kuhnt & Wiedmann, 1995; El Albani et al., 1999a,b; Kolonic et al., 2002, 2005), leading to pronounced redox cycles on orbital time scales that alternated between sulphidic and anoxic ferruginous water-column conditions (Poulton et al., 2015).

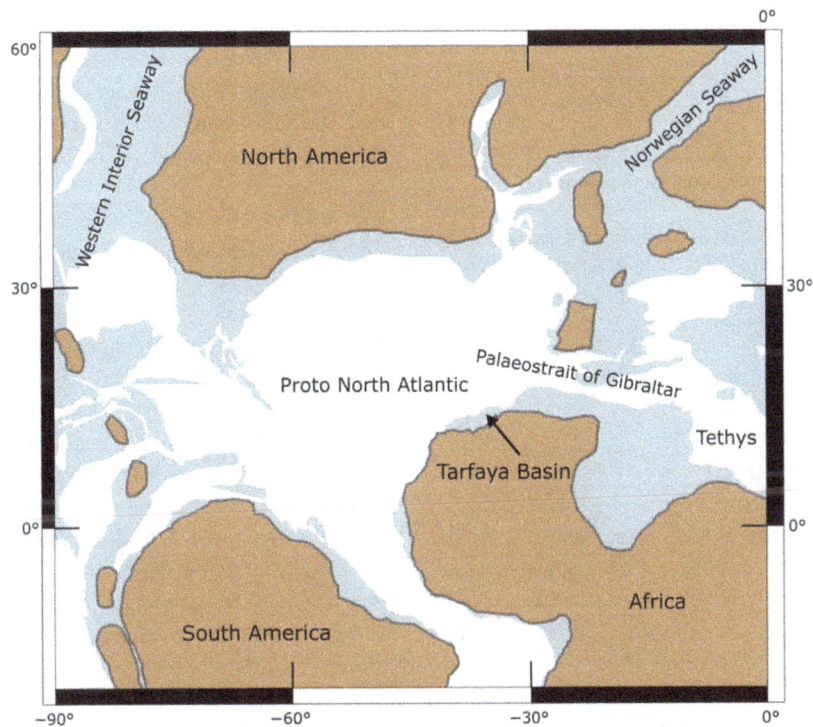

Fig. 1. Palaeogeographical map of the mid-Cretaceous (ca 94 Ma) showing the study site within the Tarfaya Basin, Morocco. Light blue areas represent continental plates covered by shallow epicontinental seas, brown areas represent dry land. Brown, thick lines delineate palaeocoastlines (modified from Kuypers et al., 2002).

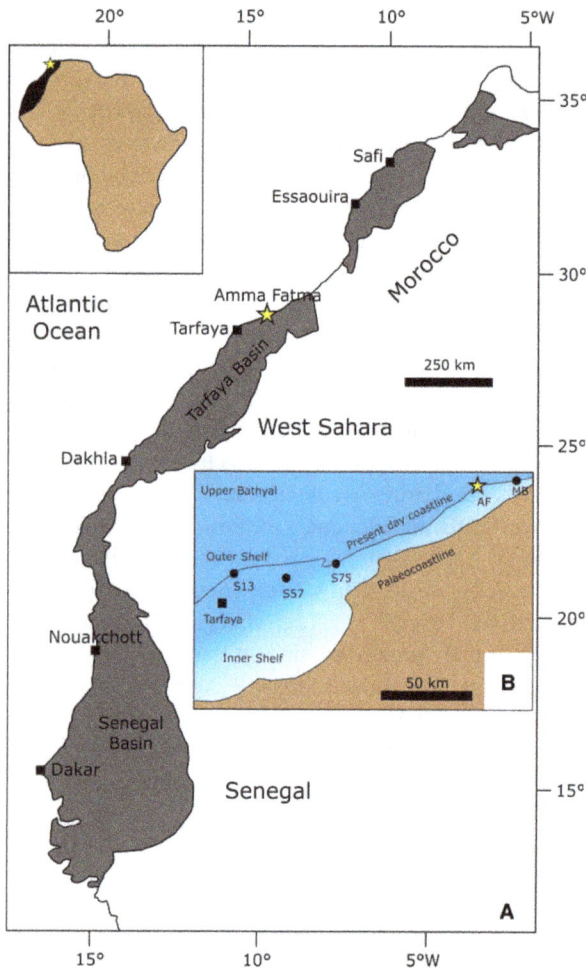

Fig. 2. (A) Locality map showing the distribution of unfolded Mesozoic and Cenozoic Basins containing black shales (grey) from the C/T boundary along the northwestern African continental margin (modified from Kolonic *et al.*, 2005). Yellow stars indicating the Amma Fatma section in (A), (B) and the small inset map in the upper left, which indicates the location of the studied section in Africa. (B) Palaeogeographic map of the Tarfaya Basin (modified from Kolonic *et al.*, 2005); AF = Amma Fatma, MB = Mohammed Plage, S13, S57 and S75 = drill cores described by Leine (1986) and Kolonic *et al.* (2005).

The carbonate body studied here crops out at Amma Fatma Plage, which is a coastal to middle shelf section at the northern edge of the Tarfaya Basin (Fig. 2). Amma Fatma Plage is located *ca* 8 km to the south-west of Mohammed Plage, the latter being a well-studied coastal section of the basin where deposition of organic-rich shales and carbonate beds occurred close to the palaeocoastline (El Albani *et al.*, 1999b; Kolonic *et al.*, 2005). Both sites experienced a series of transgressional cycles during the mid to late Cretaceous that caused repeated and widespread flooding, and initiated the deposition of laminated biogenic sediments associated with OAE 2

(Kolonic *et al.*, 2002, 2005; Gebhardt *et al.*, 2004; Kuhnt *et al.*, 2009; Gertsch *et al.*, 2010). Amma Fatma Plage lies along a 175 km long palaeotransect through the Tarfaya Basin, together with shallow exploration wells S13, S57, S75 and Mohammed Plage (cf. Kolonic *et al.*, 2002, 2005). The transect runs in SW to NE direction and connects its deepest and most distal part at wells S13 and S57 to its most proximal part at Mohammed Plage (Fig. 2B). Exploration well S13 is interpreted as an open-marine shelf setting with water depths of 200 to 300 m, whereas Mohammed Plage and Amma Fatma Plage represent shallower, proximal successions deposited at the northwestern margin of the Sahara platform (Kolonic *et al.*, 2002, 2005; Kuhnt *et al.*, 2009).

MATERIAL AND METHODS

Petrography, mineralogy and palaeontology

Rock samples from the carbonate body and from a vertical profile at Amma Fatma Plage were collected during a field trip in 2003. Samples from the studied sedimentary succession were recovered from a fresh coastal exposure in direct proximity to the carbonate body, and taken at regular intervals of 5 to 10 cm across the profile. A total of 20 samples were collected from various zones of the carbonate body, two further samples were collected from a carbonate bank covering the structure (hereafter referred to as cover bed). Thin sections (150 × 100 mm, 100 × 75 mm) were made at the University of Bremen and the University of Vienna. Thin section microscopy was conducted using a Leica DM4500 P polarization microscope (Wetzlar, Germany). Photographs were taken with a Leica DFC 450 C camera using the Leica Application Suite 4.4.0 software. Additionally, a Nikon SMZ 1500 stereomicroscope coupled to a Prog.Res Speed XT core 5 camera, as well as a Nikon Optiphot-2 optical microscope (Tokyo, Japan) were used for petrographic observations using the software Prog.Res Capture pro 2.8 for image analysis and camera control. To discriminate the different carbonate minerals under the microscope, several thin sections were stained with a mixture of potassium ferricyanide and alizarin red S solution dissolved in 0·1% hydrochloric acid (cf. Dickson, 1966). Powdered samples for X-ray diffraction (XRD) and carbon and oxygen stable isotope analyses were obtained from polished rock slabs using a hand-held microdrill at low to medium rotational speed. For XRD analysis, powdered samples were analysed with a Panalytical PW 3040/60 X'Pert PRO (CuKα radiation, 40 kV, 40 mA, step size 0·0167·5 s per step) diffractometer at the University of Vienna. X-ray diffraction patterns were interpreted using the Panalytical software "X'Pert High score plus". Calcareous nannoplankton assemblages were studied in 150 samples with a high-resolution spacing of about 4 to 7 cm within the slabbed intervals. For the calcareous nannoplankton, smear slides were prepared using techniques described in Bramlette & Sullivan (1961) and Hay (1961, 1965). The slides were examined at a magnification of about 1250 times under both cross-polarized and phase-contrast illumination.

Inorganic geochemistry – Total organic carbon, carbon and oxygen stable isotopes and trace elements

Total carbon (TC) and total organic carbon (TOC) contents of the Amma Fatma carbonate body were determined using a LECO RC-612 Carbon Analyzer (St. Joseph, MI, USA), equipped with a solid-state infrared detector at the Department of Environmental Geosciences at the University of Vienna. All samples were ground to fine powder in an agate mortar prior to measurements. To measure TC, carbon dioxide was released at a temperature of 1000°C. Prior to sample measurements a pure $CaCO_3$ standard (Co. Merck, Kenilworth, NJ, USA) was measured, revealing a relative standard deviation of 0·06%. The TOC contents were determined at 550°C. A Synthetic Carbon Leco 502-029 (1·01 ± 0·02 carbon%) standard was measured prior to sample analyses and a standard deviation of 0·005% was calculated. Carbonate ($CaCO_3$%) contents were calculated according to the equation: $CaCO_3$% = (TC − TOC) × 8·333. All black shale samples were carefully dried and ground in an agate mortar. The TIC and TOC contents of black shales were measured using a LECO CS-300 carbon-sulphur analyser at the Geoscience Department, University of Bremen. Before TOC determination, inorganic carbon was carefully removed by repetitive addition of 0·25 N HCl. The $CaCO_3$ contents were calculated as above (precision of measurements ±3%). The $\delta^{13}C_{TOC}$ values (±0·1‰ vs. Vienna Pee Dee belemnite, V-PDB) were determined on decalcified sediment samples using automated on-line combustion followed by conventional isotope-ratio mass spectrometry.

A total of 48 sample powders from 12 different samples throughout the Amma Fatma carbonate body were analysed for their stable carbon and oxygen isotope content. Sample powders were obtained from polished rock slabs using a hand-held microdrill. Stable carbon and oxygen isotope analyses were performed at the light stable isotope laboratory of the Institute of Earth Sciences, Karl-Franzens University, Graz. Effective accuracy was ±0·05 for $\delta^{13}C$ values and ±0·11 for $\delta^{18}O$ values. The $\delta^{13}C$ and $\delta^{18}O$ values are reported relative to V-PDB (standard deviation smaller than 0·04‰), and appropriate correction factors were applied.

Trace metal compositions of black shales and limestones were determined using X-ray fluorescence (XRF) spectrometry on fused-borate glass beads, and by inductively coupled plasma atomic emission spectrometry (ICP-AES). The XRF calibration curves were based on international standard reference material. For ICP-AES, about 50 mg of sample powder was digested in a mixture of 3 ml HNO_3 (65%), 2 ml HF (40%) and 2 ml HCl (36%) of suprapure quality at 200°C and 30 kbar in closed Teflon vessels (Heinrichs et al., 1986). Sample powders were then dried by evaporation, and the residue was re-dissolved in 0·5 ml HNO_3 (65%) and 4·5 ml deionized water. The solutions were analysed with ICP-AES, and the results were checked with international standard reference material. Relative standard deviations in duplicate measurement were below 3%. Enrichment factors (EFs) of trace elements were calculated according to Brumsack (2006):

$$EF_{element} = (element/Al)_{sample}/(element/Al)_{average\ shale}$$

Average shale refers to the composition of collected shale samples compiled by Wedepohl (1969). An element with an EF either above or below 1 is considered to be enriched or depleted compared to average shale, respectively.

Organic geochemistry – Lipid biomarkers

Four carbonate rock samples were chosen for biomarker analysis, one from the cover bed and one from each of the zones 4, 3 and 1. The samples were cleaned repeatedly by washing with 10% hydrochloric acid (HCl) and then dichloromethane (DCM) to avoid any contamination, and then ground to fine powder. All carbonates were treated with a 'two-step extraction procedure' to discriminate preferentially intercrystalline-bound from preferentially intracrystalline-bound molecular fossils (cf. Thiel et al., 1999; Guido et al., 2013), and to separate compounds tightly associated with the carbonate matrix from easily extractable compounds. To obtain preferentially intercrystalline-bound compounds, the ground samples were extracted by ultrasonication in DCM : MeOH (3 : 1) nine times until the solvents became colourless to get the first total lipid extract (TLE). This extract is referred to as the extract before decalcification below. To obtain preferentially intracrystalline-bound molecular fossils, the residual sediment of the first extraction step was decalcified and then extracted again. In detail, doubly distilled water was added and 10% HCl was slowly poured onto the samples. To avoid transesterification reactions at low pH, HCl addition was stopped when ca 75% of the carbonate was dissolved. The remaining sample powder was centrifuged and extracted as described above to obtain a second TLE. This extract is referred to as the extract after decalcification below. From here on, the extracts before and the extract after decalcification are treated in the same fashion. After washing with clean water, the samples were saponified with 6% potassium hydroxide (KOH) in methanol (MeOH) to release bond carboxylic acids. The supernatants were decanted and the residues were subsequently extracted via ultrasonication in DCM/MeOH (3 : 1; v : v) four times until the solvents became colourless. The combined extracts were then treated with 10% HCl to transfer the free fatty acids (FAs) to the organic solvent phase. For gas chromatography (GC) analyses the extracts were dried with sodium sulphate and the TLE of both extraction procedures were separated by column chromatography. Alcohols were derivatized by adding 100 µl pyridine and 100 µl N,O-bis(trimethylsilyl) trifluoracetamide (BSTFA) to the alcohol fraction at 70°C for 30 min. The derivatized fraction was dried under a stream of nitrogen (N$_2$) and dissolved in n-hexane prior to injection. The alcohol fractions are not described below due to the lack of pristine compounds. Free carboxylic acids were reacted with 1 ml 14% boron trifluoride (BF$_3$) in MeOH at 70°C for 1 h to form FA methyl esters. After cooling, the mixture was extracted four times with 2 ml n-hexane. Combined extracts were evaporated under a stream of N$_2$ and redissolved in n-hexane before injection. The derivatized carboxylic acid fractions were analysed using GC-mass spectrometry (GC-MS) with an Agilent 7890 A GC system coupled to an Agilent 5975C inert MSD mass spectrometer. Quantification was done using GC-flame ionization detection (GC-FID) with an Agilent 7820 A GC system. Internal standards used were 5α-cholestane for the hydrocarbon and 2-Me-C$_{18}$ FA for the carboxylic acid fractions. Both GC systems were equipped with HP-5 MS UI fused silica columns (30 m × 0·25 mm i.d., 0·25 µm film thickness). Helium was used as the carrier gas. The GC temperature program for hydrocarbons and carboxylic acids was 60°C (1 min) to 150°C at 10°C min^{-1}, then to 320°C (held 25 min) at 4°C min^{-1}. Compound assignment was based on retention times and published mass spectra.

Compound-specific stable carbon isotope analysis was performed using a Thermo Fisher Trace GC Ultra connected via a Thermo Fisher GC Isolink interface to a Thermo Fisher Delta V Advantage spectrometer at the Department of Terrestrial Ecosystem Research, University of Vienna. Compound-specific carbon isotope values are given as δ values in per mil (‰) relative to V-PDB. The δ^{13}C values of carboxylic acids (methyl esters) were corrected for additional carbon introduced after derivatization. Each measurement was calibrated using several pulses of carbon dioxide (CO$_2$) of known isotopic composition prior to and after the run. Precision was checked

with an n-alkane mixture (C$_{14}$ to C$_{38}$) of known isotopic composition. Analytical standard deviation was <0·7‰.

RESULTS

Amma Fatma section – Lithology, biostratigraphy and inorganic geochemistry

The complete section at Amma Fatma Plage measures ca 14 m from top to bottom and features cyclic alternations of black shales, siltstones and organic-rich, marly limestone beds (Figs 3 and 4). The limestones are represented by (1) carbonate nodules mainly composed of mudstones and wackestones that are aligned parallel to stratification and (2) discontinuous bioclastic storm beds with hummocky cross-stratification (cf. El Albani et al., 1999b). The limestones contain Foraminifera, calcispheres, molluscs including ammonites, brachiopods, variable amounts of quartz grains and are partly bioturbated. A detailed microfacies analysis is included in El Albani et al. (1997, 1999a). The succession of 22 beds is shown by the stratigraphic chart of the outcrop in Figs 3 and 5. Limestone and black shale beds alternate regularly, whereby their thicknesses measure 20 to 80 cm and 60 to 300 cm, respectively. Beds comprising limestone nodules with chert measure up to 20 cm in thickness. One exception is a 1 m thick bed 3 m above the base of the profile featuring large carbonate nodules with a corona made of chert. Calcium carbonate and TOC contents of the black shale beds vary from 60 to 90% and 4 to 9%, respectively. Calcium carbonate and TOC values throughout the section are shown in Table 1. The characteristic positive δ^{13}C$_{TOC}$ excursion that is prominently observed across the C/T boundary in adjacent sites is not found at Amma Fatma Plage, supporting a post OAE period of deposition. The measured δ^{13}C$_{TOC}$ values vary slightly between $-26·2$ and $-27·8$‰.

Biostratigraphic data confirm the positioning relative to the OAE 2 carbon isotope excursion, placing the investigated sedimentary succession within the $H. helvetica$ planktonic foraminiferal zone and the $coloradoense$ and $nodosoides$ ammonite zone of the upper part of the Lower Turonian stage (cf. Marzouk & Lüning, 2005). The position of the nannofossil zonal datum marker CC11/CC12 is placed at about 9·5 m above the $helvetica$ and $nodosoides$ zones. Planktonic Foraminifera are present in very high numbers and benthonic foraminiferal faunas are of low diversity. A highly diverse and highly abundant calcareous nannofossil flora was observed in all samples studied. Preservation is moderate, except in the uppermost part of the section (from 13·5 m onwards), where preservation is good. $Watznaueria barnesae$ and $Eiffelithus turriseffeilii$ are abundant throughout the section (except

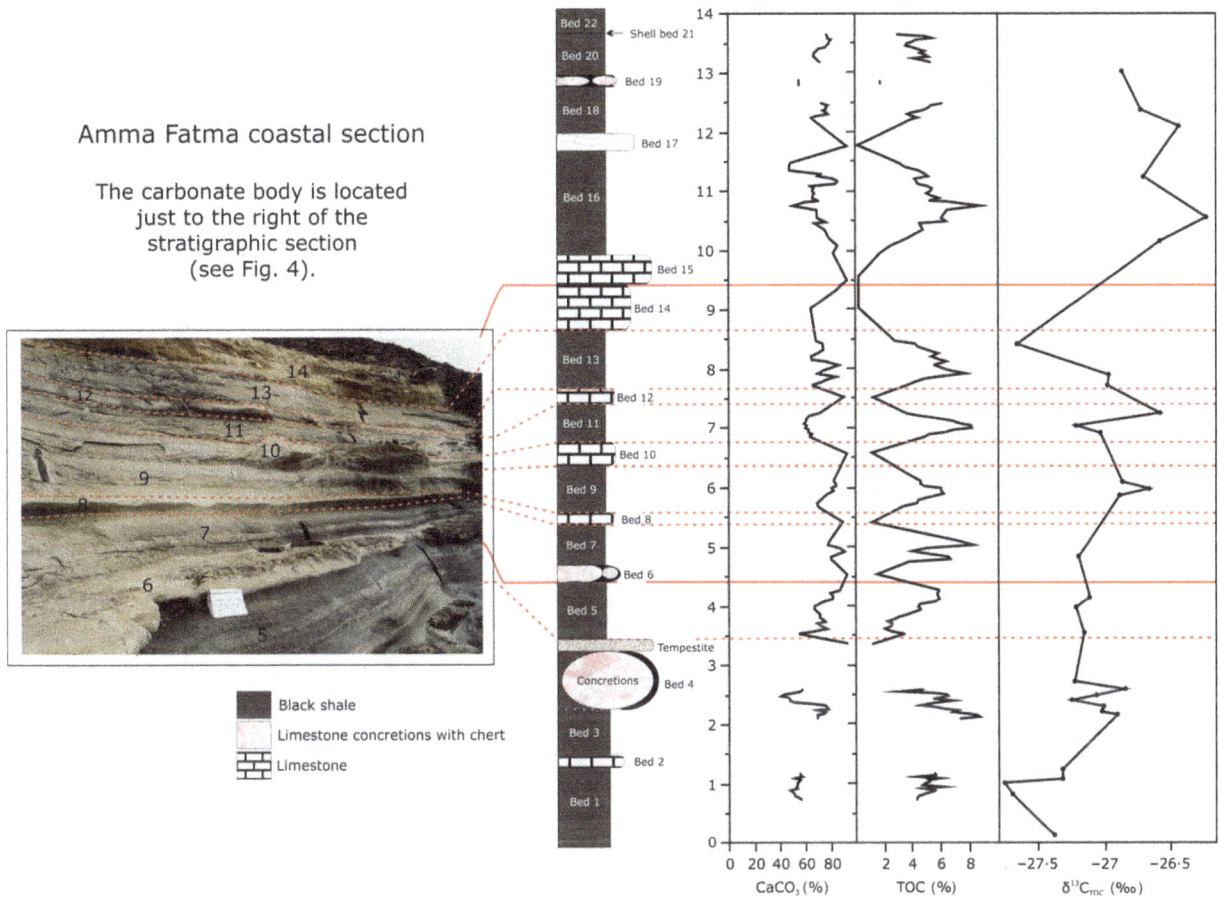

Fig. 3. (Left) Photograph of the Amma Fatma outcrop section showing beds 5 to 14 separated by dashed red lines; the Amma Fatma carbonate body lies within beds 6 to 14 delimited by the thick red lines (cf. Fig. 4). (Centre) Lithostratigraphic chart of the Amma Fatma coastal section showing lithologies and their corresponding thickness. (Right) CaCO₃ content, total organic carbon (TOC) content and its carbon stable isotope composition within the corresponding beds.

in two carbonate beds), and are known to be very resistant to solution. Although frequent, the abundance of *Watznaueria barnesae* does not surpass 40%, a cut-off value proposed by Roth & Krumbach (1986) to indicate that nannofossil distribution patterns can still largely be used for palaeoecological interpretations, despite diagenetic and syndepositional dissolution. Notably, *Cyclagelosphaera margerelii* shows a similarly abundant distribution and, therefore, appears to be also solution resistant.

The nannofossil distribution in the Amma Fatma section is characterized by many species having longer term, gradual changes, with only some species exhibiting stronger, high-frequency abundance fluctuations over a few centimetres (*Tranolithus orionatus*, *Zeugrhabdus diplogrammus*). Abundant species throughout the section include *Zygodiscus erectus*, *Rhagodiscus splendens* and *Eprolithus floralis*. *R. splendens* occurs in moderate and high abundances in the lowermost and uppermost parts of the studied section, and is absent in the interval in-

between. Notably, the distribution of *E. floralis* in the section is characterized by similar abundance peaks at the base and top of the section. Other identified species *Broisonia* spp. (except *B. enormis*) and *Gartnerago* spp. (Thierstein, 1976; Hattner *et al.*, 1980) are notably rare throughout the section. An exception is a *Gartnerago segmentatum* abundance peak between 7.5 and 8 m. Abundance and distribution of all identified nannofossil species and phosphorus content throughout the Amma Fatma section are shown in Fig. 5.

Total contents of all measured elements are shown in Table 1. Prominent trace metal enrichment is restricted to stratigraphic intervals that correspond to the organic carbon-rich black shales. Trace metal patterns vary across the section between alternating black shales and limestone beds, and also within the black shale successions themselves. Calculated EFs of trace metals whose deposition is influenced by bottom and pore water redox conditions are plotted with depth in Fig. 6. Mean trace metal

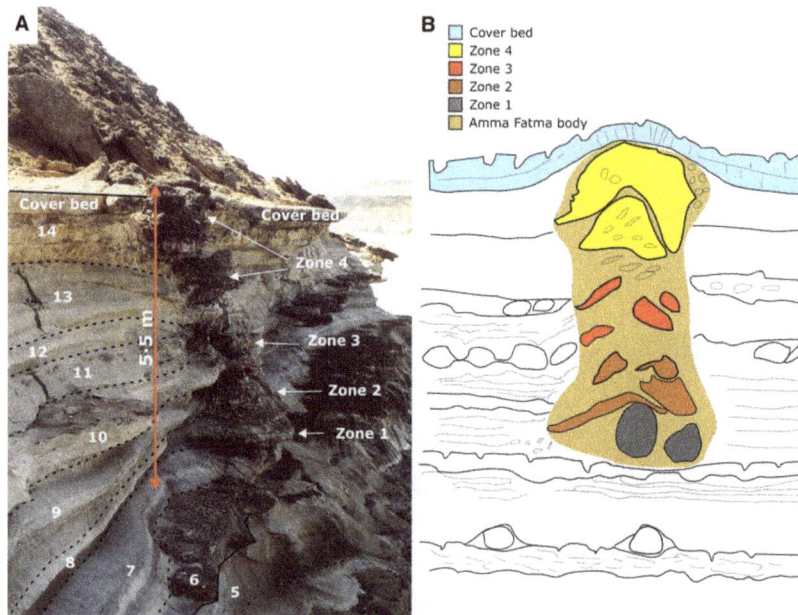

Fig. 4. (A) Photograph of the Amma Fatma carbonate body with its four zones highlighted by arrows amongst the beds shown in Fig. 3. (B) Schematic sketch of the Amma Fatma body highlighting the four zones from bottom to top and the cover bed capping the body.

concentrations and EFs of black shale and limestone beds that correspond to zones 1 to 4 and the cover bed are shown in Tables 2 and 3, respectively. Zone 1 shows the lowest concentration of several major elements including Al, Mg, Fe and Mn (Tables 1 and 2). Particularly, zone 1 has an Al content that is half that of zone 4, which has a considerable effect on calculated EFs. Zone 4 reveals relatively high Al contents, and all EFs are higher in the lower zones 1 and 2 compared to zone 4 (Table 3). As EFs may be exaggerated due to varying Al content, they must be considered together with absolute contents of the respective elements. This is particularly the case for bed 14, which consists mainly of marly limestone and exhibits low Al concentrations. Chromium, Co and U are the only trace metals whose concentrations increase from zone 1 to 4 (Table 2), whereas Mo concentrations in zone 1 and 4 are almost equal. There is a distinct peak of all trace metals except Cr and Co in zone 3, which is not as apparent in respective EFs due to high Al contents in this zone. Bed 14 exhibits the lowest contents of most trace metals with the exception of U. Molybdenum/U ratios decrease steadily from bottom to top and fall below unity in bed 14. Correlation coefficients between trace metals and Al, as well as between trace metals themselves were calculated and are shown in Table 4. The only trace metals that exhibit correlation with Al are Cr and Co, with the latter correlated better than the former. Trace metal correlations can be distinguished into three groups of elements that correlate well among each other, these being

(1) Cr with Cu, (2) Mo with Ni, Zn and V and (3) Zn with Ni and Cd (Table 4).

The Amma Fatma carbonate body

The carbonate body associated with black shales (Figs 3 and 4) is enclosed by beds 6 to 14 of the Amma Fatma section, and measures 5·5 m in height and 3·6 m in width. The carbonate body was divided into 4 zones, illustrated in a schematic sketch (Fig. 4B). Zones 1 and 2 comprise the base and lowermost part of the structure and feature concretionary and nodular carbonates penetrated by numerous veins made of secondary precipitates (Fig. 7). A carbonate bank measuring 17 to 19 cm in thickness separates zone 1 from 2 (Fig. 7B). Zone 3 comprises the middle part of the body, and features massive limestone blocks and nodules. Zone 4 is the topmost part of the carbonate body and yielded abundant molluscan fossils, as well as cylindrical to tubular trace fossils (Fig. 8C). The top of the body is capped by a 15 to 20 cm thick carbonate bank, referred to as a cover bed (Figs 4 and 8A to C), which corresponds to bed 14 of the section (Figs 3 and 4A).

The Amma Fatma carbonate body – Petrography and mineralogy

Micrite (i.e. microcrystalline calcite with a crystal size below 10 μm) is the dominant primary mineral that

Table 1. Total concentrations of all measured elements of the Amma Fatma black shale section. All data are given in ppm except depth (in m), $CaCO_3$ and total organic carbon (TOC) content (both in %)

Depth [m]	CaCO₃ [%]	TOC [%]	Ti	Al	Ba	Ca	Fe	K	Mg	Mn	P	Cr	Mo	Ni	Cu	Zn	V	Pb	As	Cd	Co	Re	U
13·5	10·0	3·62	342	6342	664	293 386	2946	3567	9149	41·7	716	28·6	10·3	44·6	10·6	40·2	93·1	2·07	7·19	1·36	1·97	0·050	9·93
13·4	34·6	5·66	426	6792	43·7	279 035	3381	4069	8796	35·0	528	35·3	22·6	67·5	13·7	139	154	2·04	7·13	2·55	2·00	0·100	9·05
13·3	18·9	4·13	501	8446	52·3	284 508	4474	4613	10 626	46·1	438	34·5	20·7	53·8	10·7	75·5	183	1·76	7·38	2·14	2·12	0·040	8·73
13·1	19·9	5·47	614	11 928	60·6	252 778	6346	5416	9189	44·7	1419	40·8	23·9	63·1	11·6	79·8	206	2·42	9·41	2·10	2·51	0·060	9·15
13·0	27·7	5·59	421	9278	54·3	264 626	4559	4146	7795	38·4	1103	46·6	23·4	62·4	12·7	111	192	2·32	8·87	3·03	1·95	0·050	8·01
12·9	19·5	2·51	1158	15 361	81·5	211 050	8242	7054	16 448	81·5	735	43·4	23·3	59·8	11·3	78·3	246	3·12	9·04	1·76	3·13	0·020	6·14
12·4	36·4	6·28	550	9047	57·7	263 207	4937	4656	9411	42·4	724	40·6	26·4	78·0	13·4	146	161	2·59	7·31	3·12	2·27	0·110	8·47
12·3	33·3	5·32	473	7043	60·0	276 795	2912	4008	8186	36·3	613	39·7	13·2	55·8	11·9	133	106	2·11	7·11	2·80	1·97	0·060	8·99
12·1	29·7	4·37	833	11 564	64·2	243 988	5638	5432	14 622	58·9	522	46·1	22·7	68·8	15·4	119	160	2·95	9·22	2·39	2·49	0·090	9·39
11·7	2·7	0·231	60·0	967	85·2	363 527	523	513	3699	38·4	122	4·61	3·00	6·89	1·90	10·7	38	1·04	3·06	1·02	1·06	0·020	0·756
11·2	28·8	3·78	1786	22 793	108	188 394	11 925	9355	18 824	102	612	58·8	18·6	56·0	15·5	115	185	4·30	15·1	2·67	4·58	0·030	5·54
11·1	37·6	5·43	1065	13 377	82·4	240 279	6123	6743	14 798	70·0	633	53·9	17·1	54·5	17·5	150	170	3·31	11·4	4·96	2·72	0·040	7·96
10·9	21·2	5·67	1066	14 505	88·5	253 042	6787	6767	12 459	66·6	830	62·2	13·2	45·7	18·0	84·9	127	3·22	11·9	1·70	3·06	0·050	6·34
10·8	28·4	6·07	1006	13 631	85·0	251 781	5493	6760	13 487	65·9	616	59·3	8·54	52·3	15·3	114	97·7	2·73	10·3	2·25	3·05	0·040	5·07
10·7	49·1	6·54	793	12 000	72·1	234 427	6510	5709	12 893	65·3	505	63·9	18·9	82·6	23·0	197	160	4·00	13·3	3·89	3·13	0·060	9·04
10·6	47·4	6·46	647	9490	63·3	244 942	4457	4428	10 286	49·0	557	51·9	15·9	73·2	18·4	190	135	3·06	10·8	3·96	2·48	0·030	9·58
10·4	34·0	5·09	785	10 620	76·2	229 982	4745	5950	17 944	60·0	469	46·9	9·46	56·7	12·1	136	86·2	2·89	12·4	2·97	2·45	0·030	7·31
10·2	14·0	2·37	597	7754	53·4	245 794	3220	4770	30 725	70·2	262	21·7	3·57	26·5	8·11	55·9	31·8	1·54	7·68	1·37	2·13	0·020	6·39
9·91	3·48	0·068	116	1695	76·5	346 508	781	1229	6814	31·5	131	6·61	3·44	5·30	3·08	13·9	13·8	1·17	11·5	1·38	1·13	0·040	1·86
9·35	3·62	0·093	170	2556	37·7	303 607	548	1959	8324	27·7	142	18·2	2·27	5·86	5·66	14·5	48·3	1·56	5·67	1·83	1·14	0·030	5·20
8·67	11·3	0·064	498	7031	55·6	279 359	3260	3950	14 793	54·0	609	31·6	5·18	19·2	5·28	45·2	127	3·27	9·52	3·07	2·90	0·030	6·37
8·39	21·9	3·45	918	13 497	80·8	239 388	6821	5831	23 383	75·9	613	41·1	13·0	42·5	11·9	87·7	150	2·77	13·0	1·47	2·67	0·020	6·41
8·20	31·7	6·03	779	11 637	76·1	242 917	5315	5620	19 351	59·6	683	59·5	19·1	52·9	24·6	127	148	2·99	11·6	1·62	2·29	0·050	7·67
8·07	21·8	6·51	602	9544	79·3	268 301	3827	4640	9738	43·7	825	51·0	19·2	47·4	16·5	87·2	135	2·69	8·86	1·60	1·90	0·040	9·77
7·87	25·1	7·63	526	10 743	96·0	259 113	4782	4902	8169	41·0	688	59·7	23·6	55·2	19·2	101	167	3·14	10·7	1·69	2·09	0·040	10·1
7·8	17·9	5·20	452	6305	55·8	289 013	2794	3134	7930	36·4	583	38·2	13·1	40·2	13·2	71·6	98·8	2·10	7·82	1·64	1·74	0·020	7·87
7·69	15·4	4·15	995	14 670	81·4	236 752	6152	5843	18 177	68·2	426	53·7	10·8	32·1	13·0	61·7	107	3·10	9·98	1·52	2·56	0·040	5·61
7·45	10·8	1·17	485	7054	134	301 984	2927	3015	9045	43·0	197	19·1	6·01	14·0	6·67	43·3	57·5	2·01	7·05	1·32	1·55	0·002	2·40
7·22	21·3	4·08	846	11 967	84·7	277 504	4594	4909	10 053	52·9	695	46·0	10·6	44·9	11·8	85·2	121	2·48	8·63	5·10	2·87	0·030	5·95
7·13	43·4	5·94	1029	14 959	85·1	245 875	7350	5747	9414	54·0	930	55·8	25·1	85·5	19·8	174	247	4·03	11·9	4·87	3·62	0·080	9·35
7·00	90·3	8·00	969	13 749	79·2	243 523	6853	5450	9233	49·5	1122	59·3	36·7	122·0	25·0	361	403	4·27	12·3	8·24	3·55	0·150	12·2
6·94	70·9	6·45	972	14 450	80·3	246 417	7215	5654	11 185	54·8	857	55·0	31·4	99·2	20·3	284	322	4·21	12·6	6·89	3·42	0·220	9·57
6·88	44·3	5·26	1057	15 136	85·1	247 997	7506	5896	13 140	59·3	1083	55·5	25·4	78·6	17·4	177	254	4·46	13·0	4·32	3·23	0·160	9·70
6·5	5·5	1·00	220	3212	39·5	354 767	1272	1392	5525	40·7	199	11·9	3·74	12·6	4·29	22·1	64·6	1·54	4·88	1·42	1·38	0·020	1·97
6·05	23·2	4·96	422	6315	58·2	303 839	2701	3090	7409	35·6	610	38·2	13·9	52·6	12·5	92·9	88·2	2·53	7·17	2·56	2·16	0·060	7·66
5·93	33·0	6·34	570	8509	61·6	291 161	3829	3824	7684	41·8	550	36·4	15·7	63·4	14·7	132	108	3·14	7·80	3·29	2·31	0·050	6·48
5·82	29·5	4·77	920	13 799	82·0	260 758	6191	5599	14 255	66·6	515	49·1	16·8	53·3	13·8	118	121	3·19	11·1	3·01	3·04	0·030	5·20
5·44	11·7	1·12	509	7378	107	312 712	3433	3224	9423	50·9	250	19·5	5·40	17·2	5·33	46·8	40·4	2·13	6·66	1·90	2·27	0·020	1·70
4·92	61·2	8·09	351	5401	55·4	300 061	2429	2891	8959	32·0	753	48·6	22·7	96·5	16·0	245	173	3·25	7·90	5·75	2·18	0·060	8·54
4·85	38·0	4·48	251	4097	44·5	330 172	1950	2190	7617	31·7	574	38·8	16·9	62·5	11·2	152	161	2·72	5·43	4·80	1·96	0·050	6·80
4·78	72·8	6·73	467	7096	58·7	296 750	3893	3377	10 794	45·8	956	45·1	28·3	94·6	15·9	292	231	3·34	8·76	7·55	2·50	0·050	7·76

(Continued)

Table 1. Continued.

Depth [m]	CaCO₃ [%]	TOC [%]	Ti	Al	Ba	Ca	Fe	K	Mg	Mn	P	Cr	Mo	Ni	Zn	Cu	V	Pb	As	Cd	Co	Re	U
4·75	37·1	4·29	506	7904	61·0	304 941	3895	3588	12 375	53·6	662	37·1	19·8	64·1	149	11·9	194	3·08	6·95	5·29	2·39	0·050	5·96
4·55	14·7	1·46	193	3026	56·0	336 650	1400	1634	6612	41·6	358	15·2	3·67	18·5	59·0	4·98	83·2	2·24	4·18	1·72	1·51	0·020	2·54
4·23	37·4	6·10	423	6121	66·4	307 471	2695	2839	7314	33·8	704	41·1	14·2	63·9	150	13·2	124	2·79	7·10	4·57	2·14	0·050	8·24
4·08	29·1	6·10	549	8718	63·0	294 824	3456	3196	7015	33·8	482	40·6	19·0	51·0	117	13·5	150	3·79	8·14	3·02	2·02	0·020	7·12
3·91	36·0	5·01	918	13 538	77·5	258 689	6963	4864	14 672	61·5	604	46·6	21·4	61·4	144	13·8	199	3·67	11·2	4·18	2·93	0·050	6·93
3·62	17·9	3·20	1139	16 723	92·1	280 049	8198	5923	13 293	81·8	504	41·1	14·0	39·9	71·5	12·1	157	3·20	12·9	2·04	3·02	0·010	5·16
3·48	20·6	4·02	1623	23 868	124	209 749	11 962	8254	16 204	93·5	501	80·6	16·2	44·7	82·5	15·4	190	4·00	17·0	2·13	4·05	0·040	5·41
3·14	5·70	1·05	335	4713	124	348 899	2432	1764	7141	40·0	198	13·3	4·46	15·0	22·8	4·10	78·4	1·81	5·71	0·98	1·56	0·020	1·64
2·65	37·8	5·11	1130	16 500	94·2	222 161	8856	6632	15 091	68·3	730	43·7	27·0	73·3	151	16·1	362	4·11	13·8	3·60	3·32	0·030	7·12
2·62	37·9	3·75	888	13 315	81·0	220 796	6738	5278	12 784	58·9	628	35·1	21·4	58·7	152	12·2	296	3·05	10·4	2·79	2·82	0·060	5·86
2·52	57·1	6·61	1557	22 446	114	169 364	12 514	8565	17 283	82·3	974	55·7	35·1	84·4	229	19·3	428	4·76	18·7	6·54	4·18	0·110	9·49
2·47	56·9	5·75	1508	21 271	117	180 611	11 811	8226	16 347	80·2	846	71·7	37·5	84·5	228	18·6	414	4·61	18·4	4·93	4·20	0·030	9·19
2·42	59·3	6·90	1405	20 537	112	180 636	10 620	7893	15 405	70·7	850	55·3	36·7	84·6	238	20·2	389	5·03	17·5	4·93	3·90	0·030	8·98
2·33	29·6	4·99	652	8907	92·3	272 670	4273	4067	10 505	47·3	706	35·4	18·3	44·1	118	11·3	198	2·90	8·26	2·23	2·05	0·050	6·57
2·23	41·7	6·90	624	9153	75·6	268 239	4562	3946	7994	44·6	849	37·5	29·6	67·6	167	16·7	222	3·49	9·78	3·75	2·06	0·070	9·17
2·14	52·1	8·47	649	10 269	74·0	257 396	4016	4293	6995	35·9	510	47·5	30·8	88·5	208	18·6	267	3·22	10·0	4·38	2·34	0·070	8·37
2·09	52·1	7·33	672	10 629	76·6	257 256	39	4515	7951	38·8	484	48·6	22·4	78·4	209	16·6	244	3·21	10·7	3·75	2·27	0·090	8·40
1·24	59·0	5·83	1333	20 823	101	199 540	10 035	6974	19 995	71·9	621	62·8	31·4	67·9	236	18·1	332	4·53	14·6	5·21	3·49	0·010	5·46
1·18	42·6	4·81	1424	22 475	103	203 908	10 222	7184	18 097	65·7	466	59·1	25·1	53·5	170	17·0	299	4·92	17·4	2·80	3·40	0·002	4·49
1·15	87·1	6·10	1479	23 513	108	186 148	11 410	7741	22 088	80·7	722	65·3	36·9	85·6	349	20·0	380	5·02	16·6	7·03	4·18	0·020	6·54
1·02	53·9	5·09	1534	24 136	109	192 277	11 533	7682	19 396	75·2	513	65·4	27·6	71·5	216	18·0	306	4·88	16·9	5·01	4·24	0·020	5·31
0·990	71·5	6·32	1459	23 156	105	186 532	10 952	7669	21 246	75·9	651	63·8	34·9	82·0	286	19·9	343	4·98	16·2	6·46	3·86	0·030	6·20
0·960	42·0	5·15	1538	23 965	109	193 622	10 423	7714	19 272	73·0	473	66·2	30·6	72·6	168	19·1	299	4·99	17·7	3·38	4·39	0·030	5·48
0·920	35·5	5·87	1708	26 651	116	156 833	12 810	8534	20 359	80·0	554	53·4	25·2	68·0	142	16·5	274	4·24	16·3	3·22	3·83	0·040	7·99
0·730	21·5	4·75	1428	21 727	107	184 326	11 082	6854	13 574	72·0	1329	54·8	10·0	47·4	86·0	14·8	106	3·11	12·2	1·70	3·08	0·020	9·73
0·030	44·9	4·43	1148	16 750	89·4	207 076	6587	5357	9817	57·9	440	67·20	28·76	86·6	180	23·2	302	5·74	22·3	4·27	4·70	0·040	6·71

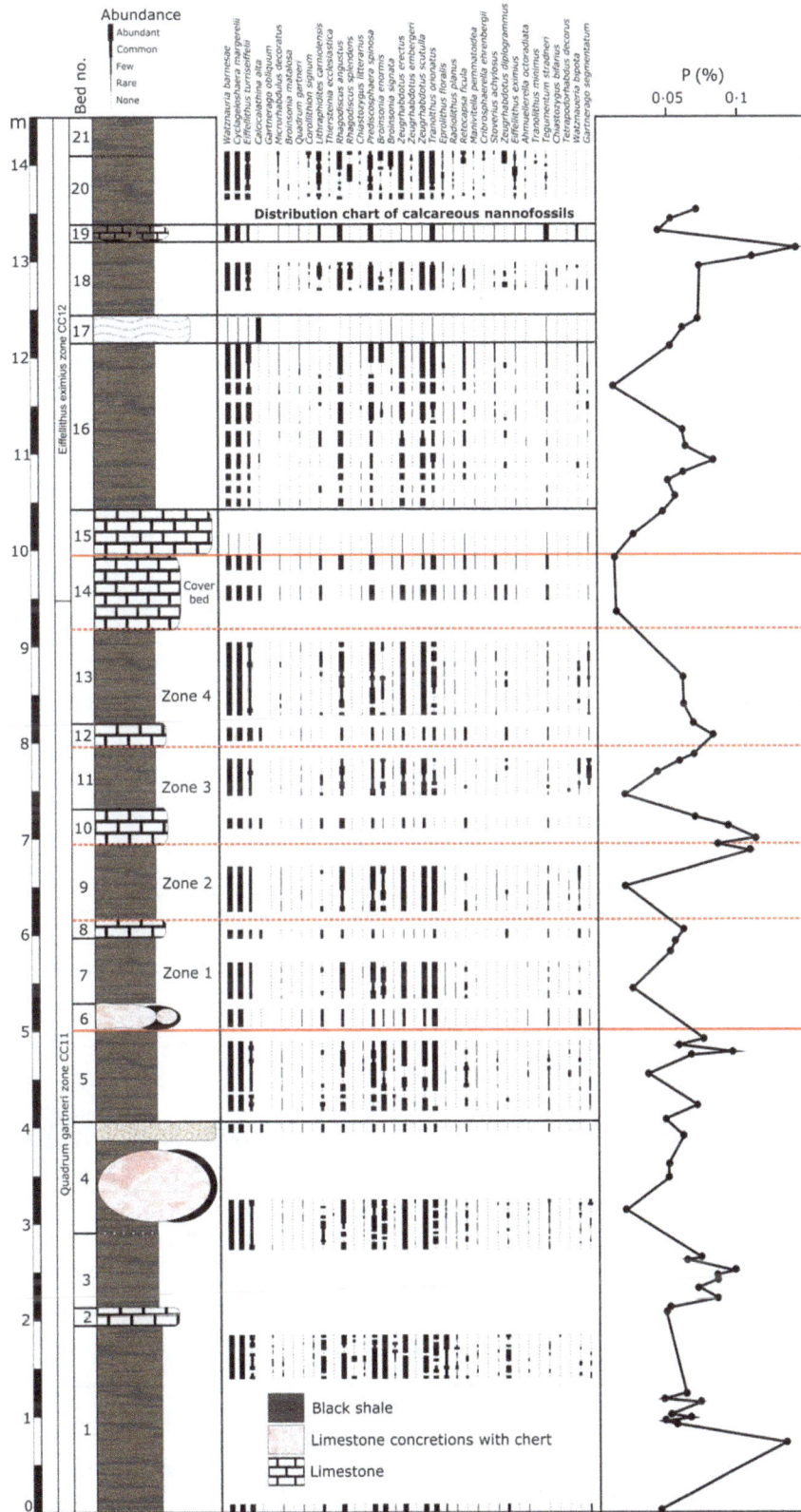

Fig. 5. Distribution chart of nannofossils and phosphorus content across the Amma Fatma section. Solid red lines delineate upper and lower boundaries of the Amma Fatma carbonate body; dashed red lines outline zones 1 to 4 and the cover bed (cf. Fig. 4).

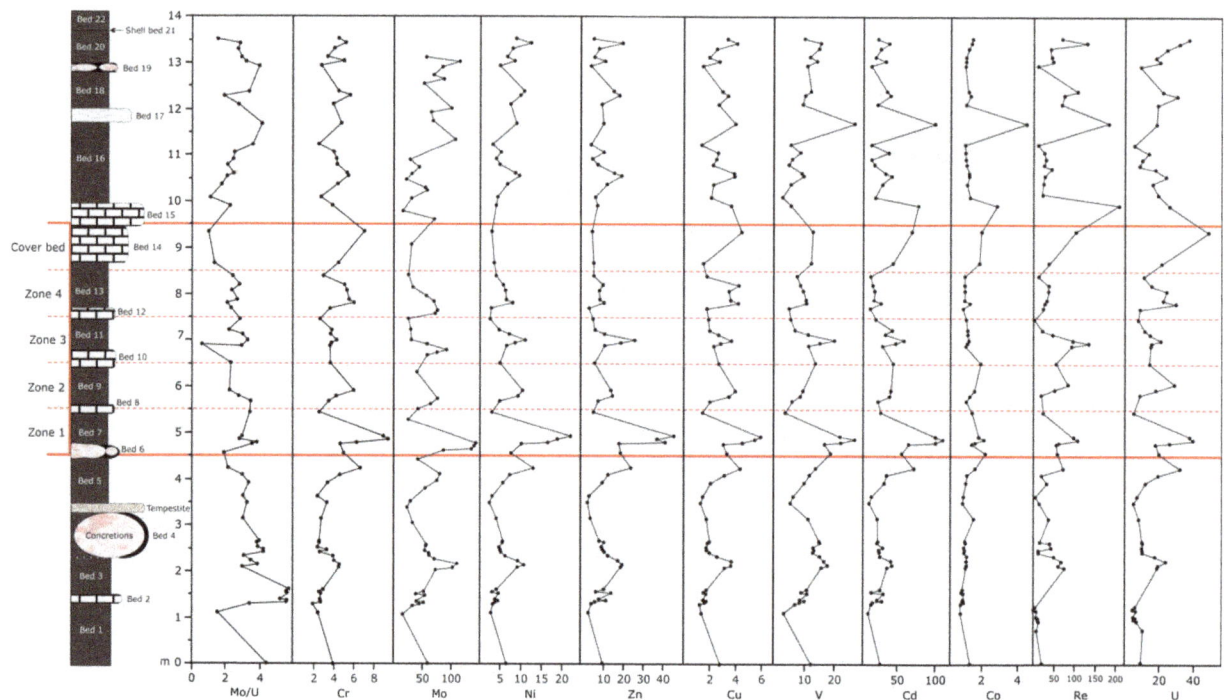

Fig. 6. Trace metal enrichment factors and Mo/U ratios across the Amma Fatma section. Solid red lines delineate upper and lower boundaries of the Amma Fatma carbonate body; dashed red lines outline zones 1 to 4 and the cover bed (cf. Fig. 4, 5).

Table 2. Mean element concentrations in ppm of black shale beds. The stratigraphically corresponding zones of the carbonate body are indicated for comparison

Beds	Zone	Al	Cr	Mo	Ni	Zn	Cu	V	Cd	Co	Re	U	P	Mo/U
14	Cover bed	4794	24·9	3·72	12·5	29·9	5·47	87·4	2·45	2·02	0·030	5·78	376	0·625
13	4	11 066	50·5	16·5	45·1	89·3	16·4	134	1·59	2·21	0·035	7·91	636	2·07
10 to 12	3	11 504	43·2	19·9	65·2	164	15·0	210	4·60	2·81	0·095	7·30	726	2·54
9	2	9541	41·2	15·5	56·5	114	13·6	106	2·96	2·50	0·047	6·45	558	2·49
6 to 8	1	5817	34·0	16·1	58·9	157	10·9	147	4·50	2·13	0·042	5·55	592	2·79

Table 3. Mean major and trace element enrichment factors of black shale beds. The stratigraphically corresponding zones of the carbonate body are indicated for comparison

Beds	Zone	Cr	Mo	Ni	Zn	Cu	V	Cd	Co	Re	U
14	Cover bed	5·67	27·5	3·25	5·61	2·90	12·5	63·4	1·99	70·4	35·0
13	4	4·69	54·3	5·69	7·95	3·09	8·63	17·2	0·956	28·7	18·7
10 to 12	3	3·60	53·7	6·71	11·8	2·50	11·9	43·1	1·22	64·2	14·5
9	2	4·52	59·4	8·48	12·0	3·06	8·00	37·1	1·29	51·5	18·6
6 to 8	1	6·06	94·6	13·6	26·0	3·85	18·2	87·9	1·81	67·5	24·4

makes up the matrix of the Amma Fatma carbonate body. Matrix micrite and peloidal micrite occur, whereby the former is the most abundant carbonate phase in zones 1 and 2, making up at least 80% of the bulk rock volume (Figs 9 and 10A). Matrix micrite is light to dark brown

in colour, exhibits a very heterogeneous and partly clotted texture and is rich in detrital material comprising large amounts of recrystallized bivalve and gastropod shell fragments (Fig. 9A), benthonic Foraminifera (Fig. 9B), *Palaxius* faecal pellets (Fig. 10B), radiolarians (Fig. 11A) and

Table 4. Mean correlation coefficients of major and trace elements of black shale beds. The stratigraphically corresponding zones of the carbonate body are indicated for comparison

Beds	Zone	Al:Cr	Al:Mo	Al:Ni	Al:Zn	Al:Cu	Al:V	Al:Cd	Al:Co	Al:Re	Al:U	Cr:Cu	Mo:Ni	Mo:Zn	Mo:V	Ni:Zn	Zn:V	Zn:Cd
14	Cover bed	0.95	0.85	0.99	0.99	0.51	0.99	0.99	0.99	-0.62	0.80	0.76	0.90	0.91	0.75	1.00	0.96	0.97
13	4	0.35	-0.25	-0.25	0.01	-0.06	0.27	-0.67	0.95	0.29	-0.58	0.77	0.94	0.70	0.75	0.86	0.74	0.38
10 to 12	3	0.98	0.84	0.85	0.74	0.89	0.79	0.79	0.96	0.73	0.91	0.94	0.99	0.97	0.99	0.97	0.99	0.93
9	2	0.91	0.93	-0.18	0.44	0.41	0.93	0.41	0.99	-1.00	-0.98	0.00	0.20	0.74	1.00	0.81	0.73	1.00
6 to 8	1	0.27	0.39	0.24	0.24	0.29	0.26	0.37	0.92	0.19	0.10	0.98	0.97	0.96	0.95	0.97	0.89	0.96

clay minerals. Peloidal micrite is particularly abundant in zones 3 and 4, where it accounts for around 70% of the rock volume (Figs 10C and 11B). This micrite is characterized by abundant oval to spherically shaped peloids, which measure 100 to 500 µm in diameter. The peloids do not show any obvious internal structure and exhibit a cloudy and heterogeneous fabric. The space between the peloids is occupied by fine, dispersed equant calcite cement. Similar to the matrix micrite, the peloidal micrite exhibits scattered occurrences of *Palaxius* faecal pellets, gastropod, bivalve and ammonite shell fragments, as well as Foraminifera and radiolarian tests. The peloids and *Palaxius* faecal pellets are fluorescent under UV light, suggesting that they are enriched in organic matter compared to the surrounding micrite. *Palaxius* faecal pellets are present in all zones but are particularly common in zones 1 and 2 (Fig. 10B). The crescent-shaped canals are arranged in two groups around a symmetry plane, and are oriented in different directions towards and away from the symmetry plane. They are oval to spherical in shape and measure up to 500 µm in diameter. Their internal structure clearly distinguishes them from the aforementioned micritic peloids. Pyrite occurs as very small and finely dispersed grains measuring only several microns in size. In some rare cases pyrite also forms aggregates and clusters of idiomorphic crystals (Fig. 9B). Dolomite occurs within zones 1 and 2. The dolomite crystals are pale grey in colour and show well-defined crystal faces (Figs 9B and 10B).

The Amma Fatma limestone is penetrated by numerous veins, cracks and large cavities throughout zones 1 to 4. Cavities and veins show the same paragenetic mineral sequence in all zones: Matrix micrite or peloidal micrite represents the primary fabric of the rock. Cavities and veins are rimmed by microquartz, followed by equant calcite spar and megaquartz (Figs 10C, D, 11B, C, 12A and B). Microquartz (*sensu* Knauth, 1994) is pale brown to beige in colour (Figs 10C and 11B, C), and typified by very small grain sizes reaching no more than several microns in diameter. In places it can also be found finely dispersed among the equant calcite crystals. Microquartz occasionally forms crystal fans showing oscillatory extinction patterns under crossed polarized light. Volumetrically, equant calcite is the dominant void filling phase in all four zones (Figs 10A, 11C, 12A and B); its crystals measure up to 1 mm in diameter. The last phase in the paragenetic phase sequence is megaquartz (*sensu* Knauth, 1994). Megaquartz crystals reach up to 1 mm or more in diameter (Fig. 10C and D). An accessory peculiar carbonate phase present within fractures and voids is an authigenic rim of microcrystalline calcite referred to as seam micrite (Fig. 12). It is present in all zones and occurs exclusively along the outer rims of veins and cavities,

Fig. 7. (A) Bottom view of the Amma Fatma body. The width of the image is 4 m at the base of the carbonate body. (B) A carbonate bank separating zone 2 from zone 1, hammer for scale. (C) Concretionary carbonate from zone 2 penetrated by numerous carbonate veins, hammer for scale.

Fig. 8. Photographs from the topmost part of the body. (A) View from northeast, the cover bed capping the body bends upwards, hammer for scale (cf. Fig. 4A). (B) Same as (A), view from south-west, hammer (arrow) for scale. (C) Tubular fossils from zone 4 of the carbonate body, coin for scale.

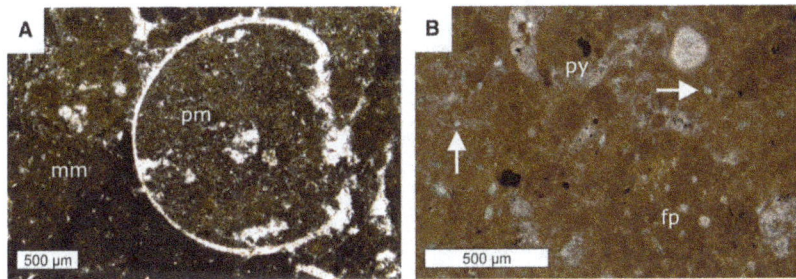

Fig. 9. Photomicrographs of zone 1 limestones; plane-polarized light. (A) Detritus-rich matrix micrite (mm) surrounding a gastropod shell filled by peloidal micrite (pm). (B) Stained thin section showing peloidal micrite with faecal pellets (fp), pyrite aggregates (py) and grey, unstained dolomite crystals (arrows).

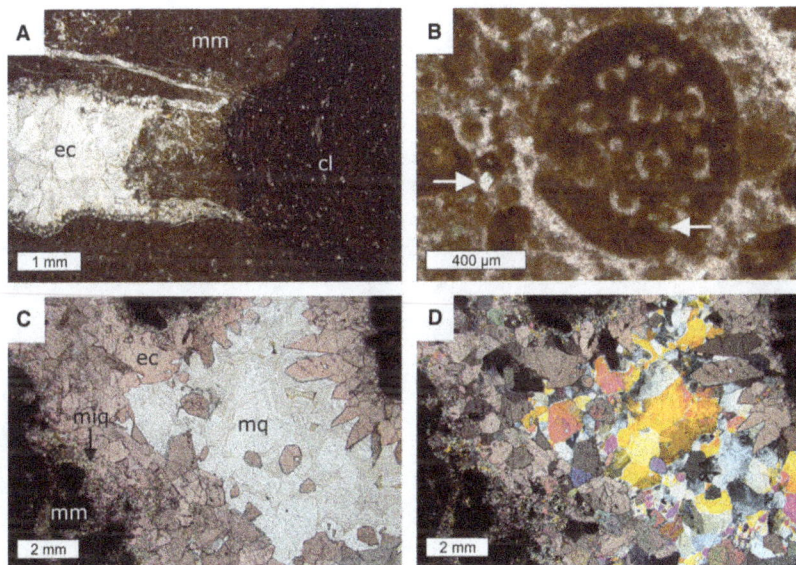

Fig. 10. Photomicrographs of zone 2 limestones. (A) Matrix micrite (mm) surrounding a large cavity filled by equant calcite (ec), several mm thick clay layers (cl) to the left; plane-polarized light. (B) Detailed view of a *Palaxius* faecal pellet with rhombohedral dolomite crystals (arrows); stained thin section, plane-polarized light. (C) Large cavity surrounded by matrix micrite (mm), rimmed by microquartz (miq) and equant calcite (ec); centre of the cavity is filled by large megaquartz crystals (mq); stained thin section, plane-polarized light. (D) Same photograph as (C) in crossed-polarized light.

where it forms highly irregular patches and fringes (Fig. 12A and B). Seam micrite is dark brown, almost black in colour and is very homogenous, showing no detrital inclusions. It shows strong epifluorescence when excited with UV light (Fig. 12C).

The carbonate cover bed (Fig. 4) shows a quite similar paragenetic phase sequence to zones 1 to 4 of the carbonate body (Fig. 4). The matrix comprises micrite, which, compared to the matrix micrite of zones 1 to 4, contains less detrital material. It is relatively homogeneous showing no clotted or peloidal textures and contains minor dolomite crystals (Fig. 13). The fossil content is low, represented by mainly radiolarian and foraminiferan tests (Fig. 13A). Similar to the carbonate body, the cover bed is penetrated by numerous cracks and veins, which have been filled by secondary precipitates. Cavities contain microquartz and later formed equant calcite spar (Fig. 13B and C).

Macrofauna

The macrofossils collected from Zone 4 of the carbonate body comprise a low diversity molluscan assemblage of gastropods, bivalves and ammonites, together with *Palaxius* faecal pellets and burrows. The gastropods comprise 23 specimens of the aporrhaid *Drepanocheilus* sp. (Fig. 14A and B), mostly juveniles, but also a few adults up to 21 mm high, and a single, partially fragmented specimen that might represent a neogastropod belonging to the genera *Drilluta*, *Bellifusus* or *Paleopsephaea* (Fig. 14C). The bivalves belong to three taxa: a single articulated lucinid specimen (Fig. 14E), juvenile

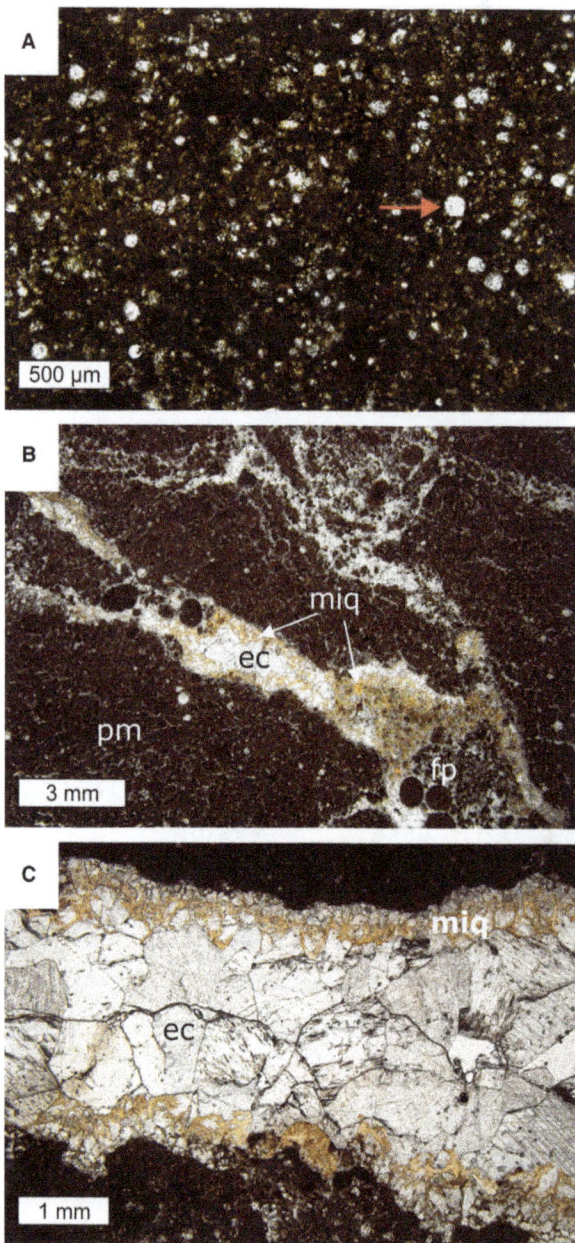

Fig. 11. Photomicrographs of zone 3 limestones; plane-polarized light. (A) Microcrystalline calcite matrix featuring abundant radiolarian tests (arrow). (B) Peloidal micrite with faecal pellets (fp) surrounding veins and cracks rimmed by microquartz (miq) and filled by equant calcite (ec). (C) Close-up view of a vein rimmed by microquartz (miq) and filled by equant calcite (ec).

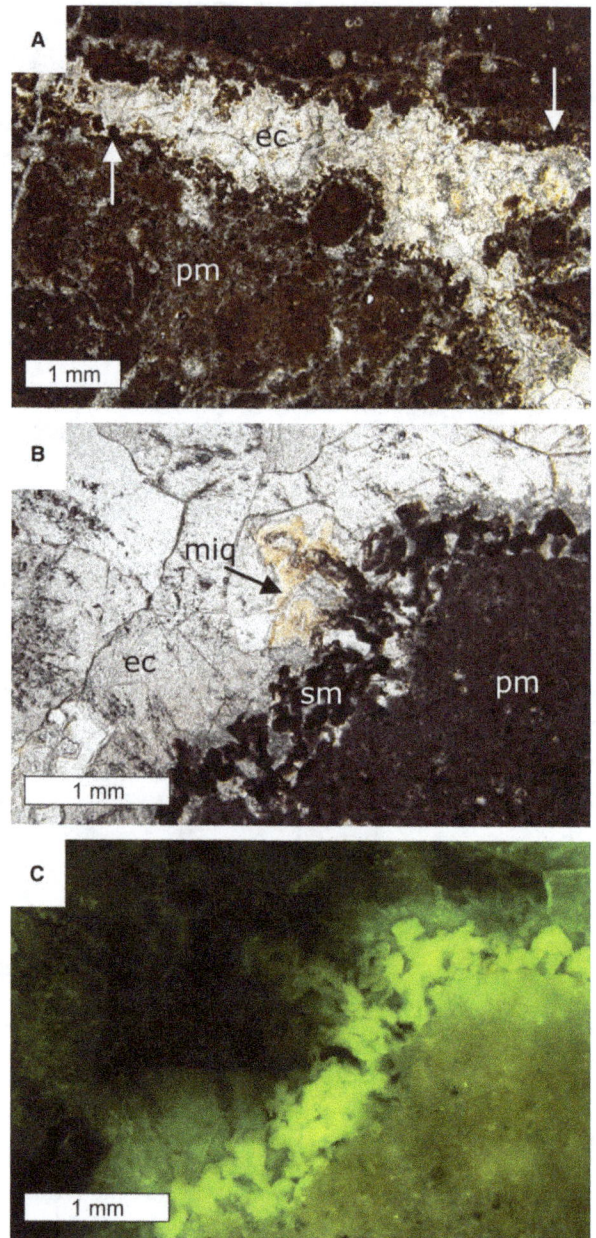

Fig. 12. Photomicrographs of zone 4 limestones. (A) Peloidal micrite (pm) surrounding a void filled by equant calcite (ec), arrows indicate seam micrite (sm) lining the cavity; plane-polarized light. (B) Close-up view of the seam micrite (sm), pm = peloidal micrite, miq = microquartz, ec = equant calcite; plane-polarized light. (C) Same detail as (B) under fluorescent light.

inoceramids ($n = 4$; Fig. 14D) and gryphaeid oysters ($n = 6$), which could belong to the genus *Pycnodonte*. The small specimen sizes (<10 mm) of the two latter taxa and the articulation of the lucinid precludes further identification. The ammonites ($n = 5$) are juveniles and adults belonging to the genus *Benueites*, including *Benueites* cf. *benueensis*.

Isotope geochemistry, TC and TOC

Results of carbon and oxygen stable isotope analyses are shown in Fig. 15. Matrix micrite is ^{13}C-depleted to different degrees throughout all zones of the carbonate body, including the cover bed. It revealed the lowest δ^{13}C values

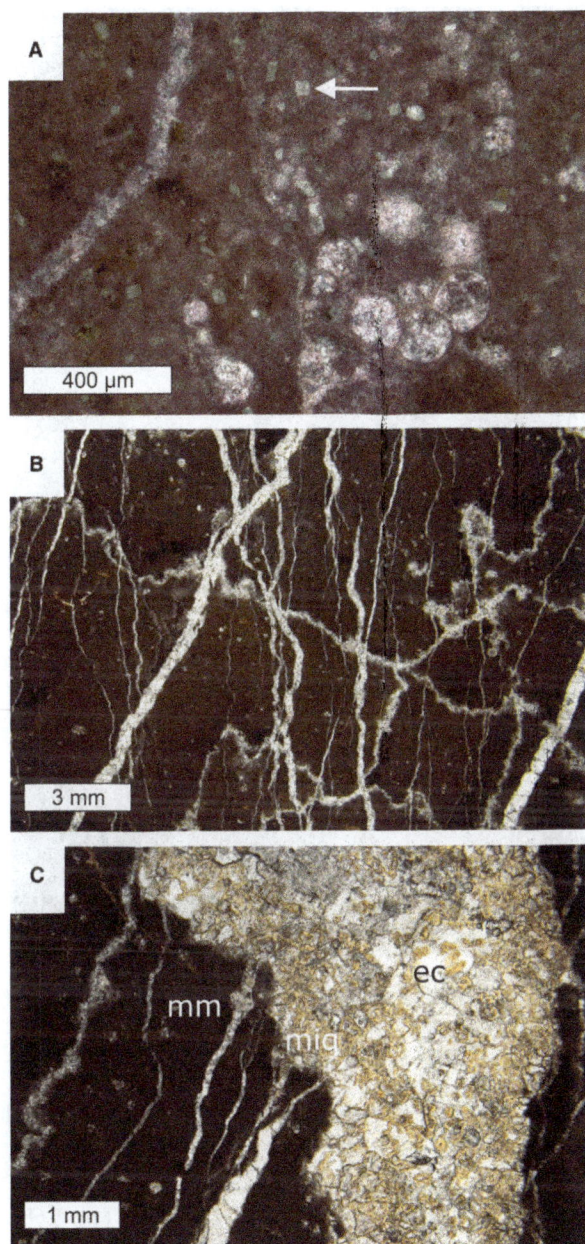

Fig. 13. Photomicrographs of the cover bed; plane-polarized light. (A) Close-up view of matrix micrite featuring dolomite crystals (arrow) and foraminifera tests, stained thin section. (B) Dark brown, fine-grained micrite pervaded by several veins made of secondary equant calcite. (C) Large cavity surrounded by matrix micrite (mm) and filled by microquartz (miq) and equant calcite (ec).

-2.3 to $-0.4‰$. The $\delta^{13}C$ values of equant calcite were analysed from zones 3, 4 and the cover bed. Generally, equant calcite is less ^{13}C-depleted than matrix micrite. A sample from zone 3 revealed the only positive $\delta^{13}C$ value of $1.6‰$. Samples from zone 4 and the cover bed show little heterogeneity and fall between -3.7 to $-1.5‰$ and -0.9 to $-0.7‰$, respectively.

The $\delta^{18}O$ values of matrix micrite and equant calcite from all sections of the carbonate body are exclusively negative. Matrix micrites from zones 1, 2 and 3 exhibit values from -2.9 to $-2.3‰$, -3.2 to $-2.1‰$ and -3.4 to $-1.8‰$, respectively. Zone 4 matrix micrite displays more scattered $\delta^{18}O$ values from -4.3 to $-1.7‰$. Matrix micrite of the cover bed yielded values from -3.3 to $-2.6‰$. The $\delta^{18}O$ value of equant calcite from zone 3 and 2 is $-3.4‰$. Equant calcite from zone 4 shows the most negative $\delta^{18}O$ values from -6.2 to -2.6. Equant calcite of the cover bed is less ^{18}O depleted (-2.3 and $-1.7‰$).

A total of six samples from the Amma Fatma carbonate body were analysed for their TC and TOC contents (Table 5). The TOC content increases significantly from top to bottom, from 0.1% in zone 4 to 0.7% in zone 1. The cover bed exhibits a similar TOC content as the topmost zone 4 of 0.1%.

Biomarkers

Hydrocarbons

Organic matter enclosed in the sediments of the Tarfaya Basin has been shown to be of low maturity (Kolonic *et al.*, 2002; Sachse *et al.*, 2012). The hydrocarbon fraction before decalcification from the cover bed is characterized by short and long chain *n*-alkanes that range from C_{16} to C_{34}. The shorter-chain C_{16} to C_{20} *n*-alkanes dominate over longer chain *n*-alkanes, whereby *n*-alkanes peak at *n*-C_{20}. Apart from *n*-alkanes, the only compounds detected are minor amounts of the regular isoprenoids pristane and phytane, with phytane more abundant than pristane. The hydrocarbon fraction after decalcification from the cover bed shows a very similar pattern, except that the contents are much reduced compared to the fraction before decalcification.

Hydrocarbons in the extract before decalcification from zone 1 are characterized by the dominance of *n*-alkanes from C_{16} to C_{30}. In contrast with the cover bed, C_{16} and C_{18} dominate over C_{20}. Pristane and phytane are the only non-alkane compounds detected in zone 1, with phytane slightly more abundant than pristane. The hydrocarbon fraction after decalcification is similar to the fraction before decalcification except that the overall intensities of all compounds are greatly reduced. Very similar results

in zone 1, ranging from -23.5 to $-21.3‰$. Zone 2 micrite yielded a wider range of values from -23.1 to $-7.9‰$, and zone 3 micrite more positive values between -11.0 to $-0.3‰$. Zone 4 micrite shows the largest scatter of $\delta^{13}C$ values, falling between -22.9 and $-1.3‰$. The matrix micrite of the cover bed yielded $\delta^{13}C$ values from

Fig. 14. Representative macrofossils from Amma Fatma Plage. Carbonate body: (A to E); bed 2 Amma Fatma section: (F to I) from collections of the Oxford University Museum, U.K. (KX codes). (A) Adult *Drepanocheilus* sp. (B) Juvenile *Drepanocheilus* sp. (C) Possible neogastropod; specimen at outcrop had siphon broken from (but still closely associated with) the main part of the body whorl. (D) Juvenile inoceramid. (E) Lucinid; right valve with partially fractured anterior margin. (F) Lucinid; right valve (KX.15683). (G) Lucinid; detail of dorsal margin of articulated specimen, anterior to right. Note well-developed lunule and calcified ligament (KX.15684). (H) Inoceramid, oblique dorsal view of right valve (KX.15696). (I) Adult *Drepanocheilus* sp. (KX.15683). All scales = 5 mm.

were obtained for two carbonate samples of zones 3 and 4, and are not further detailed here.

Carboxylic acids

Carboxylic acid fractions before and after decalcification from the cover bed and zone 1 limestone are shown in Fig. 16. The carboxylic acid fraction from the cover bed is dominated by saturated short chain *n*-fatty acids ranging from C_{12} to C_{18}. Most abundant *n*-fatty acids are C_{16} and C_{18}, which dominate over C_{12}. Unsaturated *n*-fatty acids and long-chained saturated *n*-fatty acids are present in very small amounts (Fig. 16A). The carboxylic acid fraction after decalcification varies only slightly from the fraction before decalcification, with medium-chain *n*-fatty

acids up to C_{23} more abundant in the former (Fig. 16B). The contents of short chained *n*-fatty acids are markedly decreased (*ca* 50% of all carboxylic acids) compared to the fraction before decalcification (Table 6), however, the distribution of C_{12}, C_{16} and C_{18} is similar.

The carboxylic acid fraction prior to decalcification of zone 1 limestone (Fig. 16C) is characterized by short-chained, C_{12} to C_{18} *n*-fatty acids. The C_{12}, C_{16} and C_{18} *n*-fatty acids are dominant over all other compounds. Additionally, αβ- and ββ-hopanoic acids with 31 and 32 carbon atoms are accessory compounds. The carboxylic acid fraction after decalcification of zone 1 limestone (Fig. 16D) differs markedly from all other samples. Apart from short-chain *n*-fatty acids numerous middle- to long-chain compounds are abundant. These compounds

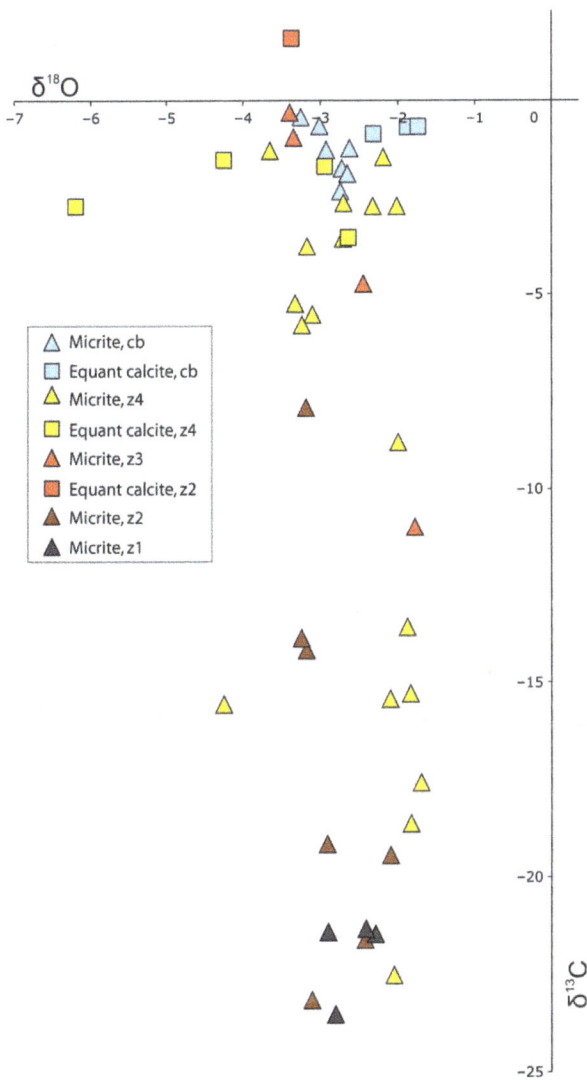

Fig. 15. Carbon and oxygen stable isotope cross-plot of carbonate phases from zones 1 to 4 (z1 to z4) and the cover bed (cb). All values are relative to V-PDB.

Table 5. Total organic carbon (TOC) content of the zones of the carbonate body including the cover bed

Sample	Zone	TOC%	TIC%	TC%
M-01	Cover Bed	0·12	8·36	8·48
M-03	Zone 4	0·13	8·23	8·36
M-06		0·13	8·02	8·15
M-20	Zone 3	0·45	7·75	8·20
M-11	Zone 1	0·69	7·59	8·28
M-17		0·59	7·65	8·24

biphytanic diacid (BP-2) dominates over acyclic- (BP-0), monocyclic- (BP-1) and tricyclic- (BP-3) biphytanic diacids. BP-2 is strongly ^{13}C-depleted ($-96‰$), while BP-3 shows a higher δ^{13}C value of $-22‰$.

The total amount of carboxylic acids varies strongly between the samples. For the cover bed the amount of acids in the extract after decalcification is half that of the fraction before decalcification. For the samples from zones 4, 3 and 1 the ratio is reversed, where the carboxylic acids were found to be more tightly bound to the carbonate lattice. The amount of acids increases from zone 4 to zone 1. Contents and phase-specific stable carbon isotope values of the carboxylic acids are shown in Table 6.

INTERPRETATION AND DISCUSSION

The Amma Fatma carbonate body – A hydrocarbon seep deposit

The comprehensive data set obtained for the Amma Fatma carbonate body allows constraining of the geological setting and the processes that triggered its formation. Although this deposit is unlike many hydrocarbon-seep deposits in terms of its faunal content (cf. Campbell, 2006), there is multiple evidence that it formed at a shallow-water seep in a high productivity realm. The strongest indication for methane seepage is the presence of ^{13}C-depleted biphytanic diacids, which are molecular fossils of anaerobic methane oxidizing archaea (ANMEs; Birgel et al., 2008a). These compounds are putative diagenetic breakdown products of glycerol dibiphytanyl glycerol tetraethers (GDGTs; Liu et al., 2016) and are membrane lipids of archaea including ANMEs (see Schouten et al., 2013 for a review). Biphytanic diacids with extreme ^{13}C-depletions have been reported from various methane-rich environments (Schouten et al., 2003;

include C_{16}, C_{18} and C_{22} α,ω-diacids and minor hopanoic acids, namely C_{32}-17β(H),21β(H)-hopanoic acid. Acyclic to tricyclic biphytanic diacids were only found in zone 1 limestone. The mono- and bicyclic biphytanic diacids only contain cyclopentane rings, whereas the tricyclic biphytanic diacid contains two cyclopentane rings and one cyclohexane ring, according to its mass spectrum and retention time. Of all biphytanic diacids, bicyclic

Fig. 16. Gas chromatograms (total ion currents) of the carboxylic acid fractions from total rock samples before decalcification (intercrystalline compounds) and from the residual powders after decalcification (intracrystalline compounds), IS = internal standard. (A) Carboxylic acids from the cover bed before decalcification. (B) Carboxylic acids from the cover bed after decalcification. (C) Carboxylic acids from zone 1 before decalcification. (D) Carboxylic acids from zone 1 after decalcification; numbers 0 to 3 refer to the biphytanic diacids BP-0 to BP-3, their δ^{13}C values are indicated in parentheses and are versus V-PDB.

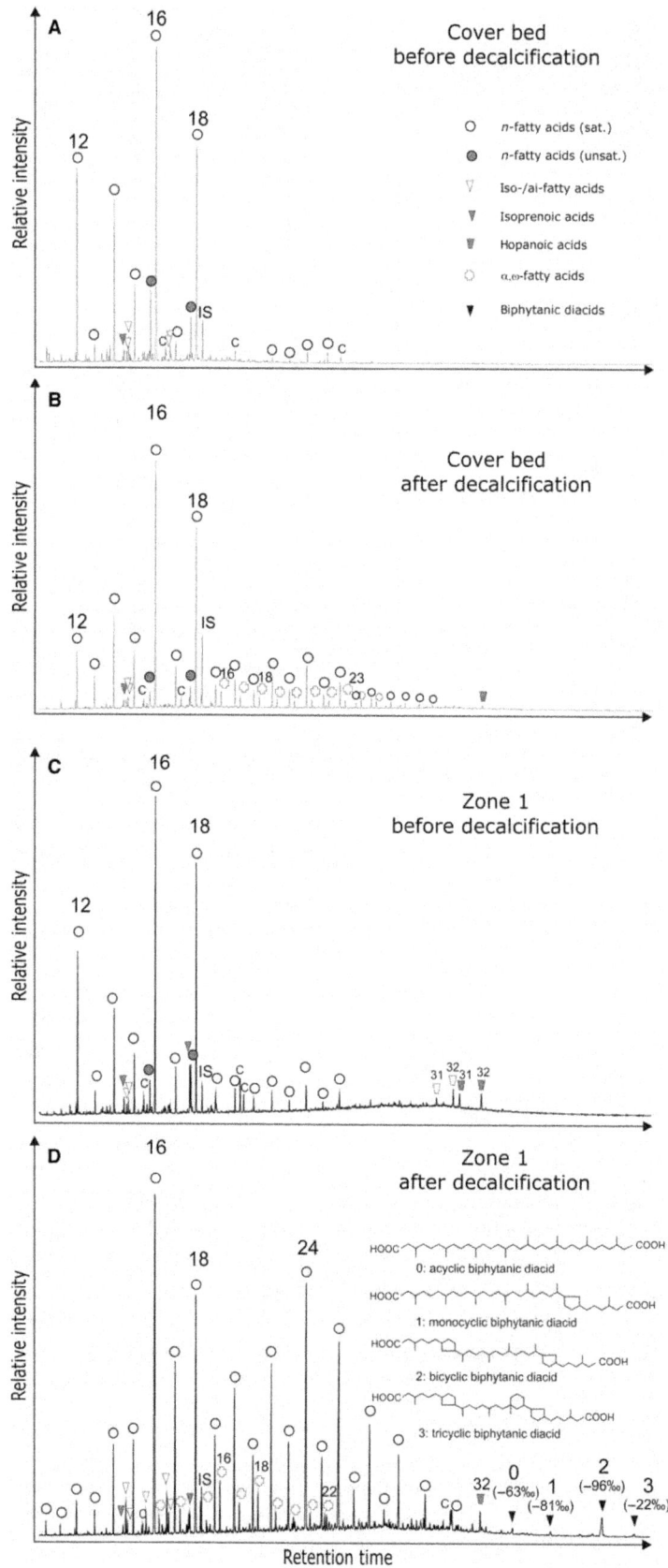

Birgel *et al.*, 2008a; De Boever *et al.*, 2009; Natalicchio *et al.*, 2012). GDGTs and biphytanic diacids are uncommon in Mesozoic and Palaeozoic seep limestones, and the more persistent irregular isoprenoids crocetane and pentamethylicosane (PMI) have been used to track methane oxidation in the rock record (Birgel *et al.*, 2008b), but neither of these were found in the Amma Fatma deposit. The most abundant biphytanic diacid in the Amma Fatma limestone is BP-2. Values as low as $-96‰$ for BP-2 in the Amma Fatma deposit have only been reported from seep-dwelling ANMEs. A similar predominance of ^{13}C-depleted BP-2 among the biphytanic diacids was also reported for other Cenozoic and modern methane-seep carbonates (Birgel *et al.*, 2008a). In contrast with BP-2, BP-3 is derived from other sources, as indicated by its $\delta^{13}C$ value similar to that of crenarchaeol-derived tricyclic biphytanes of planktonic Thaumarchaeota (Könneke *et al.*, 2012; Schouten *et al.*, 2013; Pearson *et al.*, 2016).

Interestingly, biphytanic diacids are only present in the lowermost zone 1, suggesting a heterogeneous composition of the Amma Fatma carbonate body. In zones 2 to 4, however, biphytanic diacids and other molecular fossils of hydrocarbon seepage such as isoprenoid hydrocarbons are possibly obscured by hydrocarbons and unresolvable complex mixtures, similar to some modern seep carbonates, particularly oil-seep carbonates (Naehr *et al.*, 2009; Feng *et al.*, 2014). Since biphytanic diacids were only detected within the carboxylic acids fraction after decalcification, these compounds are tightly bound to the carbonate lattice and consequently reflect environmental conditions during authigenesis, consistent with anaerobic methanotrophy contributing to carbonate formation.

Many ancient seep limestones are typified by a distinct paragenetic sequence of mineral phases that enables the reconstruction of processes that governed their formation (Beauchamp & Savard, 1992; Campbell *et al.*, 2002; Peckmann *et al.*, 2002; Hagemann *et al.*, 2013; Zwicker *et al.*, 2015). Microcrystalline calcite (i.e. micrite) forms the matrix of most seep limestones, and several types can be distinguished including clotted (Peckmann *et al.*, 2002), peloidal (Roberts & Aharon, 1994) and detritus-rich matrix micrite (Kiel *et al.*, 2013; Tong *et al.*, 2016). Micrite of the Amma Fatma deposit exhibit a wide range of $\delta^{13}C$ values, a pattern commonly found for seep limestones (cf. Greinert *et al.*, 2001). The composition of the carbon pool from which carbonates precipitate at seeps varies over time and depends on the degree of mixing between various carbon sources (Suess & Whiticar, 1989; Peckmann & Thiel, 2004). Biogenic and thermogenic methane, higher hydrocarbon compounds including crude oil and dissolved inorganic carbon from sea water all contribute to the carbon pool of the parent fluid, and determining the contribution of each source to carbonate

precipitation at ancient seeps thus remains problematic. Sulphate-dependent AOM is considered to be the main process triggering carbonate precipitation at seeps, whereby the carbonates inherit the ^{13}C-depletion of the organic carbon sources (Ritger *et al.*, 1987; Peckmann & Thiel, 2004; Campbell, 2006). Regardless of the carbon source controlling the isotopic composition of the carbon pool, the carbonates that precipitate at seeps are usually more ^{13}C-enriched than the organic carbon sources due to an admixture of marine carbonate (Peckmann & Thiel, 2004).

The lowest $\delta^{13}C$ value of the Amma Fatma matrix micrite of $-23·5‰$ therefore suggests that AOM did not necessarily contribute substantially to carbonate precipitation, although ^{13}C-depleted biphytanic diacids testify to the presence of microbial communities performing AOM. Slightly lower $\delta^{13}C$ values were obtained from seep deposits associated with Jurassic black shales from Kilve, UK, and were attributed to biogenic methane production in the host sediments and its oxidation close to the sea floor (Xu *et al.*, 2016). The oxidation of higher hydrocarbon compounds may have contributed to carbonate formation at Amma Fatma. This notion, however, is rather speculative as evidence for the presence of oil at the ancient seep is missing, and it is only supported by $\delta^{13}C$ values that are within the range of carbonates precipitating at modern oil seeps (cf. Anderson *et al.*, 1983; Sassen *et al.*, 1999; Formolo *et al.*, 2004). A new approach using trace and rare earth metal contents may shed light on the composition of fluids of the Amma Fatma seep (cf. Smrzka *et al.*, 2016), and future work may assess the impact of crude oil at ancient seeps with more confidence. Finally, the Amma Fatma black shales themselves might have acted as a readily available source of organic carbon, causing microbially remineralization-driven carbonate formation. The $\delta^{13}C_{TOC}$ values of the Amma Fatma succession are slightly lower than the $\delta^{13}C$ values of the authigenic carbonates. Remineralization of large amounts of organic carbon buried during OAE 2 may have consequently contributed to carbonate precipitation, but it does not explain the presence of ^{13}C-depleted biphytanic diacids.

Seam micrite is a peculiar carbonate phase of the Amma Fatma deposit. Peckmann *et al.* (2003) described a similar phase in Eocene seep deposits from Washington State that also lines cavity walls and predates later-stage calcite cements. The purity and microfabric of the Amma Fatma seam micrite are indicative of an *in situ* precipitation, and its intense epifluorescence (Fig. 12C) distinguishes it from matrix and peloidal micrite. The latter observation suggests an increased content of organic residue, agreeing with an involvement of microbial activity in its formation (cf. Peckmann & Thiel, 2004). Equant

Table 6. Contents and compound-specific stable carbon isotope values of carboxylic acids. Summed up values are in ng g⁻¹ rock, grouped lipid contents are in relative percentages of all carboxylic acids, δ¹³C values in parentheses are in ‰ relative to V-PDB

Sample	Cover Bed		Zone 4		Zone 3		Zone 1	
Before/after decalcification	Before	After	Before	After	Before	After	Before	After
Short-chain n-fatty acids (14 to 18)	87% (−30‰)	58% (−30‰)	88% (−30‰)	58% (−31‰)	76% (−31‰)	28% (−30‰)	68% (−30‰)	35% (−30‰)
Long-chain n-fatty acids (26 to 32)	0	4% (n.m.)	2% (n.m.)	9% (−31‰)	1% (n.m.)	31% (−31‰)	1% (n.m.)	17% (−31‰)
Iso-& anteiso-fatty acids (15 and 17)	5% (n.m.)	2% (n.m.)	0	1% (n.m.)	0	0	2% (n.m.)	3% (n.m.)
Short-chain α,ω-diacids (14 to 24)	0	4% (n.m.)	0	3% (n.m.)	0	0	0	5% (n.m.)
Long-chain α,ω-diacids (25 to 28)	0	3% (n.m.)	0	3% (n.m.)	0	0	0	0
Hopanoic acids (αβ 31/32 and ββ 31/32)	0	<1% (n.m.)	0	2% (n.m.)	5% (−32‰)	0	6% (−31‰)	1% (n.m.)
Biphytanic diacids (BP 0 to 3)	0	0	0	0	0	0	0	3% (−66‰)
Others	8%	29%	10%	25%	18%	40%	23%	36%
Sum of all carboxylic acids [ng g⁻¹ rock]	2322·2	1117·3	2235·4	4640·5	3149·1	4886·9	3191·8	5925·5

calcite is a typical, later-stage mineral phase of seeps limestones (Campbell et al., 2002; Peckmann et al., 2003), and is interpreted as an abiotically formed cement that precipitated during increasing burial and progressive diagenesis, which agrees with the only moderate ¹³C-depletion of the Amma Fatma equant calcite.

Authigenic silica minerals typify many Mesozoic and Cenozoic seep limestones (Goedert et al., 2000; Himmler et al., 2008; Peckmann et al., 2011; Kuechler et al., 2012; Kiel et al., 2013; Smrzka et al., 2015). Most of the dissolved silica required for quartz precipitation in seep and non-seep environments is derived from the dissolution of radiolarian tests, diatom frustules and sponge spicules (Lancelot, 1973; Von Rad et al., 1977; Bohrmann et al., 1994; DeMaster, 2002; Kuechler et al., 2012), although dissolution of volcanic glass and clay minerals may also become relevant during late burial diagenesis (Keene, 1975). The abundance of radiolarian tests within the Amma Fatma carbonate body (Fig. 11A) suggests a biogenic source of silica. The sequence of microquartz occurring as thin linings along crack and cavity walls (Figs 11B, C and 12) and later megaquartz occluding large cavities (Fig. 10C and D) is typical of seep limestones (Smrzka et al., 2015). The presence of silica polymorphs suggests that changing silicic acid concentrations controlled the mineralogy of the precipitate over time (Heany, 1993), whereby microquartz precipitated from fluids with higher concentrations of silicic acid and megaquartz formed during a later stage when silica concentrations had decreased.

Although the Amma Fatma deposit features mineral phases that are common in ancient and modern seep limestones, it lacks a carbonate phase that typifies many seep deposits, namely banded and botryoidal aggregates of fibrous aragonite cement. This type of cement typically exhibits low δ¹³C values, reflecting its origin from AOM (Beauchamp & Savard, 1992; Roberts et al., 1993; Savard et al., 1996). The lack of this phase together with the dominance of micrite suggests that the Amma Fatma seep was characterized by diffusive rather than advective seepage of hydrocarbon-rich fluids (cf. Peckmann et al., 2009).

Benthonic fauna

The benthonic macrofossil assemblage of the top part of the Amma Fatma carbonate body includes aporrhaid (Drepanocheilus sp.) and possible neogastropod gastropods, lucinid, inoceramid and gryphaeid (Pycnodonte sp.) bivalves. The small inoceramids and gryphaeids were epifaunal filter feeders, and the lucinid was an infaunal taxon that almost certainly had sulphide-oxidizing chemosymbionts, by comparison with modern members of this family (Dando *et al.*, 1986). The aporrhaid was an epifaunal/shallow infaunal suspension or deposit feeder. The possible neogastropod was most likely epifaunal, but without further taxonomic information its diet cannot be inferred. The presence of this fauna suggests that oxygenated conditions were at least episodically present in the water column and sediment during the formation of at least the upper part of the carbonate body, although the absence of large inoceramid and oyster specimens may hint at challenging environmental conditions. A taxonomically very similar low diversity fauna is present in shell-rich micritic carbonate concretions from below the carbonate body in bed 2 at Amma Fatma Plage, collected by Andrew Gale in 2003 and now housed in the collections of the Oxford University Museum, UK. This includes the same species of lucinid (Fig. 14F and G) and *Drepanocheilus* sp. (Fig. 14I), but also much larger inoceramids (up to 35 mm long, Fig. 14H). Furthermore, species of *Pycnodonte*, inoceramids, lucinids and aporrhaid gastropods are common elements in Late Cretaceous faunas from across North Africa (Busson *et al.*, 1999; El Qot, 2006; Ayoub-Hannaa *et al.*, 2014). This indicates that the Amma Fatma seep fauna contains mostly elements of normal marine background faunas, and lacks specialized obligate taxa, possibly apart from the lucinid, which might belong to the seep obligate genus *Nymphalucina* (Kiel, 2013), although is also similar to the lucind figured in El Qot (2006, plate 14, figs 7 to 10) as *Lucina fallax* Forbes, 1846 from the Late Cenomanian of Egypt. This paucity of obligate seep taxa is a common feature of shallow water seep faunas in the Mesozoic and Cenozoic (Kiel, 2010; Little *et al.*, 2015). Other examples include the slightly older Cenomanian Tropic Shale seep faunas from southern Utah, USA (Kiel *et al.*, 2012), which are dominated by mostly small specimens of the bivalve genera *Inoceramus* and *Nymphalucina*, the gastropod genus *Drepanochilus*, rarer examples of the neogastropod *Paleopsephaea* and *Callianassa* decapod claws. A younger example is the Paleocene Panoche Hills seep deposit from California, USA (Schwartz *et al.*, 2003), which contains numerous specimens of *Nymphalucina* and less common aporrhaid gastropods (Kiel, 2013).

The *Palaxius* faecal pellets found throughout the Amma Fatma carbonate body are microcoprolites, which are today produced by callianassid decapod crustaceans. *Palaxius* microcoprolites are common in Palaeozoic and Mesozoic shallow water limestones (Senowbari-Daryan *et al.*, 1992; Schweigert *et al.*, 1997) and other methane-seep deposits (Peckmann *et al.*, 2007; Senowbari-Daryan *et al.*, 2007; Zwicker *et al.*, 2015). The number and shape of the crescent shaped canals in the Amma Fatma microcoprolites strongly resemble *Palaxius decemlunulatus*, which has previously been found in Cretaceous shallow

water carbonate deposits of northern Africa (Senowbari-Daryan & Kuss, 1992). The high density of *Palaxius* microcoprolites in the Amma Fatma body suggests high productivity and organic matter availability, sustaining dense populations of crustaceans.

Mode of formation of the Amma Fatma carbonate deposit

The Amma Fatma carbonate body is not older than Lower Turonian in age. The presence of ammonites belonging to the genus *Benueites* in zone 4 shows that the upper part of the seep deposit belongs to the *nodosoides* Zone, the youngest zone of the Lower Turonian, as elsewhere in the Tarfaya Basin. *Benueites* species, including *benueensis*, are associated with *Mammites nodosoides* (Collignon, 1966). El Albani *et al.* (2001) described the laminated marls at Amma Fatma and deduced an early Turonian age based on the ammonite fauna and planktonic Foraminifera. The macrofaunal assemblage of the upper part of the Amma Fatma deposit suggests that the entire water column was at least episodically oxygenated at that time. The absence of the diagnostic $\delta^{13}C$ isotope excursion observed for OAEs confirms that the black shales and the carbonate body formed in the aftermath of OAE 2, and is consistent with biostratigraphic evidence.

The Amma Fatma deposit is a tall, concretionary carbonate body, and in this respect resembles some of the exhumed Oxfordian seep deposits from Beauvoisin (Peckmann *et al.*, 1999) and Miocene deposits of Italy (Conti *et al.*, 2004). Concretions of various shapes and sizes are common in Mesozoic strata, and typically stand out from the host sediment after erosion (Coleman, 1993; Coleman & Raiswell, 1995; Sellés-Martínez, 1996; Seilacher, 2001; Marshall & Pirrie, 2013; Liang *et al.*, 2016). What is most indicative of concretionary growth is that the Amma Fatma carbonate body clearly stands out from the host successions. Apart from this large carbonate deposit, laminated strata from Amma Fatma and Mohammed Plage feature isolated, nodular, diagenetic concretions that precipitated within the organic carbon-rich limestones and black shales (El Albani *et al.*, 1997, 1999a). El Albani *et al.* (2001) described lower Turonian, elongated carbonate concretions in direct proximity to the Amma Fatma carbonate body. These concretions formed during early diagenesis before significant compaction occurred, as fossils were well-preserved and surrounding beds of the host marls wrap above and below the outside of the concretions. The Amma Fatma deposit crosses almost the entire section of laminated shales and limestones, making it difficult to ascertain if the carbonates precipitated during or after their deposition.

The cover bed capping the Amma Fatma deposit sheds light on the mode of accretion. It thins out around the top of the carbonate body (Fig. 8A and B), which agrees with the presence of a relief on the sea floor during the time of deposition. Such a scenario is in contrast with a thickening of a cover bed as seen during pockmark formation (cf. Agirrezabala et al., 2013). The circumstance that the cover bed is much thinner directly on top of the seep deposit, and thickens readily away from it, indicates that the deposit was elevated above the sea floor. Given the changing environmental conditions, the carbonate body may have projected into the water column during episodes of sulphidic bottom waters as observed today in the Black Sea (Peckmann et al., 2001). However, there is no evidence for an ancient hardground or firmground, as cementing epifauna are absent. It is conceivable that concretionary growth accompanied by compaction of the host sediment were the main modes of accretion of the Amma Fatma carbonate body, and that carbonate precipitation occurred in the subsurface but close to the sea floor. Large concretionary bodies require a constant and large supply of pore water that transports the necessary solutes to the locus of precipitation (Raiswell & Fischer, 2004; Mozley & Davis, 2005). Diffusive and advective transport of solutes by seeping fluids is the most likely mechanism that sustained concretionary growth over long timescales that subsequently fostered compaction and cementation of the Turonian black shales and marly limestones.

Mean $\delta^{18}O$ values of the predominantly authigenic matrix micrite ranging between -2.6 and $-2.9‰$ also point towards precipitation close to the sea floor, as they are within the range of values for limestones that precipitated in equilibrium with Cretaceous sea water at Mohammed Plage (cf. Kuhnt et al., 2009). They show no appreciable deviation throughout the body, suggesting that pore waters did not experience large temperature shifts or meteoric water input that would have produced more negative $\delta^{18}O$ values (Hudson, 1978; Astin & Scotchman, 1988). Such fairly homogeneous oxygen isotope compositions are common for concretions that grew in shallow sediments close to the sea floor (Hudson & Friedman, 1974; Raiswell & Fisher, 2000). The $\delta^{18}O$ values of the later-stage equant calcite do not differ substantially from those of the matrix micrite, suggesting that neither phase experienced much elevated temperatures during burial diagenesis (cf. Campbell et al., 2002).

In contrast, the trend towards lower $\delta^{13}C$ values from the topmost zone 4 to the lowermost zone 1 (Fig. 15) is interesting. The bottom zone 1 not only shows the highest ^{13}C depletion, but also the sole occurrence of biphytanic diacids and the highest TOC content. Explaining these patterns is challenging, as the preferential growth

direction of the Amma Fatma deposit is not obvious. Neither pervasive growth, outward, downward, nor upward growth can be excluded, and assumptions on the mode of accretion remain speculative. Seep carbonate formation is fuelled by a source of reduced, hydrocarbon-rich fluids from below. Zone 1 was possibly affected most by these fluids, which drove microbial activity as evidenced by the occurrence of ^{13}C-depleted biphytanic diacids. Fluid seepage apparently continued during ongoing sedimentation of organic material, and zone 1 remained the hotspot of hydrocarbon-driven authigenesis that clogged up the pore space and reduced permeability. The impact of methane-bearing fluids on carbonate precipitation in the upper zones 3 and 4 was apparently lower, resulting in gradually less negative $\delta^{13}C_{carbonate}$ values as the dominant carbon source changed from isotopically lighter, seepage-derived hydrocarbons to heavier organic matter derived from the host black shales associated with OAE 2.

Methane seepage and carbonate precipitation in the aftermath of OAE 2

The depositional environment of Amma Fatma Plage was most likely an outer shelf setting, as suggested by the foraminiferal assemblages and smectite-dominated clay mineral assemblages (Holbourn et al., 1999) as well as the macrofossil content of the seep deposit (see above). The three palaeoecologically significant Foraminifera species identified at Amma Fatma are Zygodiscus (=Zeugrhabdotus) erectus, Rhagodiscus splendens and Eprolithus floralis. Generally, temperature and trophic resources are considered the most important factors affecting the distribution of calcareous nannoplankton, whereby ecological preferences of some species are already partly known (Eshet & Almogi-Labin, 1996; Mattioli, 1997; Lüning et al., 1998; Herrle, 2003). The foraminiferal assemblages in combination with the occurrence of radiolarians and diatoms suggest high surface water productivity (El Albani et al., 1999b; Holbourn et al., 1999). Specifically, Zygodiscus (=Zeugrhabdotus) erectus is interpreted as an indicator for high productivity (Roth, 1981; Roth & Krumbach, 1986; Erba et al., 1992; Herrle, 2003). This species is abundant throughout the Amma Fatma section and lacks any systematic distributional changes, suggesting persistent high productivity conditions throughout the entire section consistent with overall high carbon burial. Another species identified as high productivity-prone is Rhagodiscus splendens. This may indicate further intensification of upwelling conditions, possibly marking high productivity phases at the base and the top of the section within a generally highly productive setting. An additional argument for upwelling conditions is the presence of

Eprolithus floralis, which is also considered to be an indicator for cooler waters (Kuhnt *et al.*, 2001). The co-occurrence of these three species suggests that intensified upwelling of relatively cool, intermediate water occurred during the lower Turonian along the Tarfaya shelf. This interpretation is supported by pronounced peaks in phosphorus content that coincide with higher abundances of *Eprolithus floralis* and *Rhagodiscus splendens* at the bottom and top parts of the Amma Fatma section (Fig. 5). These observations, together with the characterization of the macrofaunal assemblage, help to reconstruct the environmental conditions in which the Amma Fatma seep deposit formed.

Carbonates from zone 4 and the cover bed formed under the influence of more prolonged periods of oxygenated conditions. The occurrence of marine background fauna suggests that it dominated and outcompeted any specialized seep-specific taxa in this shallow water and high productivity environment. The absence of benthonic macrofauna in the lower part of the body might suggest lower levels of oxygen compared to the upper part. The presence of *Palaxius* in the lower part does not contradict this notion, as callianassid decapod crustaceans can tolerate extended periods of anoxia (Forster & Graf, 1992; Anderson *et al.*, 1994; Bromley, 1996). The carbonate body formed either during or shortly after deposition of the host strata, and vital information on the environmental conditions can be derived by analysing the trace metal content of host black shales and limestones. Organic-rich shales are distinctly enriched in trace metals including Cr, Mo, U, Ni, Zn, Cu, V, Cd, Co and Re, whereby the modes and mechanisms of enrichment differ somewhat between these elements (Calvert & Pedersen, 1993; Algeo & Maynard, 2004; Brumsack, 2006). The fact that all trace metals – with the exception of Cr and Co – do not correlate with Al confirms that their authigenic enrichment is due to geochemical changes in bottom and pore waters. The use of Cr in palaeoenvironmental analysis has been cautioned, as it is to a large degree of detrital provenance and is highly mobile in marine sediments (Tribovillard *et al.*, 2006). Cobalt mainly resides within the alumosilicate fraction of the black shales as it shows the strongest correlation with Al of all trace metals, and will be consequently omitted from the discussion below. Black shale beds 7 to 13 show EFs of varying magnitudes, which argue for changing environmental conditions during deposition (Table 3). Molybdenum, Cd, Re and U exhibit the highest EFs. Molybdenum behaviour in sedimentary pore waters is linked to the presence of hydrogen sulphide that causes a speciation change from oxidized Mo to reduced thiomolybdate, which can be much more readily deposited either on organic species, or be incorporated and co-precipitated by iron sulphides (Helz *et al.*,

1996). Cadmium exhibits a similar behaviour towards dissolved hydrogen sulphide, except that it accumulates as a separate sulphide phase and can be just as easily enriched in barely reducing conditions at very low hydrogen sulphide concentrations (Rosenthal *et al.*, 1995). The ease of Cd enrichment in anoxic sediments and its high sensitivity towards hydrogen sulphide leads to high EFs, as well as to the large spread in EFs and total concentrations between zones 1 and 4 (Tables 2 and 3). Rhenium EFs are the highest on average of any trace metal. This is due to its very low crustal concentrations and the ease with which it accumulates in anoxic sediments (Crusius *et al.*, 1996). Rhenium enrichment does not require the presence of free hydrogen sulphide or Fe or Mn (oxy)hydroxides, and is solely controlled by the redox potential of the ambient water (Morford *et al.*, 2005). This suggests that the sediments in which carbonates of zones 1 to 4 formed were all characterized by a very shallow oxygen penetration depth or, more likely, periodically anoxic bottom waters.

In modern sediments U enrichment is observed at redox potentials present within the Fe reduction zone, before Mo accumulation occurs under more reducing conditions (Morford & Emerson, 1999). It has been argued that preferential U over Mo uptake is an indicator for suboxic conditions, whereas the reverse is the case when pore waters are strongly sulphidic (Algeo & Tribovillard, 2009). The decrease in Mo/U ratios, total Mo concentrations and Mo, Cd and Re EFs, combined with the concomitant increase in U content from zone 1 to 4 suggests a trend from euxinic conditions in zone 1 to more mildly anoxic conditions in zone 4. Nickel and Zn accumulate most effectively under euxinic conditions and reside in Ni- and Fe-sulphides, respectively (Algeo & Maynard, 2004). Vanadium can also accumulate efficiently within the nitrate reduction zone of marine sediments, and does not necessarily require dissolved sulphide. However, the combined enrichment of V, Zn, Ni and Mo suggest at least temporal euxinia at the sediment-water interface throughout zones 1 to 4 (cf. Hetzel *et al.*, 2009). The notion that ambient water mass chemistry lead to authigenic enrichment of Mo, Ni, Zn, V and Cd is supported by the strong positive correlation between these elements, and their lack of correlation to Al (Table 4).

Trace metal contents of black shales and limestones are governed by bottom water redox conditions, as well as sedimentary redox conditions. Redox-sensitive metals precipitate at redox boundaries in the sediments that often lie substantially below the sediment surface, which may lead to a strong offset between the age of authigenic precipitates and the age of the sediment in which the elements are enriched. It can therefore be difficult to discriminate between enrichment of trace metals via direct precipitation from bottom waters, and enrichment via the precipitation of solid phases within redox zones in the sediment (cf. Kasten et al., 2003; Reitz et al., 2004). However, this age offset between authigenic minerals and the host sediment decreases the more the redox zonation is compressed in the sediment like in the case of anoxic bottom waters. Analyses of trace metal enrichment argue for a formation of black shale and marly limestone beds 7 to 14 under episodic euxinic bottom waters. More importantly, the distribution of Mo, U, Cd and Re implies changes in hydrogen sulphide content of the water mass, from higher contents in zone 1 to lower contents in zone 4. This questions the proposition that the carbonate body grew in its entirety and simultaneously within the shales and limestones. Zone 4 carbonates formed under at least episodically oxygenated conditions, which would have hampered the formation of black shales in bed 13.

Despite the occurrence of macrofossils in zone 4, the Amma Fatma body is poor in fossils compared to most Phanerozoic seep deposits, which are usually replete in shelly macrofauna (Campbell & Bottjer, 1995; Kiel & Peckmann, 2007; Little et al., 2015). Most of the shells identified in thin sections are severely fragmented and were most likely transported to the site over some distance, indicating that most of these molluscs did not live at the seep. Environmental conditions for macrofaunal species at the Amma Fatma seep were at the very least challenging, if not precluding their survival altogether. Euxinic conditions occurred during all stages of carbonate precipitation, interrupted by sporadic oxygenation during formation of zone 4 that enabled macrofaunal assemblages to temporarily colonize the seep. Bottom water conditions were temporarily sulphidic, as were the shallow sediments where carbonate precipitation occurred, consistent with reconstructions for the wider Tarfaya palaeo-shelf area (Kolonic et al., 2005; Poulton et al., 2015). The carbonate mineralogy of the Amma Fatma seep deposits is almost exclusively calcite, lacking any banded and botryoidal aragonite cement. Calcite precipitation at seeps is favoured over aragonite precipitation under the influence of strongly reducing pore waters with low sulphate concentrations (Haas et al., 2010; Nöthen & Kasten, 2011). The environmental conditions as constrained by trace metal analyses of the black shales did consequently not only govern the mode of accretion, but apparently influenced the mineralogy of authigenic carbonate at the Amma Fatma seep as well.

At passive continental margins various processes control the mechanism of fluid flow from sea floor sediments, including tectonic processes, thermohaline convection, sea-level changes and changes in sedimentation rates (Judd & Hovland, 2007; Suess, 2014). Relevant factors that may have influenced the mode of seepage in the Cretaceous Tarfaya Basin include differential compaction and pore water overpressure, both of which are in turn controlled by sedimentation rates and changes in sea-level. Low sea levels promote seepage of fluids through the sediment column (Teichert et al., 2003), as the hydraulic pressure needs to exceed the pressure of the water column to enable upward fluid flow towards the sea floor (Carson & Screaton, 1998). High-frequency sea-level changes at Amma Fatma Plage are indicated by intercalations of bioclastic storm beds into the marly, hemipelagic background sediments (cf. El Albani et al., 1999b). The Amma Fatma coastal section is early Turonian in age, and studies from the Tarfaya Basin have demonstrated that the sea-level rose continuously during the Cenomanian due to a combination of eustatic sea-level rise and shelf subsidence (Gebhardt et al., 2004; Mort et al., 2008; Kuhnt et al., 2009). This sea-level rise culminated during the earliest Turonian where sea-levels were possibly up to 250 m higher than today, the highest during the Cretaceous (Haq, 2014). However, Kuhnt et al. (2009) inferred that the lowermost Turonian at Mohammed Plage was characterized by a sea-level lowstand, linked to the ephemeral build-up of Antarctic ice sheets during that time. This is in accord with the inferred outer shelf water depths at Amma Fatma, which should have promoted seepage of hydrocarbon-rich fluids due to the comparatively low pressure of the water column. The presence of shallow water species Broisonia enormis and the abundance peak of Gartnerago segmentatum (Thierstein, 1976; Hattner et al., 1980) further suggests that a short-term shallowing event may have occurred. Finally, El Albani et al. (1997) noted an upward increase in the frequency and thickness of the storm beds, and interpreted this observation as evidence for a falling sea-level through the section corresponding to the eustatic sea-level development during the late Early Turonian also suggested by Hardenbol et al. (1998).

Apart from sea-level changes, sedimentation rates also affect seepage dynamics in continental margin sediments. Sedimentation rates in the Tarfaya Basin during OAE 2 were estimated to have been above 10 cm kyr^{-1} (Kuhnt et al., 2005), or correspond to 4·8 cm kyr^{-1} at well S13 (see Fig. 2) and 1·6 cm kyr^{-1} at Mohammed Plage (Kolonic et al., 2005). Despite the differences in these reconstructions, all estimates reflect high sedimentation rates. Such high sedimentation rates increase sediment loading on the continental slope that can lead to pore water overpressure, which squeezes reduced fluids from the underlying organic carbon-rich source rocks towards the sea floor (cf. Suess, 2014). Taken together, favourable conditions for seepage over long timescales were apparently favoured by a relative sea-level fall and high sedimentation rates, which facilitated the precipitation of large amounts of authigenic carbonate at the Amma Fatma seep.

CONCLUSIONS

The Amma Fatma carbonate body is an unusual seep deposit, as seep-endemic macrofauna that characterizes many other seep deposits is absent. Nonetheless, lipid biomarkers testify to the presence of microbial communities that performed AOM and contributed to authigenic carbonate formation. With $\delta^{13}C$ values of matrix micrite as low as $-23.5\permil$ it appears that anaerobic oxidation of methane was not the only trigger for carbonate precipitation. Remineralization of organic matter residing in the host black shales may have been an additional source of carbon for limestone formation. Persistent seepage was probably favoured by a sea-level lowstand and high sedimentation rates. The Amma Fatma deposit formed during the aftermath of OAE 2 over a considerable length of time. High productivity and partly euxinic conditions resulted in the deposition and preservation of black shales rich in organic matter. The Amma Fatma deposit is poor in macrofossils, and those present in the uppermost zone 4 are all Cretaceous shallow water species. Their presence testifies to the sporadic occurrence of oxygenated conditions in the bottom water. However, nannofossil distribution and pronounced enrichments of trace metals within the host strata point towards a high productivity setting with recurrent episodes of euxinia that shaped the shallow marine environment. Identifying seep deposits in such a shallow water setting poses a challenge. Diagnostic signals are masked by high primary production favouring heterotrophic taxa, which take advantage of the vast amounts of organic matter derived from photosynthetic primary production and outcompete obligate, endemic chemosymbiotic seep fauna. The Amma Fatma carbonate body may serve as an example on how to identify seep deposits in high productivity, shallow water settings. In such an environment, only a combination of methods will allow to ascertain that authigenic carbonate deposits formed at hydrocarbon seeps, and to constrain the modes of carbonate formation.

ACKNOWLEDGEMENTS

We thank Lukas Gerdenits for analytical support, Leopold Slawek for the preparation of thin sections, Susanne Gier for XRD measurements, Petra Körner for LECO measurements, Sabine Kasten and Matthias Zabel for help with element analysis and Sylvain Richoz for stable isotope analyses of carbonate samples. CTSL thanks Louella Saul, Annie D'Hondt, Franz Fürsich, Wagih Hannaa, Dick Squires, Andrew Gale, Steffen Kiel and Jim Kennedy for advice about macrofossil identifications, and Eliza Howlett for access to specimens curated in the Oxford University Natural History Museum. Comments by Sabine Kasten and the two journal reviewers Nicolas Tribovillard and an anonymous referee are gratefully acknowledged.

References

Agirrezabala, L.M., Kiel, S., Blumenberg, M., Schäfer, N. and **Reitner, J.** (2013) Outcrop analogues of pockmarks and associated methane-seep carbonates: a case study from the Lower Cretaceous (Albian) of the Basque-Cantabrian Basin, western Pyrenees. *Palaeogeogr. Palaeoclimatol. Palaeoecol.*, **390**, 94–115.

Algeo, T.J. and **Maynard, J.B.** (2004) Trace-element behavior and redox facies in core shales of Upper Pennsylvanian Kansas-type cyclothems. *Chem. Geol.*, **206**, 289–318.

Algeo, T.J. and **Tribovillard, N.** (2009) Environmental analysis of paleoceanographic systems based on molybdenum-uranium covariation. *Chem. Geol.*, **268**, 211–225.

Amblés, A., Halim, M., Jacquesy, J.C., Vitorovic, D. and **Ziyad, M.** (1994) Characterization of kerogen from Timahdit shale (Y-layer) based on multistage alkaline permanganate degradation. *Fuel*, **73**, 17–24.

Anderson, R.K., Scalan, R.S., Parker, R.L. and **Behrens, E.W.** (1983) Seep oil and gas in Gulf of Mexico slope sediments. *Science*, **222**, 619–622.

Anderson, S.J., Taylor, A.C. and **Atkinson, R.J.A.** (1994) Anaerobic metabolism during anoxia in the burrowing shrimp *Calocaris macandraea* Bell (Crustacea: Thalassinidea). *Comp. Biochem. Phys. A*, **108**, 515–522.

Arthur, M.A. and **Sageman, B.** (1994) Marine black shales, depositional mechanisms and environments of ancient deposits. *Annu. Rev. Earth Pl. Sc.*, **22**, 499–551.

Arthur, M.A., Dean, W.E. and **Schlanger, S.O.** (1985) Variations in the global carbon cycle during the Cretaceous related to climate, volcanism, and changes in atmospheric CO_2. In: *The Carbon Cycle and Atmospheric CO_2: Natural Variations Archean to Present* (Eds E.T. Sundquist and W.S. Broecker), *Geophys. Monogr. Ser.*, **32**, 504–529. AGU, Washington, DC.

Astin, T.R. and **Scotchman, T.C.** (1988) The diagenetic history of some septarian concretions from the Kimmeridge clay, England. *Sedimentology*, **35**, 349–368.

Ayoub-Hannaa, W., Fürsich, F.T. and **El Qot, G.M.** (2014) Cenomanian-Turonian bivalves from eastern Sinai, Egypt. *Palaeontogr. Abt. A*, **301**, 63–168.

Barron, E.J. (1983) A warm, equable Cretaceous: the nature of the problem. *Earth Sci. Rev.*, **19**, 305–338.

Baudin, F. (1995) Depositional controls on Mesozoic source rocks in the Tethys. In: *Paleogeography, Paleoclimate, and Source Rock* (Ed. A.Y. Huc), *AAPG Stud. Geol.*, **40**, 191–211.

Beauchamp, B. and Savard, M. (1992) Cretaceous chemosynthetic carbonate mounds in the Canadian Arctic. *Palaios*, **7**, 434–450.

Birgel, D., Peckmann, J., Klautzsch, S., Thiel, V. and Reitner, J. (2006) Anaerobic and aerobic oxidation of methane at late Cretaceous seeps in the Western Interior Seaway, USA. *Geomicrobiol J.*, **23**, 565–577.

Birgel, D., Elvert, M., Han, X. and Peckmann, J. (2008a) ^{13}C-depleted biphytanic acids as tracers of past anaerobic oxidation of methane. *Org. Geochem.*, **39**, 152–156.

Birgel, D., Himmler, T., Freiwald, A. and Peckmann, J. (2008b) A new constraint on the antiquity of anaerobic oxidation of methane: Late Pennsylvanian seep limestones from southern Namibia. *Geology*, **36**, 543–546.

Bohrmann, G., Abelmann, A., Gersonde, R., Hubberten, H. and Kuhn, G. (1994) Pure siliceous ooze, a diagenetic environment for early chert formation. *Geology*, **22**, 207–210.

Bonarelli, G. (1891) *Il territorio di Gubbio*. Notizie Geologiche: Tipografia Economica, Roma, 38 pp.

Bramlette, M.N. and Sullivan, F.R. (1961) Coccolithophorids and related nannoplankton of the early Tertiary in California. *Micropaleontology*, **7**, 129–174.

Bromley, R.G. (1996) *Trace Fossils: Biology, Taphonomy and Applications*, 2nd edn. Chapman and Hall, London, 361 pp.

Brumsack, H.J. (1980) Geochemistry of Cretaceous black shales from the Atlantic Ocean (DSDP Legs 11, 14, 36, and 41). *Chem. Geol.*, **31**, 1–25.

Brumsack, H.J. (2006) The trace metal content of recent organic carbon-rich sediments: implications for Cretaceous black shale formation. *Palaeogeogr. Palaeoclimatol. Palaeoecol.*, **232**, 344–361.

Busson, G., Dhondt, A., Amédro, F., Néraudeau, D. and Cornée, A. (1999) La grande transgression du Cénomanien supérieur-Turonien inférieur sur la Hamada de Tinrhert (Sahara algérien): datations biostratigraphiques, environnement de dépôt et comparaison d'un témoin épicratonique avec les séries contemporaines à matière organique du Maghreb. *Cretaceous Res.*, **20**, 29–46.

Calvert, S.E. and Pedersen, T.F. (1993) Geochemistry of recent oxic and anoxic marine sediments – implications for the geological record. *Mar. Geol.*, **113**, 67–88.

Campbell, K.A. (2006) Hydrocarbon seep and hydrothermal vent paleoenvironments and paleontology: past developments and future directions. *Palaeogeogr. Palaeoclimatol. Palaeoecol.*, **232**, 362–407.

Campbell, K.A. and Bottjer, D.J. (1995) Brachiopods and chemosynbiotic bivalves in Phanerozoic hydrothermal vent and cold seep environments. *Geology*, **23**, 321–324.

Campbell, K.A., Farmer, J.D. and Des Marais, D. (2002) Ancient hydrocarbon seeps from the Mesozoic convergent margin of California: carbonate geochemistry, fluids and paleoenvionments. *Geofluids*, **2**, 63–94.

Carson, B. and Screaton, E.J. (1998) Fluid flow in accretionary prisms: evidence for focused, time-variable discharge. *Rev. Geophys.*, **36**, 329–351.

Coleman, M.L. (1993) Microbial processes: controls on the shape and composition of carbonate concretions. *Mar. Geol.*, **113**, 127–140.

Coleman, M.L. and Raiswell, R. (1995) Source of carbonate and origin of zonation in pyritiferous carbonate concretions: evaluation of a dynamic model. *Am. J. Sci.*, **295**, 282–308.

Collignon, M. (1966) Les céphalopodes crétacés du bassin côtier de Tarfaya. *Service Geol. Maroc, Notes et Mem.*, **175**, 1–35.

Conti, S., Fontana, D., Gubertini, A., Sighinolfi, G., Tateo, F., Fioroni, C. and Fregni, P. (2004) A multidisciplinary study of middle Miocene seep-carbonates from the northern Apennine foredeep (Italy). *Sed. Geol.*, **169**, 1–19.

Crusius, J., Calvert, S., Pedersen, T. and Sage, D. (1996) Rhenium and molybdenum enrichments in sediments as indicators of oxic, suboxic and sulfidic conditions of deposition. *Earth Planet. Sci. Lett.*, **145**, 65–78.

Dando, P.R., Southward, A.J. and Southward, E.C. (1986) Chemoautotrophic symbionts in the gills of the bivalve mollusc *Lucinoma borealis* and the sediment chemistry of its habitat. *P. R. Soc. London B. Biol.*, **227**, 227–247.

De Boever, E., Birgel, D., Thiel, V., Muchez, P., Peckmann, J., Dimitrov, L. and Swennen, R. (2009) The formation of giant tubular concretions triggered by anaerobic oxidation of methane as revealed by archaeal molecular fossils (Lower Eocene, Varna, Bulgaria). *Palaeogeogr. Palaeoclimatol. Palaeoecol.*, **280**, 23–36.

DeMaster, D.J. (2002) The accumulation and cycling of biogenic silica in the Southern Ocean: revisiting the marine silica budget. *Deep-Sea Res. Pt. II*, **49**, 3155–3167.

Dickson, J.A.D. (1966) Carbonate identification and genesis as revealed by staining. *J. Sed. Petrol.*, **36**, 491–505.

El Albani, A., Caron, M., Deconinck, J.-F., Robazynski, F., Amédro, F., Daoudi, L., Ezaidi, A., Terrab, S. and Thurow, J. (1997) Origin of nodules in Lower Turonian organic-rich sediments in Tarfaya Basin (Southwest Morocco). *CR Acad. Sci. Ser. IIA Earth Planet. Sci.*, **324**, 9–16.

El Albani, A., Kuhnt, W., Luderer, F., Herbin, J.P. and Caron, M. (1999a) Palaeoenvironmental evolution of the Late Cretaceous sequence in the Tarfaya Basin (Southwest of Morocco). In: *The Oil and Gas Habitats of the South Atlantic* (Eds N.R. Cameron, R.H. Bate and V.S. Clure), *Geol. Soc. London. Spec. Publ.*, **153**, 223–240.

El Albani, A., Vachard, D., Kuhnt, W. and Chellai, H. (1999b) Signature of hydrodynamic activity caused by rapid sea level changes in pelagic organic-rich sediments, Tarfaya Basin (southern Morocco). *CR Acad. Sci. Ser. IIA Earth Planet. Sci.*, **329**, 397–404.

El Albani, A., Vachard, D., Kuhnt, W. and Thurow, J. (2001) The role of diagenetic carbonate concretions in the preservation of the original sedimentary record. *Sedimentology*, **48**, 875–886.

El Qot, G.M. (2006) Late Cretaceous macrofossils from Sinai, Egypt. *Beringeria*, **36**, 3–163.

Erba, E., Castradori, D., Guasti, G. and Ripepe, M. (1992) Calcareous nannofossils and Milankovitch cycles: the example of the Albian Gault Clay Formation (Southern England). *Palaeogeogr. Palaeoclimatol. Palaeoecol.*, **93**, 47–69.

Eshet, Y. and Almogi-Labin, A. (1996) Calcareous nannofossils as paleoproductivity indicators in Upper Cretaceous organic-rich sequences in Israel. *Mar. Micropaleontol.*, **29**, 37–61.

Feng, D., Birgel, D., Peckmann, J., Roberts, H.H., Joye, S.B., Sassen, R., Liu, X.-L., Hinrichs, K.-U. and Chen, D. (2014) Time integrated variation of sources of fluids and seepage dynamics archived in authigenic carbonates from Gulf of Mexico Gas Hydrate Seafloor Observatory. *Chem. Geol.*, **385**, 129–139.

Formolo, M.J., Lyons, T.W., Zhang, C., Kelley, C., Sassen, R., Horita, J. and Cole, D.R. (2004) Quantifying carbon sources in the formation of authigenic carbonates at gas hydrate sites in the Gulf of Mexico. *Chem. Geol.*, **205**, 253–264.

Forster, S. and Graf, G. (1992) Continuously measured changes in redox potential influenced by oxygen penetrating from burrows of *Callianassa subterranea*. *Hydrobiologia*, **235**, 527–532.

Forster, A., Schouten, S., Moriya, K., Wilson, P.A. and Sinninghe Damsté, J.S. (2007) Tropical warming and intermittent cooling during the Cenomanian/Turonian oceanic anoxic event 2: sea surface temperature records from the equatorial Atlantic. *Paleoceanography*, **22**, PA1219. doi:10.1029/2006PA001349.

Gale, A.S., Smith, A.B., Monks, N.E.A., Young, J.A., Howard, A., Wray, D.S. and Huggett, J.M. (2000) Marine biodiversity through the Late Cenomanian-Early Turonian: paleoceanographic controls and sequence stratigraphic biases. *J. Geol. Soc. London*, **157**, 745–757.

Gebhardt, H., Kuhnt, W. and Holbourn, A. (2004) Foraminiferal response to sea level change, organic flux and oxygen deficiency in the Cenomanian of the Tarfaya Basin, southern Morocco. *Mar. Micropaleontol.*, **53**, 133–157.

Gertsch, B., Adatte, T., Keller, G., Tantawy, A.A.A., Berner, Z., Mort, H.P. and Fleitmann, D. (2010) Middle and late Cenomanian oceanic anoxic events in shallow and deeper shelf environments of western Morocco. *Sedimentology*, **57**, 1430–1462.

Goedert, J.L., Peckmann, J. and Reitner, J. (2000) Worm tubes in an allochtonous cold-seep carbonate from lower Oligocene rocks of western Washington. *J. Paleontol.*, **74**, 992–999.

Greinert, J., Bohrmann, G. and Suess, E. (2001) Gas hydrate-associated carbonates and methane-venting at Hydrate

Ridge: classification, distribution, and origin of authigenic lithologies. *Geoph. Monog.*, **124**, 99–113.

Guido, A., Heindel, K., Birgel, D., Rosso, A., Mastandrea, A., Sanfilippo, R., Russo, F. and Peckmann, J. (2013) Pendant bioconstructions cemented by microbial carbonate in submerged marine caves (Holocene, SE Sicily). *Palaeogeogr. Palaeoclimatol. Palaeoecol.*, **388**, 166–180.

Haas, A., Peckmann, J., Elvert, M., Sahling, H. and Bohrmann, G. (2010) Patterns of carbonate authigenesis at the Kouilou pockmarks on the Congo deep-sea fan. *Mar. Geol.*, **268**, 129–136.

Hagemann, A., Leefmann, T., Peckmann, J., Hoffmann, V.-E. and Thiel, V. (2013) Biomarkers from individual carbonate phases of an Oligocene cold-seep deposit, Washington State, USA. *Lethaia*, **46**, 7–18.

Hallam, A. and Bradshaw, A. (1977) Bituminous shales and oolitic ironstones as indicators of transgressions and regressions. *J. Geol. Soc. London*, **36**, 57–64.

Handoh, I.C., Bigg, G.R., Jones, E.J.W. and Inoue, M. (1999) An ocean modeling study of the Cenomanian Atlantic: equatorial paleoupwelling, organic-rich sediments and the consequences for a connection between the proto-North and South Atlantic. *Geophys. Res. Lett.*, **26**, 223–226.

Haq, B.U. (2014) Cretaceous eustasy revisited. *Global Planet. Change*, **113**, 44–58.

Hardenbol, J., Thierry, J., Farley, M.B., Jacquin, T., De Graciansky, P.C. and Vail, P.R. (1998) Mesozoic and Cenozoic sequence chronostratigraphic framework of European Basins. *SEPM Spec. Publ.*, **60**, 3–13.

Hattner, J.G., Wind, F.H. and Wise, S.W. (1980) The Santonian-Campanian boundary. Comparison of nearshore-offshore calcareous nannofossil assemblages. *Cah. Micropaléontol.*, **3**, 9–26.

Hay, W.W. (1961) Note on the preparation of samples for discoasterids. *J. Paleontol.*, **35**, 873.

Hay, W.W. (1965) Calcareous nannofossils. In: *Handbook of Paleonological Techniques*(Eds. B. Kummel and D. Raup), pp. 3–7. Freeman, San Francisco, CA.

Heany, P.J. (1993) A proposed mechanism for the growth of chalcedony. *Contrib. Mineral. Petrol*, **115**, 66–74.

Heinrichs, H., Brumsack, H.-J., Loftfield, N. and König, N. (1986) Verbessertes Druckaufschlußsystem für biologische und anorganische Materialen. *Z. Pflanzenernähr. Bodenk.*, **149**, 350–353.

Helz, G.R., Miller, C.V., Charnick, J.M., Mosselmans, J.F.W., Pattrick, R.A.D., Garner, C.D. and Vaughan, D.J. (1996) Mechanism of molybdenum removal from the sea and its concentration in black shales: EXAFS evidence. *Geochim. Cosmochim. Acta*, **60**, 3631–3642.

Herbin, J.P., Montadert, L., Mueller, C., Gomez, R., Thurow, J. and Wiedmann, J. (1986) Organic-rich sedimentation at the Cenomanian–Turonian Boundary in oceanic and coastal basins in the North Atlantic and Tethys. In: *North Atlantic*

Palaeoceanography (Eds C. Summerhayes and N.J. Shackleton), *Geol. Soc. London*, **21**, 389–422.

Herrle, J.O. (2003) Reconstructing nutricline dynamics of mid-Cretaceous oceans: evidence from calcareous nannofossils from the Niveau Paquier black shale (SE France). *Mar. Micropalaeontol.*, **47**, 307–321.

Hetzel, A., Böttcher, M.E., Wortmann, U.G. and Brumsack, H.-J. (2009) Paleo-redox conditions during OAE 2 reflected in Demarara Rise sediment geochemistry (ODP Leg 207). *Palaeogeogr. Palaeoclimatol. Palaeoecol.*, **273**, 302–328.

Himmler, T., Freiwald, A., Stollhofen, H. and Peckmann, J. (2008) Late Carboniferous hydrocarbon-seep carbonates from the glaciomarine Dwyka Group, southern Namibia. *Palaeogeogr. Palaeoclimatol. Palaeoecol.*, **257**, 185–197.

Holbourn, A., Kuhnt, W., El Albani, A., Pletsch, T., Luderer, F. and Wagner, T. (1999) Upper Cretaceous palaeoenvironments and benthonic foraminiferal assemblages of potential source rocks from the western African margin, Central Atlantic. In: *The Oil and Gas Habitats of the South Atlantic* (Eds N.R. Cameron, R.H. Bate and V.S. Clure), *Geol. Soc. London. Spec. Publ.*, **153**, 195–222.

Hudson, J.D. (1978) Concretions, isotopes and the diagenetic history of the Oxford Clay (Jurassic) of central England. *Sedimentology*, **25**, 339–370.

Hudson, J.D. and Friedman, I. (1974) Carbon and oxygen isotopes in concretions: relationship to pore water changes during diagenesis. In: *Proceedings of the International Symposium on Water-Rock Interaction*(Eds. J. Cadetm and T. Paces), pp. 331–339. Geological Survey, Prague.

Ingall, E.D., Bustin, R.M. and Van Cappellen, P. (1993) Influence of water column anoxia on the burial and preservation of carbon and phosphorus in marine shales. *Geochim. Cosmochim. Acta*, **57**, 303–316.

Jenkyns, H.C. (2003) Evidence for rapid climate change in the Mesozoic-Palaeogene greenhouse world. *Philos. T. R. Soc. London*, **361**, 1885–1916.

Jenkyns, H.C. (2010) Geochemistry of oceanic anoxic events. *Geochem. Geophys. Geosys.*, **11**, 1–30.

Jenkyns, H.C., Matthews, A., Tsikos, H. and Erel, Y. (2007) Nitrate reduction, sulfate reduction, and sedimentary iron isotope evolution durint the Cenomanian – Turonian oceanic anoxic event. *Palaeoceanography*, **22**, 1–17.

Jones, C.E. and Jenkyns, H.C. (2001) Seawater strontium isotopes, oceanic anoxic events, and seafloor hydrothermal activity in the Jurassic and Cretaceous. *Am. J. Sci.*, **301**, 112–149.

Judd, A.J. and Hovland, M. (2007) *Submarine Fluid Flow, the Impact on Geology, Biology, and the Marine Environment.* Cambridge University Press, Cambridge.

Kaiho, K. (1994) Benthic foraminiferal dissolved-oxygen index and dissolved-oxygen levels in the modern ocean. *Geology*, **22**, 719–722.

Kasten, S., Zabel, M., Heuer, V. and Hensen, C. (2003) Processes and signals of nonsteady-state diagenesis in deep-

sea sediments and their pore waters. *The South Atlantic in the Late Quaternary*, pp. 431–459. Springer, Berlin, Heidelberg.

Kauffman, E.G., Arthur, M.A., Howe, B. and Scholle, P.A. (1996) Widespread venting of methane-rich fluids in Late Cretaceous (Campanian) submarine springs (Tepee Buttes), Western Interior seaway, U.S.A. *Geology*, **24**, 799–802.

Keene, J.B. (1975) Cherts and porcellanites from the North Pacific. *Initial Rep. Deep Sea.*, **32**, 429–507.

Kerr, A.C. (1998) Oceanic plateau formation: a cause of mass extinction and black shale deposition around the Cenomanian-Turonian boundary? *J. Geol. Soc. London*, **155**, 619–626.

Kiel, S. (2010) On the potential generality of depth-related ecologic structure in cold-seep communities: evidence from Cenozoic and Mesozoic examples. *Palaeogeogr. Palaeoclimatol. Palaeoecol.*, **295**, 245–257.

Kiel, S. (2013) Lucinid bivalves from ancient methane seeps. *J. Malacol. Soc. London*, **79**, 346–363.

Kiel, S. and Peckmann, J. (2007) Chemosymbiotic bivalves and stable carbon isotopes indicate hydrocarbon seepage at four unusual Cenozoic fossil localities. *Lethaia*, **40**, 345–357.

Kiel, S., Wiese, F. and Titus, A.L. (2012) Shallow-water methane-seep faunas in the Cenomanian Western Interior Seaway: no evidence for onshore-offshore adaptations to deep-sea vents. *Geology*, **40**, 839–842.

Kiel, S., Birgel, D., Campbell, K.A., Crampton, J.S., Schiøler, P. and Peckmann, J. (2013) Crecateous methane-seep deposits from New Zealand and their fauna. *Palaeogeogr. Palaeoclimatol. Palaeoecol.*, **390**, 17–34.

Knauth, P.L. (1994) Petrogenesis of chert. In: *Reviews in Mineralogy, vol. 29: Silica – Physical Behavior, Geochemistry and Materials Applications*(Eds. P.W. Heany, C.T. Prewitt and G.V. Gibbs), pp. 233–258. Mineralogical Society of America, Washington, DC.

Kolonic, S., Sinninghe Damsté, J.S., Böttcher, M.E., Kuypers, M.M.M., Kuhnt, W., Beckmann, B., Scheeder, G. and Wagner, T. (2002) Geochemical characterization of Cenomanian/Turonian black shales from the Tarfaya Basin (SW Morocco): relationships between paleoenvironmental conditions and early sulfurization of sedimentary organic matter. *J. Petrol. Geol.*, **25**, 325–350.

Kolonic, S., Wagner, T., Forster, A., Sinninghe Damsté, J.S., Walsworth-Bell, B., Erba, E., Turgeon, S., Brumsack, H.J., Chellai, E.H., Tsikos, H., Kuhnt, W. and Kuypers, M.M.M. (2005) Black shale deposition on the northwest African shelf during the Cenomanian-Turonian oceanic anoxic event: climate coupling and global organic carbon burial. *Paleoceanography*, **20**, 1–18.

Könneke, M., Lipp, J.S. and Hinrichs, K.-U. (2012) Carbon isotope fractionation by the marine ammonia-oxidizing archaeon *Nitrosopumilus maritimus. Org. Geochem.*, **48**, 12–24.

Kuechler, R.R., Birgel, D., Kiel, S., Freiwald, A., Goedert, J.L., Thiel, V. and Peckmann, J. (2012) Miocene methane-derived carbonates from southwestern Washington (USA) and a model for silicification at seeps. *Lethaia*, **45**, 259–273.

Kuhnt, W. and **Wiedmann, J.** (1995) Cenomanian –Turonian source rocks: paleobiogeographic and paleoenvironmental aspects. In: *Paleogeography, Paleoclimate and Source Rocks* (Ed. A.Y. Huc), *Stud. Geol.*, **40**, 213–232. AAPG, Tulsa, OK.

Kuhnt, W., Herbin, J.P., Thurow, J. and **Wiedmann, J.** (1990) Distribution of Cenomanian–Turonian organic facies in the western Mediterranean and along the adjacent Atlantic margin. In: *Deposition of Organic Facies* (Ed. A.Y. Huc), *AAPG Stud. Geol.*, **30**, 133–160.

Kuhnt, W., El Chellai, H., Holbourn, A., Luderer, F., Thurow, J., Wagner, T., El Albani, A., Beckmann, B., Herbin, J.P., Kawamura, H., Kolonic, S., Nederbragt, S., Street, C. and **Ravilious, K.** (2001) Morocco basin's sedimentary record may provide correlations for Cretaceous paleoceanographic events worldwide. *EOS Trans. Am. Geophys. Union*, **82**, 361–364.

Kuhnt, W., Luderer, F., Nederbragt, A., Thurow, J. and **Wagner, T.** (2005) Millennial resolution record of the Late Cenomanian Oceanic Anoxic Event (OAE2) in the Tarfaya Basin (Morocco). *Int. J. Earth Sci.*, **94**, 147–159.

Kuhnt, W., Holbourn, A., Gale, A., Chellai, E.H. and **Kennedy, W.J.** (2009) Cenomanian sequence stratigraphy and sea-level fluctuations in the Tarfaya Basin (SW Morocco). *Geol. Soc. Am. Bull.*, **121**, 1695–1710.

Kuypers, M.M.M., Pancost, R.D., Nijenhuis, I.A. and **Sinninghe Damsté, J.S.** (2002) Enhanced productivity led to increased organic carbon burial in the euxinic North Atlantic basin during the late Cenomanian oceanic anoxic event. *Paleoceanography*, **17**, 1–13.

Lancelot, Y. (1973) Chert and silica diagenesis in sediments from the central Pacific. *Initial Rep. Deep Sea*, **32**, 429–507.

Larson, R.L. (1991) Latest pulse of Earth; evidence for a mid-Cretaceous super-plume. *Geology*, **19**, 547–550.

Larson, R.L. and **Erba, E.** (1999) Onset of the Mid-Cretaceous greenhouse in the Barremian-Aptian: igneous events and the biological, sedimentary, and geochemical responses. *Paleoceanography*, **14**, 663–678.

Layeb, M., Fadhel, M.B. and **Youssef, M.B.** (2012) Thrombolitic and coral buildups in the Upper Albian of the Fahdene basin (North Tunisia): stratigraphy, sedimentology and genesis. *Bull. Soc. Géol. France*, **183**, 217–231.

Layeb, M., Fadhel, M.B., Layeb-Tounsi, Y. and **Youssef, M.B.** (2014) First microbialites associated to organic-rich facies of the Oceanic Anoxic Event 2 (Northern Tunisia, Cenomanian-Turonian transition). *Arab. J. Geosc.*, **7**, 3349–3363.

Leine, L. (1986) Geology of the Tarfaya oil shale deposit, Morocco. *Geol. Mijnbouw*, **65**, 57–74.

Liang, H., Chen, X., Wang, C., Zhao, D. and **Weissert, H.** (2016) Methane-derived authigenic carbonates of mid-Cretaceous age in southern Tibet: types of carbonate concretions, carbon sources, and formation processes. *J. Asian Earth Sci.*, **115**, 153–169.

Little, C.T.S., Birgel, D., Boyce, A.J., Crame, A.J., Francis, J.E., Kiel, S., Peckmann, J., Pirrie, D., Rollinson, G.K. and **Witts, J.D.** (2015) Late Cretaceous (Maastrichtian) shallow water hydrocarbon seeps from Snow Hill and Seymour Islands, James Ross Basin, Antarctica. *Palaeogeogr. Palaeoclimatol. Palaeoecol.*, **418**, 213–228.

Liu, X.-L., Birgel, D., Elling, F.J., Sutton, P.A., Lipp, J.S., Zhu, R., Zhang, C., Könneke, M., Peckmann, J., Rowland, S., Summons, R.E. and **Hinrichs, K.-U.** (2016) From ether to acid: a plausible degradation pathway of glycerol dialkyl glycerol tetraethers. *Geochim. Cosmochim. Acta*, **183**, 138–152.

Lüning, S., Marzouk, A.M. and **Kuss, J.** (1998) Latest Maastrichtian high frequency litho- and ecocycles from the hemipelagic of Eastern Sinai, Egypt. *J. Afr. Earth Sci.*, **27**, 373–395.

Lüning, S., Kolonic, S., Belhadj, E.M., Belhadj, Z., Cota, L., Baric, G. and **Wagner, T.** (2004) Integrated depositional model for the Cenomanian-Turonian organic-rich strata in North Africa. *Earth Sci. Rev.*, **64**, 51–117.

Marshall, J.D. and **Pirrie, D.** (2013) Carbonate concretions – explained. *Geol. Today*, **29**, 53–62.

Marzouk, A.M. and **Lüning, S.** (2005) Calcareous nannofossil biostratigraphy and distribution patterns in the Cenomanian-Turonian of North Africa. *Grmena*, **1**, 107–122.

Mascle, J., Lohmann, G.P., Clift, P.D. and **Party, S.S.** (1997) Development of a passive transform margin: Cote d'Ivoire–Ghana transform margin — ODP Leg 159 preliminary results. *Geo-Mar. Lett.*, **17**, 4–11.

Mattioli, E. (1997) Nannoplankton productivity and diagenesis in the rhythmically bedded Toarcian-Aalenian Fiuminata sequence (Umbria-Marche Apennine, Central Italy). *Palaeogeogr. Palaeoclimatol. Palaeoecol.*, **130**, 113–134.

McAnena, A., Floegel, S., Hofmann, P., Herrle, J.O., Griesand, A., Pross, J., Talbot, H.M., Rethemeyer, L., Wallmann, K. and **Wagner, T.** (2013) Atlantic opening, million-year cooling and coupled marine biotic crises in the Cretaceous. *Nat. Geosci.*, **6**, 558–561.

Metz, C.L. (2010) Tectonic controls on the genesis and distribution of Late Cretaceous Western Interior Basin hydrocarbon-seep mounds (Tepee Buttes) of North America. *J. Geol.*, **118**, 201–213.

Meyer, K.M. and **Kump, L.R.** (2008) Oceanic euxinia in Earth history: causes and consequences. *Ann. Rev. Earth Planet. Sci.*, **36**, 251–288.

Morford, J.L. and **Emerson, S.R.** (1999) The geochemistry of redox sensitive trace metals in sediments. *Geochim. Cosmochim. Acta*, **63**, 1735–1750.

Morford, J.L., Emerson, S.R., Breckel, E.J. and **Kim, S.H.** (2005) Diagenesis of oxyanions (V, U, R, and Mo) in pore waters and sediments from a continental margin. *Geochim. Cosmochim. Acta*, **69**, 5021–5032.

Mort, H.P., Adatte, T., Keller, G., Bartels, D., Föllmi, K.B., Steinmann, P., Berner, Z. and Chellai, E.H. (2008) Organic carbon deposition and phosphorus accumulation during Oceanic Anoxic Event 2 in Tarfaya. *Morocco. Cretaceous Res.*, **29**, 1008–1023.

Mozley, P.S. and Davis, J.M. (2005) Internal structure and mode of growth of elongate calcite concretions: evidence for small-scale, microbially induced, chemical heterogeneity in groundwater. *Geol. Sco. Am. Bull.*, **117**, 1400–1412.

Naehr, T.H., Birgel, D., Bohrmann, G., MacDonald, I.R. and Kasten, S. (2009) Biogeochemical controls on authigenic carbonate formation at the Chaopopote "asphalt volcano", Bay of Campeche. *Chem. Geol.*, **266**, 390–402.

Natalicchio, M., Birgel, D., Dela Pierre, F., Martire, L., Clari, P., Spötl, C. and Peckmann, J. (2012) Polyphasic carbonate precipitation in the shallow subsurface: insights from microbially-formed authigenic carbonate beds in upper Miocene sediments of the Tertiary Piedmont Basin (NW Italy). *Palaeogeogr. Palaeoclimatol. Palaeoecol.*, **329–330**, 158–172.

Nederbragt, A.J. and Fiorentino, A. (1999) Stratigraphy and palaeoeoceanography of the Cenomanian-Turonian Boundary Event in Oued Mellegue, north– western Tunisia. *Cretaceous Res.*, **20**, 47–62.

Nöthen, K. and Kasten, S. (2011) Reconstructing changes in seep activity by means of pore water and solid phase Sr/Ca and Mg/Ca ratios in pockmark sediments of the Northern Congo Fan. *Mar. Geol.*, **287**, 1–13.

Owens, J.D., Gill, B.C., Jenkyns, H.C., Bates, S.M., Severmann, S., Kuypers, M.M.M., Woodfine, R.G. and Lyons, T.W. (2013) Sulfur isotopes track the global extent and dynamics of euxinia during Cretaceous Oceanic Anoxic Event 2. *Proc. Natl Acad. Sci. USA*, **110**, 18407–18412.

Pearson, A., Hurley, S.J., Walter, S.R.S., Kusch, S., Lichtin, S. and Zhang, Y.G. (2016) Stable carbon isotope ratios of intact GDGTs indicate heterogenus sources to marine sediments. *Geochim. Cosmochim. Acta*, **181**, 18–35.

Peckmann, J. and Thiel, V. (2004) Carbon cycling at ancient methane seeps. *Chem. Geol.*, **205**, 443–467.

Peckmann, J., Thiel, V., Michaelis, W., Clari, P., Gaillard, P., Martire, L. and Reitner, J. (1999) Cold seep deposits of Beauvoisin (Oxfordian; southeastern France) and Marmorito (Miocene; northern Italy): microbially induced authigenic carbonates. *Int. J. Earth Sci.*, **88**, 60–75.

Peckmann, J., Reimer, A., Luth, U., Luth, C., Hansen, B.T., Heinicke, C., Hoefs, J. and Reitner, J. (2001) Methane derived carbonates and authigenic pyrite from the northwestern Black Sea. *Mar. Geol.*, **177**, 129–150.

Peckmann, J., Goedert, J.L., Thiel, V., Michaelis, W. and Reitner, J. (2002) A comprehensive approach to the study of methane–seep deposits from the Lincoln Creek Formation, western Washington State USA. *Sedimentology*, **49**, 855–873.

Peckmann, J., Goedert, J.L., Heinrichs, T., Hoefs, J. and Reitner, J. (2003) The late Eocene 'Whiskey Creek' methane-seep deposit (western Washington State). Part II: petrology, stable isotopes, and biogeochemistry. *Facies*, **48**, 241–254.

Peckmann, J., Senowbari-Daryan, B., Birgel, D. and Goedert, J.L. (2007) The crustacean ichnofossil *Palaxius* associated with callianassid body fossils in an Eocene methane-seep limestone, Humptulips Formation, Olympic Peninsula, Washington. *Lethaia*, **40**, 273–280.

Peckmann, J., Birgel, D. and Kiel, S. (2009) Molecular fossils reveal fluid composition and flow intensity at a Cretaceous seep. *Geology*, **37**, 847–850.

Peckmann, J., Kiel, S., Sandy, M.R., Taylor, D.G. and Goedert, J.L. (2011) Mass occurrences of the brachiopod *Halorella* in Late Triassic methane-seep deposits, eastern Oregon. *J. Geol.*, **119**, 207–220.

Poulton, S.W., Henkel, S., März, C., Urquhart, H., Floegel, S., Kasten, S., Sinninghe Damsté, J. and Wagner, T. (2015) A continental-weathering control on orbitally driven redox-nutrient cycling during Cretaceous Oceanic Anoxic Event 2. *Geology*, **43**, 963–966.

Raiswell, R. and Fischer, Q.J. (2004) Rates of carbonate cementation associated with sulphate reduction in DSDP/ODP sediments: implications for the formation of concretions. *Chem. Geol.*, **211**, 71–85.

Raiswell, R. and Fisher, Q.J. (2000) Mudrock-hosted carbonate concretions: a review of growth mechanisms and their influence on chemical and isotopic composition. *J. Geol. Soc. London*, **157**, 239–251.

Reitz, A., Hensen, C., Kasten, S., Funk, J.A. and De Lange, G.J. (2004) A combined geochemial and rock-magnetic investigation of a redox horizon at the last glacial/interglacial transition. *Phys. Chem. Earth*, **29**, 921–931.

Ritger, S., Carson, B. and Suess, E. (1987) Methane derived authigenic carbonates formed by subduction-induced pore-water expulsion along the Oregon/Washington margin. *Geol. Soc. Am. Bull.*, **98**, 147–156.

Roberts, H.H. and Aharon, P. (1994) Hydrocarbon-derived carbonate buildups of the northern Gulf of Mexico contintental slope: a review of submersible investigations. *Geo-Mar. Lett.*, **14**, 135–148.

Roberts, H.H., Aharon, P. and Walsh, M.M. (1993) Cold-seep carbonates of the Louisiana continental slope-to-basin floor. In: *Carbonate Microfabrics* (Eds R. Rezak and D.L. Lavoie), pp. 95–104. Springer, Berlin, Heidelberg, New York.

Rosenthal, Y., Lam, P., Boyle, E.A. and Thomson, J. (1995) Authigenic cadmium enrichments in suboxic sediments: precipitation and postdepositional mobility. *Earth Planet. Sci. Lett.*, **132**, 99–111.

Roth, P.H. (1981) Mid-Cretaceous calcareous nannoplankton from the central Pacific: implication for paleoceanography. *Initial Rep. Deep Sea*, **62**, 471–489.

Roth, P.H. and Krumbach, K.R. (1986) Middle Cretaceous calcareous nannofossil biogeography and preservation in the Atlantic and Indian Oceans: implications for paleoceanography. *Mar. Micropaleontol.*, **10**, 235–266.

Sachse, V.F., Littke, R., Jabour, H., Schümann, T. and Kluth, O. (2012) Late Cretaceous (Late Turonian, Coniacian and Santonian) petroleum source rocks as part of an OAE, Tarfaya Basin, Morocco. *Mar. Petrol. Geol.*, **29**, 35–49.

Sames, B., Wagreich, M., Wendler, J.E., Haq, B.U., Conrad, C.P., Dobrinescu-Melinte, M.C., Hu, X., Wendler, I., Wolfgring, E., Yilmaz, I.Ö. and Zorina, S.O. (2016) Review: short-term sea-level changes in a greenhouse world – a view from the Cretaceous. *Palaeogeogr. Palaeoclimatol. Palaeoecol.*, **441**, 393–411.

Sassen, R., Joye, S., Sweet, S.T., DeFreitas, D.A., Milkov, A.V. and MacDonald, I.R. (1999) Thermogenic gas hydrates and hydrocarbon gases in complex chemosynthetic communities, Gulf of Mexico continental slope. *Org. Geochem.*, **30**, 485–497.

Savard, M.M., Beauchamp, B. and Veizer, J. (1996) Significance of aragonite cements around Cretaceous marine methane seeps. *J. Sed. Res.*, **66**, 430–438.

Schlanger, S.O. and Jenkyns, H.C. (1976) Cretaceous oceanic anoxic events: causes and consequences. *Geol. Mijnbouw*, **55**, 179–184.

Schlanger, S.O., Arthur, M.A., Jenkyns, H.C. and Scholle, P.A. (1987) The Cenomanian–Turonian oceanic anoxic event, I. Stratigraphy and distribution of organic carbon-rich beds and the marine $\delta^{13}C$ excursion. In: *Marine Petroleum Source Rocks* (Eds J. Brooks and A.J. Fleet), *Geol. Soc. Spec. Publ.*, **26**, 371–399.

Schouten, S., Wakeham, S.T., Hopmans, E.C. and Sinninghe Damsté, J.S. (2003) Biogeochemical evidence that thermophilic archaea mediate the anaerobic oxidation of methane. *Appl. Environ. Microb.*, **69**, 1680–1686.

Schouten, S., Hopmans, E.C. and Sinninghe Damsté, J.S. (2013) The organic geochemistry of glycerol dialkyl glycerol tetraether lipids: a review. *Org. Geochem.*, **54**, 19–61.

Schwartz, H., Sample, J., Weberling, K.D., Minisini, D. and Moore, J.C. (2003) An ancient linked fluid migration system: cold-seep deposits and sandstone intrusions in the Panoche Hills, California, USA. *Geo-Mar. Lett.*, **23**, 340–350.

Schweigert, G., Seegis, D.B., Fels, A. and Leinfelder, R.R. (1997) New internally structured decapod microcoprolites from Germany (Late Triassic/Early Miocene), Southern Spain (Early/Middle Jurassic) and Portugal (Late Jurassic): taxonomy, palaeoecology and evolutionary patterns. *Paläontol. Z.*, **71**, 51–69.

Seilacher, A. (2001) Concretion morphologies reflecting diagenetic and epigenetic pathways. *Sed. Geol.*, **143**, 41–57.

Sellés-Martínez, J. (1996) Concretion morphology, classification and genesis. *Earth Sci. Rev.*, **41**, 177–210.

Senowbari-Daryan, B. and Kuss, J. (1992) Anomuren-Koprolithen aus der Kreide von Ägypten. *Mitt. Geol.*, **73**, 129–157.

Senowbari-Daryan, B., Weidlich, O. and Flügel, E. (1992) Erster Nachweis von, Favreinen' (Crustaceen-Koprolithen) aus dem Perm: Oberperm, Oman-Berge. *Paläontol. Z.*, **6**, 187–196.

Senowbari-Daryan, B., Gaillard, C. and Peckmann, J. (2007) Crustacean microcoprolites from Jurassic (Oxfordian) hydrocarbon-seep deposits of Beauvoisin, southeastern France. *Facies*, **53**, 229–238.

Shapiro, R. and Fricke, H. (2002) Tepee Buttes: fossilized methane seep ecosystems. *Geol. Soc. Am. Field Guides*, **3**, 94–101.

Sinninghe Damsté, J.S. and Köster, J. (1998) A euxinic southern North Atlantic Ocean during the Cenomanian/Turonian oceanic anoxic event. *Earth Planet. Sci. Lett.*, **158**, 165–173.

Smrzka, D., Kraemer, S.M., Zwicker, J., Birgel, D., Fischer, D., Kasten, S., Goedert, J.L. and Peckmann, J. (2015) Constraining silica diagenesis in methane-seep deposits. *Palaeogeogr. Palaeoclimatol. Palaeoecol.*, **420**, 13–26.

Smrzka, D., Zwicker, J., Klügel, A., Monien, P., Bach, W., Bohrmann, G. and Peckmann, J. (2016) Establishing criteria to distinguish oil-seep from methane-seep carbonates. *Geology*, **44**, 667–670.

Suess, E. (2014) Marine cold seeps and their manifestations: geological control, biogeochemical criteria and evironmental conditions. *Int. J. Earth Sci.*, **103**, 1889–1916.

Suess, E. and Whiticar, M.J. (1989) Methane-derived CO_2 in pore fluids expelled from the Orgeon subduction zone. *Palaeogeogr. Palaeoclimatol. Palaeoecol.*, **71**, 119–136.

Teichert, B.M.A., Eisenhauer, A., Bohrmann, G., Haase-Schramm, A., Bock, B. and Linke, P. (2003) U/Th Systematics and ages of authigenic carbonates from Hydrate Ridge, Cascadia Margin: recorders of fluid flow variations. *Geochim. Cosmochim. Acta*, **67**, 3845–3857.

Thiel, V., Peckmann, J., Seifert, R., Wehrung, P., Reitner, J. and Michaelis, W. (1999) Highly isotopically depleted isoprenoids: molecular markers for ancient methane venting. *Geochim. Cosmochim. Acta*, **63**, 3959–3966.

Thierstein, H.R. (1976) Mesozoic calcareous nannoplankton biostratigraphy of marine sediments. *Mar. Micropaleontol.*, **1**, 325–362.

Tong, H., Wang, Q., Peckmann, J., Cao, Y., Chen, L., Zhou, W. and Chen, D. (2016) Diagenetic alteration affecting $\delta^{18}O$, $\delta^{13}C$ and $^{87}Sr/^{86}Sr$ signatures of carbonates: a case study on Cretaceous seep deposits from Yarlung-Zangbo Suture Zone, Tibet, China. *Chem. Geol.*, **444**, 71–82.

Tribovillard, N., Algeo, T.J., Lyons, T. and Riboulleau, A. (2006) Trace metals as paleoredox and paleoproductivity proxies: an update. *Chem. Geol.*, **232**, 12–32.

Von Rad, U., Riech, V. and Rösch, H. (1977) Silica diagenesis in contintental margin sediments off northwest Africa. *Initial Rep. Deep Sea*, **41**, 879–905.

Wagner, T. and Pletsch, T. (1999) Tectono-sedimentary controls on Cretaceous black shale deposition along the opening Equatorial Atlantic Gateway (ODP Leg 159). In: *Oil and Gas Habitats of the South Atlantic* (Eds N.R. Cameron, R.H. Bate and V.S. Clure), *Geol. Soc. London. Spec. Publ.*, **153**, 241–265.

Wagner, T., Floegel, S. and Hofmann, P. (2013) Marine black shale and Hadley Cell dynamics: a conceptual framework for the Cretaceous Atlantic Ocean. *Mar. Petrol Geol.*, **43**, 222–238.

Wedepohl, K.H. (ed.) (1969) Composition and abundance of common igneous rocks. In: *Handbook of Geochemistry*, **1**, 227–249.

Weissert, H. (1981) The environment of deposition of black shales in the Early Cretaceous: an ongoing controversy. *SEPM Spec. Publ.*, **32**, 547–560.

Weissert, H., McKenzie, J. and Hochuli, P. (1979) Cyclic anoxic events in the Early Cretaceous Tethys Ocean. *Geology*, **7**, 147–151.

Xu, W., Ruhl, M., Hesselbo, S.P., Riding, J.B. and Jenkyns, H.C. (2016) Orbital pacing of the Early Jurassic carbon cycle, black shale formation and seabed methane seepage. *Sedimentology*, **64**, 127–149.

Zwicker, J., Smrzka, D., Gier, S., Goedert, J.L. and Peckmann, J. (2015) Mineralized conduits are part of the uppermost plumbing system of Oligocene methane-seep deposits, Washington State (USA). *Mar. Petrol. Geol.*, **66**, 616–630.

6

Manganese enrichments near a large gas-hydrate and cold-seep field: a record of past redox and sedimentation events

WESLEY C. INGRAM*, STEPHEN R. MEYERS†, ZHIZHANG SHEN†, HUIFANG XU† and
CHRISTOPHER S. MARTENS*

*Department of Marine Sciences, University of North Carolina-Chapel Hill, Chapel Hill, NC 27599, USA (E-mail: MatagordaPetroleum@outlook.com)
†Department of Geoscience, University of Wisconsin-Madison, Madison, WI 53706, USA

Keywords
Biogeochemistry, cold seeps, deep-sea sediments, gas hydrates, manganese, redox boundary.

ABSTRACT

The spatial distribution, mineralogy, and origin of manganese enrichments surrounding a large gas hydrate and cold seep field (Mississippi Canyon 118, Gulf of Mexico) are investigated in this study, to better constrain their biogeochemical context in deep-sea sediments and to assess how gas hydrates may alter such records. Manganese depth profiles from 10 sediment cores, documented using centimetre-scale X-ray fluorescence core scanning, display highly-enriched 1 to 10 cm thick layers. These manganese-rich layers are more numerous, but of lower concentration, in close proximity to the field, and show no consistent relationship with sedimentology (clay vs. carbonate content) or the established chronostratigraphic framework at the site. X-ray diffraction and sequential dissolution procedures indicate that the manganese enrichments are authigenic carbonates, which formed along a palaeo redox boundary during periods of prolonged steady-state conditions. The hypothesis that spatial heterogeneity of this manganese record is linked to the nearby gas hydrate and cold seep field, by influencing redox conditions and/or sedimentation processes, is investigated here. Results are consistent with more frequent interruption of steady-state sedimentation in closer proximity to the salt-tectonic induced bathymetric mound, which contains the active cold seeps and gas hydrate deposits. Thus, spatial mapping of manganese enrichment horizons provides a tool to reconstruct sedimentation surrounding these volatile sea bed features, yielding a measure of past activity of gas hydrates and cold seeps.

INTRODUCTION

Marine gas hydrates have received much attention due in part to the vast amount of carbon contained within these deposits (Kvenvolden, 1988; Dickens, 2001; Milkov, 2004), and their potential linkages to slope destabilization and past climate events (Nisbet, 1990; Paull et al., 1996, 2003; Haq, 1998; Maslin et al., 1998, 2004; Mienert et al., 2005). These aspects have motivated detailed biogeochemical and geophysical investigations of modern gas hydrates (Sassen et al., 2004; Castellini et al., 2006; McGee, 2006; Brunner, 2007; Lapham et al., 2008, 2010; McGee et al., 2009; Macelloni et al., 2012, 2013; Simonetti et al., 2013; Feng et al., 2014; Martens et al., 2016), and the development of palaeoceanographic proxy approaches to assess the stability of hydrates in Earth's past (Dickens et al.,

1995; Dickens, 2003). Unique opportunities to understand marine gas hydrates are provided by integrated studies that seek to link modern and palaeo-perspectives, through the investigation of detailed stratigraphic, sedimentological and geochemical records surrounding present-day gas hydrates (Castellini et al., 2006; Brunner, 2007; Ingram et al., 2010, 2013).

Late Quaternary sedimentation and detailed geochemical records are considered in this study, through an analysis of deep-sea cores recovered from the first National Gas Hydrate Seafloor Observatory (McGee, 2006), located on the northern Gulf of Mexico slope within offshore Federal Lease Block Mississippi Canyon 118 (MC118). A special emphasis is placed on the genesis and interpretation of anomalous Mn enrichments ('Mn-layers') that occur at the MC118 site – documented for the first time in this study. These Mn-rich layers are investigated for their potential to reconstruct sedimentation and biogeochemical changes at the site, including linkages to the gas hydrate field. The Mn enrichments are revealed in exceptional detail through high-resolution elemental analysis (centimetre-scale) using X-ray fluorescence (XRF) scanning techniques (Richter *et al.*, 2006), applied to a network of 10 deep-sea sediment cores across the MC118 site. The Mn-layers are further characterized via chemical dissolution/extraction procedures and X-ray diffraction (XRD) analysis. This suite of analytical approaches allows a detailed assessment of the spatial distribution of the Mn-layers at the MC118 site, determination of their mineral composition, and evaluation of possible linkages to past redox cycling and/or sedimentation processes. Interpretation of the Mn record is aided by a well-established stratigraphic framework, which includes previous chronostratigraphic (Ingram *et al.*, 2010) and sedimentological (Ingram *et al.*, 2013) studies of the same suite of cores.

Based on the analyses outlined above it is postulated that the Mn-layers at MC118 record the duration of steady-state redox conditions, along a palaeo-redox boundary between oxic and post-oxic sediments, and that

movement of this boundary is linked to the gas hydrate and cold seep field through its influence on sedimentation. As will be shown, the distribution and concentration of Mn within the discrete Mn-layers is consistent with more frequent interruption of steady-state conditions closer to the field. This observation reveals a potential new method to evaluate past redox/sedimentation conditions across a wide range of time scales, through detailed characterization of solid-phase Mn profiles, by employing XRF scanning to 'map' movement of past redox boundaries. In addition, results presented here indicate that the MC118 field has influenced sea floor morphology and sedimentation, but has not contributed to a catastrophic slope failure during the last 14 000 years, consistent with prior studies (Ingram *et al.*, 2010).

The MC118 site

The offshore MC118 site includes a large sea floor mound with active cold seeps and gas hydrates (Fig. 1), and is the focus of ongoing geophysical and geochemical monitoring (McGee, 2006; Brunner, 2007; Lapham *et al.*, 2008, 2010; McGee *et al.*, 2009; Macelloni *et al.*, 2012, 2013; Simonetti *et al.*, 2013). The field itself is centred at

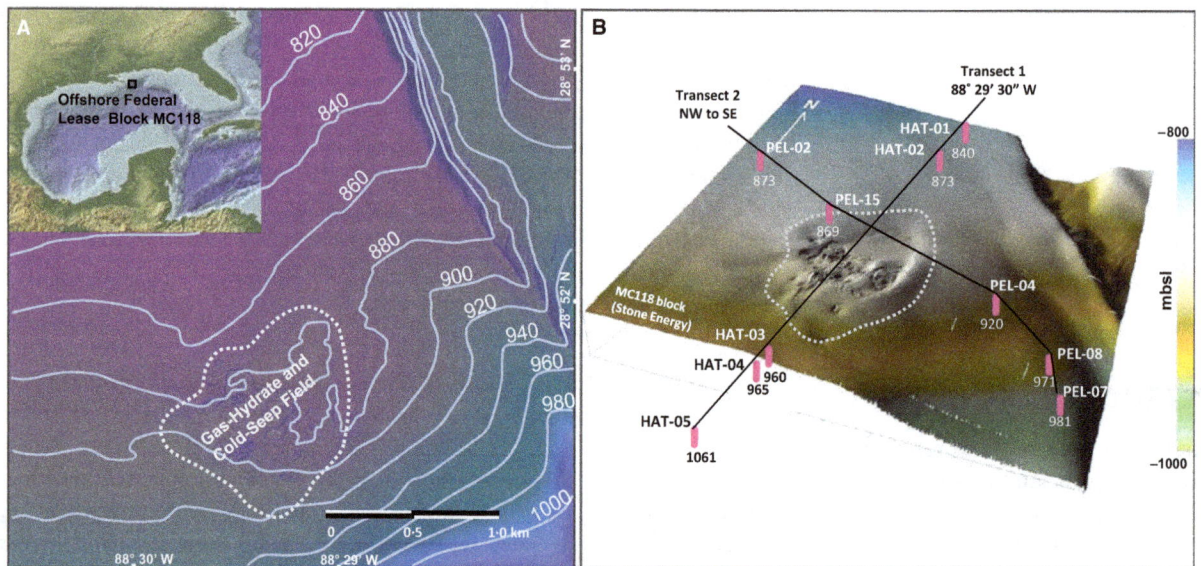

Fig. 1. (A) Bathymetric map of the study area (Block MC118) with labelled contours (light blue) in metres water depth (base map courtesy of Ken Sleeper) with inset digital elevation and bathymetry map (top left; NOAA geophysical data centre image) of the Gulf of Mexico Region. Bathymetry provided by the Gulf of Mexico Hydrate Research Consortium (modified after Sleeper *et al.*, 2006). The location of the MC118 offshore federal lease block is indicated by the red box in the inset map. The extent of the studied gas hydrate-cold seep field is outlined by the dashed line and is characterized by an area with gas vents, sea floor pockmark features, petroleum seepage, shallow faults, carbonate hardgrounds and gas hydrate deposits. (B) Bathymetric map of the study site with the location of cores collected during cruises on the R/V *Hatteras* and R/V *Pelican*; base map image is courtesy of the Gulf of Mexico Hydrate Research Consortium. Cores are indicated as vertical magenta lines, with core identification (above) and water depth (in metres) below the core symbol. The edge of the gas hydrate-cold seep field is outlined by a thin dashed white line, black lines connecting cores indicate transects taken by the R/V *Hatteras* ('Hatteras Transect') and R/V *Pelican* ('Pelican Transect'). The colour bar on the far right indicates water depth (metres below sea level).

28·8523°N and 88·4920°W and lies at approximately 890 m water depth completely within the offshore federal lease block (Fig. 1). Regionally, the northern Gulf of Mexico slope, including this study site, is influenced by salt diapirism beneath the sea floor (Diegel et al., 1995; Jackson, 1995; Galloway et al., 2000). The field is underlain by a salt diapir 200 to 300 m below the sea floor (mbsf), which contributes to the formation of the sea floor mound itself (Sassen et al., 2006; Sleeper et al., 2006). Moreover, the cold-seep mound is the epicentre for the migration and release of hydrocarbons from the sea floor. This supply of hydrocarbons (natural gas and petroleum) supports active biological seep communities (Sassen et al., 2006) and microbial chemolithotrophy in the vicinity of the active seep. The upward migrating hydrocarbons are rapidly cycled and oxidized in shallow sediments, also documented at similar sites in the Gulf of Mexico (Castellini et al., 2006). While intense redox cycling occurs immediately over the MC118 mound, it decreases markedly in sediments outside the field (Lapham et al., 2008). This is important, as the cores used in the present study are positioned mostly outside the area with active seeps, with the exception of Core PEL-15 (Fig. 1). Visible outcroppings of gas hydrates, faulted carbonate 'hardgrounds', authigenic carbonates and pockmark features are present in places across ca 1 km² of the sea floor in the vicinity of the mound (Sassen et al., 2006; Sleeper et al., 2006; Feng et al., 2014).

Marine gas hydrates are a major feature at this research site, and worldwide they represent a massive reservoir of light hydrocarbons (Milkov, 2004). The vast majority of this carbon occurs in shallow deep-sea sediments along continental slopes (Ginsburg, 1998), such as at MC118. In the Gulf of Mexico, gas hydrates often form alongside cold seeps and supply hydrocarbons to the sea floor, driving substantial changes in both pore-water and sediment chemistry through various redox processes (Sassen et al., 2004, 2006; Paull et al., 2005; Castellini et al., 2006). The present work documents authigenic Mn in deep-sea sediments complicated by nearby gas hydrates and cold seeps, and the cores investigated in this study reflect ambient sea floor conditions as well as those influenced by the mound (Fig. 1).

MATERIAL AND METHODS

Core collection and processing

A total of 10 gravity cores were collected from surface ships, five onboard the R/V Hatteras in August, 2007 and five by the R/V Pelican in April, 2008 (Sleeper & Lutken, 2008; Ingram et al., 2010; Fig. 1). Cores were transported to and processed at a shore-based laboratory at the University of North Carolina, Chapel Hill, where they were prepared for geochemical analyses (see Ingram et al., 2010, for detailed description of coring and processing methods).

X-ray fluorescence (XRF) core scanning

Cores were analysed using an Avaatech-XRF scanner (2nd generation) with a rhodium target X-ray source, and a Canberra X-PIPS detector. Continuous down-core XRF scanning was conducted at a resolution of 1 cm along the archive-half split core surface (Richter et al., 2006; Ingram et al., 2010). Manganese was scanned using a 10 kV acceleration voltage, 1000 mA, with a cellulose filter, and 90 sec measurement time, with duplicate scans every 10 cm, which yielded highly reproducible results (Table 1). The XRF scanning 'count' data are calibrated (least-squares linear fits; Table 1) to Mn concentration data for selected cores (HAT-03, PEL-04, PEL-07; 54 samples), using Inductively Coupled Plasma Atomic Emission Spectrometry (ICP-AES; SGS Laboratory method ICP-AES 40B, see Table 1 for error analysis).

Reductive and acidification dissolutions

Five Mn-rich layers were sampled from Core PEL-07 to operationally determine the mineral phases associated with the Mn enrichment. Sediment from each depth interval (Mn-layer) was first pulverized into 100-mesh powder. The procedure from Chun et al. (2010) was then

Table 1. XRF-ICP Mn Calibration Data: Calibration equations (least-squares linear fits) relating concentration from ICP-AES method to counts from X-ray fluorescence method, Pearson correlation coefficients (r^2 value) and number of samples used for calibration. Average coefficient of variation (CV) ratios for Mn XRF counts (XRF CV) are based on duplicate scans analysed every ca 10 cm, and select duplicates from ICP-ES sent to SGS Labs (SGS Laboratory method ICP-AES 40B). Average errors (CV) for XRF counts of Mn are slightly higher than other elements previously measured (Ingram et al., 2010), likely a result of lower concentration of Mn in these marine sediments.

Core	Element	Calibration Eqn.	r^2 Value	Samples
XRF-ICP manganese calibration data				
HAT-03	Mn	$y = 8{\cdot}31{*}10^{-2}(x) - 55{\cdot}268$	0·84	13
PEL-04	Mn	$y = 4{\cdot}5{*}10^{-2}(x) + 261{\cdot}57$	0·79	32
PEL-07	Mn	$y = 7{\cdot}25{*}10^{-2}(x) + 28{\cdot}554$	0·85	9

Core	XRF CV	ICP-ES CV	Core	XRF CV
Average coefficient of variation for Mn				
HAT-01	0·098	NA	PEL-02	0·116
HAT-02	0·137	NA	PEL-15	0·129
HAT-03	0·035	0·008	PEL-04	0·066
HAT-04	0·12	NA	PEL-08	0·106

used to remove oxides and oxyhydroxides by reductive dissolution, followed by an acid dissolution step (sodium acetate, pH = 3·96) to remove carbonates, with the remaining residue comprising 'insoluble' clays and/or fine-grained silicates. Manganese and Ti concentrations were measured after each step, using ICP-AES (SGS Laboratory method ICP-AES 40B) on the same split sample from each fraction of the procedure, allowing estimation of: (1) untreated (total Mn), (2) Mn oxide, (3) Mn carbonate, and (4) insoluble residue, which represents the unreactive aluminosilicates (clay content) and/or silicates (Schenau et al., 2002; Chun et al., 2010).

Manganese enrichment factors (EF), EF = (metal/Ti)$_{sample}$/(metal/Ti)$_{crust}$ were calculated to normalize Mn to Ti, a conservative detrital input. 'Excess' Mn (authigenic) was calculated using the following equation [Mn excess = Mn$_{total}$ − (Ti$_{sample}$*(Mn/Ti)$_{crust}$)] from Chun et al. (2010), and the bulk crustal Mn/Ti ratio (mol/mol) of 0·156 from Rudnick & Gao (2003). Enrichment factors and excess Mn calculations are a means to correct for inadvertent loss of mass during the sequential extraction procedures. These values are used to better constrain the fraction of Mn contained in carbonates versus oxides (Table 2).

X-ray diffraction (XRD) for select Mn-layers

A total of five Mn-layers from Core PEL-07 were analysed by X-ray diffraction. Powder samples were analysed using a Rigaku Rapid II X-ray diffractometre with a 2D detector and Molybdenum Kα radiation operated at 50 kV, 50 mA. Samples were ground to a fine powder and then mounted within capillary glass tubes, which yields improved XRD spectra with lower background compared to standard mounts.

RESULTS

Stratigraphy and XRF scanning results

Ten shallow gravity cores collected around the MC118 field comprise 38·6 m of total recovered sediment (Fig. 2), and reveal a detailed Mn record documented by centimetre-scale XRF core scans. The occurrence of Mn-layers is described in the context of the previously established chronostratigraphy at the site. Detailed sedimentological, stratigraphic and chronostratigraphic information can be found in Ingram et al. (2010, 2013) (For the chronostratigraphic framework, see figs 3–6 in Ingram

Table 2. Mn concentrations from ICP-AES determined for sediment samples from authigenic Mn-layers within Core PEL-07. Sediment samples were selected from 5 depths, and subjected to a two-step chemical dissolution to first remove oxide/oxyhydroxides, and then carbonates (see Material and Methods). Mn enrichment factor (Mn-EF) and 'excess' Mn (E. Mn) is also determined for each layer (see Methods; Chun et al., 2010), and is not applicable (NA) for insoluble clay, as 'excess' Mn is defined as the amount which exceeds the siliciclastic fraction, predominantly clay here. The insoluble (clay) fraction is the remaining material after dissolutions and is fine-grained, consisting of clay with minor silts possible. The sample was measured for Mn concentration following each step, and for untreated samples. The Mn concentrations for oxides, carbonates and insoluble (clay), are inferred using the following scheme: oxides = untreated − step 1; carbonates = step 1 − step 2; insoluble (clay) = step 2. The percentage of the total (far right column) is a ratio of Mn in each fraction, oxides, carbonates and insoluble (clay), relative to the untreated sample or total Mn.

Core	Depth	Dissolution	Mn (p.p.m.)	Ti (p.p.m.)	Mn EF	E. Mn (p.p.m.)	Composition	Mn (p.p.m.)	% Total (from p.p.m.)
Mn species data to solve for Mn mineral									
PEL-07	130	Untreated	2800	2900	5·393	2347	Oxyhydroxides	160	5·71
PEL-07	130	Step 1	2640	3100	4·756	2156	Carbonates	2373	84·75
PEL-07	130	Step 2	267	3600	0·414	NA	Insoluble (Clay)	267	9·54
PEL-07	189	Untreated	1990	2800	3·969	1553	Oxyhydroxides	60	3·02
PEL-07	189	Step 1	1930	3100	3·477	1446	Carbonates	1674	84·12
PEL-07	189	Step 2	256	3300	0·433	NA	Insoluble (Clay)	256	12·86
PEL-07	212	Untreated	1670	2800	3·331	1233	Oxyhydroxides	90	5·39
PEL-07	212	Step 1	1580	2900	3·043	1128	Carbonates	1331	79·7
PEL-07	212	Step 2	249	2900	0·48	NA	Insoluble (Clay)	249	14·91
PEL-07	577	Untreated	1610	2700	3·33	1189	Oxyhydroxides	0	0
PEL-07	577	Step 1	1610	2800	3·212	1173	Carbonates	1239	76·96
PEL-07	577	Step 2	371	3200	0·648	NA	Insoluble (Clay)	371	23·04
PEL-07	599	Untreated	2350	2700	4·861	1929	Oxyhydroxides	180	7·66
PEL-07	599	Step 1	2170	2700	4·489	1749	Carbonates	1753	74·6
PEL-07	599	Step 2	417	2700	0·863	NA	Insoluble (Clay)	417	17·74
PEL-07	Avg.	Untreated	2084	2800	4·187	1647	Oxyhydroxides	98	4·36
PEL-07	Avg.	Step 1	1986	2900	3·799	1534	Carbonates	1674	80·02
PEL-07	Avg.	Step 2	312	3100	0·555	NA	Insoluble (Clay)	312	15·62

et al., 2010). Additionally, the previously established stratigraphy is also provided here for visual context of the sedimentological units (Fig. 2).

Shallow sediments of Unit I are informally broken into Units IA and IB to better describe the occurrence of Mn-layers within the larger stratigraphic Unit I (Fig. 2). Sediments of Unit IA are late Holocene in age (ca 2300 calendar years BP to present) calcareous nannofossil silty clays, and are laterally discontinuous across the study area. Sediments of underlying Unit IB are also calcareous nannofossil silty clays, but are more nannofossil-rich than the overlying Unit IA. Across all studied cores, the more calcareous Unit IB is generally thicker than Unit IA (Fig. 2), and accumulated over a longer time period (ca 2300 to 9500 calendar years BP). Sediments of Unit IB are also lighter in colour and exhibit fewer sedimentary structures (Ingram et al., 2010).

With these sedimentological units defined, the occurrence of Mn-layers is presented in stratigraphic context (Figs 2 and 3). The shallowest Unit IA exhibits very few Mn-layers, with some exceptions. Core HAT-04 displays a shallow Mn-layer completely within Unit IA, and to a lesser extent PEL-04 shows an increased Mn concentration at the top of the core (Fig. 3). With the exception of core HAT-04, all other recovered cores lack discernable Mn-layers within this shallowest stratigraphic interval. Unit IB below also contains few authigenic Mn-layers, with one notable exception, Core HAT-03, which displays two

nearly synchronous layers at the base of the unit with Mn content well above 'background' levels (Fig. 3).

Unit II beneath the shallower carbonate-rich interval (Fig. 2) is a mottled, hemipelagic nannofossil silty clay (Ingram et al., 2010). Sediments here are markedly more clay rich with substantially lower carbonate content. This unit appears visually darker with sedimentary structures that are disturbed by burrows in places. Sediments are early Holocene to late Pleistocene in age ranging from 9500 to 14 000 calendar years from the top to the base of the unit (Fig. 2). This interval contains more Mn-layers than any other unit (at least 15 identified layers) with highly pronounced layers in Cores HAT-01, -05 and PEL-02 (Fig. 3). XRF scans of these same three cores also display less numerous Mn-layers, and in places exhibit only one or two, while cores that contain many layers generally yield lower concentrations for the individual layers. Calibration of XRF data indicates these 'more numerous' layers are ca 1000 p.p.m. Mn less concentrated than the single or double highly discrete layers (Fig. 3).

A continuation of clay-rich sediments comprises Unit III, a well-laminated hemipelagic nannofossil silty clay, which is differentiated from overlying sediments by numerous red-brown layers (Fig. 2). Sediments of this lowermost unit are chemically and lithologically similar to the unit above, yet with slightly higher lithogenic inputs (Ingram et al., 2013). The top of this unit is marked by a prominent red-brown band ('red band'),

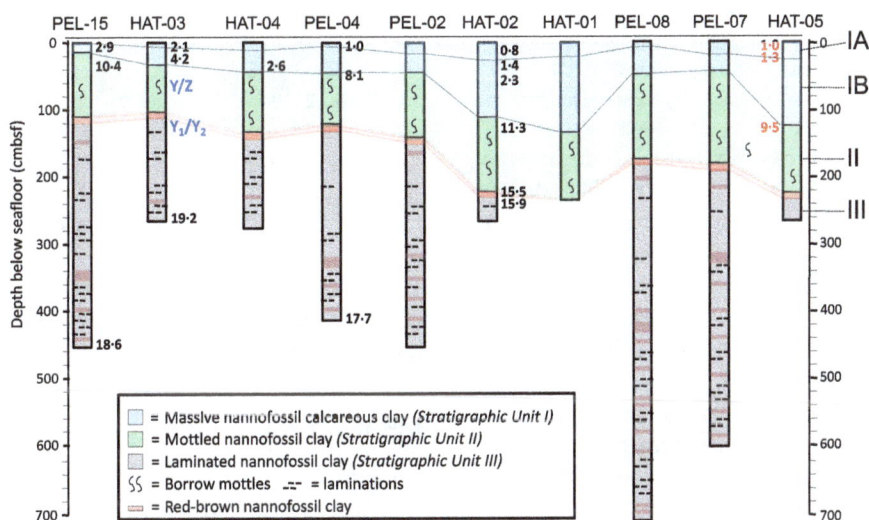

Fig. 2. Stratigraphic correlation of marine sediments over the MC118 site. Cores are arranged by increasing distance (left to right) away from the field. Dates obtained via AMS radiocarbon analysis of planktonic foraminifera (red, Ka) and foraminiferal biostratigraphic boundaries (dark blue, Y/Z ca 10 Ka and Y1/Y2 ca 15 Ka) are shown alongside their stratigraphic position; all reported ages are in calibrated calendar years BP. Three stratigraphic units are identified as follows: Unit I (light blue) = massive nannofossil calcareous silty clay; Unit II (light green) = mottled nannofossil clay; Unit III (grey) = laminated nannofossil clay containing reddish-brown nannofossil clay layers (red lines); for detailed stratigraphic description see Ingram et al. (2010). The distinct ca 5-cm thick reddish clay layer ('red band') contains highly 'reworked' (pre-Quaternary) nannofossils and defines the top of Unit III.

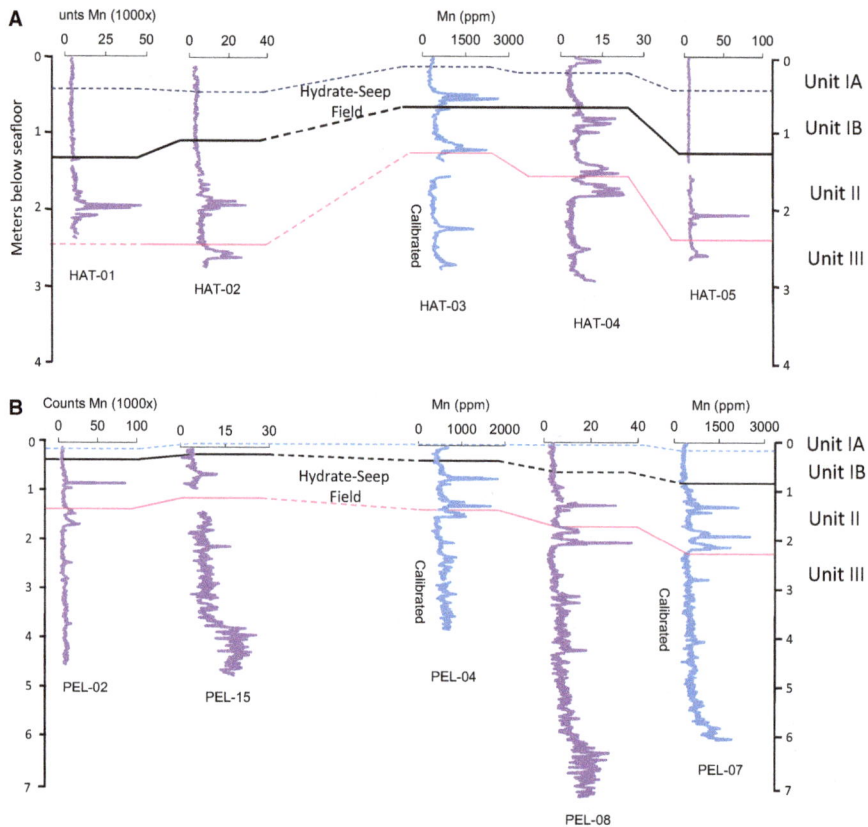

Fig. 3. Profiles of down-core Mn concentration/XRF counts across Transect 1 from the *R/V Hatteras* cruise (A) and across Transect 2 from the *R/V Pelican* cruise (B). Transect 1 is the N-S transect from HAT-01 to HAT-05 displayed in Fig. 1. Transect 2 is the NW-SE transect from PEL-02 to PEL-07 also shown in Fig. 1. Cores HAT-03, PEL-04 and PEL-07 are calibrated to Mn measured by ICP-AES, see Table 1 for calibration of XRF counts to ICP-AES Mn concentration; all other cores are presented here as XRF counts. Down-core profiles display Mn counts (1000x) with lines drawn across the profiles representing stratigraphic boundaries. The black lines separate Unit I from Unit II, which mark a substantial shift in carbonate content. The thin-dashed blue line separates informal units IA and IB, defined by a change in carbonate content (Ingram *et al.*, 2010, 2013). The light-red line is the depth to the 'red band', a chronostratigraphic marker (Ingram *et al.*, 2010), which defines the top of Unit III.

also observed by previous investigators (Lutken *et al.*, 2006; Brunner, 2007; Sleeper & Lutken, 2008; Ingram *et al.*, 2010). This marker bed is coincident with reworked pre-Quaternary nannofossils deposited during Melt Water Pulse 1A (MWP-1A; Marchitto & Wei, 1995), and is dated between 14 000 and 15 000 calendar years BP (Ingram *et al.*, 2010). The unit lacks a defined base, and thus extends to the greatest depth recovered. Based on accumulation rates and radiocarbon dating (Ingram *et al.*, 2010), sediments are latest Pleistocene in age and certainly older than the MWP-1A reworked nannofossil horizon, which defines the top of the unit (Fig. 2).

Manganese profiles spanning units II and III may appear correlative, such as in Cores PEL-04, -08 and -07, yet upon close examination, Mn layers are not chronostratigraphically equivalent (PEL-04, PEL-07 and -08; Fig. 3). It is also apparent that Unit III mostly lacks the highly concentrated Mn-layers, with one exception in Core PEL-08 (near the top of the unit;

Fig. 3), yet recovery of this unit was incomplete using shallow gravity coring. The Pelican transect recovered more of Unit III than the previous cruise and yields several metres of late Pleistocene-aged sediments in some cores. While this unit lacks the discrete highly concentrated Mn-layers, it does display many 'minor' layers (Fig. 3).

Mn-layer mineralogy

A multi-step reductive dissolution and acid digestion procedure was used to operationally determine probable mineral phases associated with the authigenic Mn (Fig. 4; Table 2). From five sediment samples taken within the Mn-layers, just over 80 percentage of the total Mn content is comprised of Mn-rich calcite ($Ca-Mn(CO_3)_2$) and/or rhodochrosite ($MnCO_3$) (Table 2). Manganese oxides account for slightly less than 10 percent, and the 'insoluble' clay fraction accounts for *ca* 10% (Table 2). Thus,

Mn within the concentrated layers is predominantly associated with carbonates, followed by aluminosilicates and minor amounts of oxides. Some change in concentration may be an artefact of the total change in sample mass associated with dissolution of carbonates. This is a concern for sediments with very high carbonate content, however, the samples investigated here (selected for sequential dissolution in PEL-07) are from clay-rich Units II and III with less than 13 wt.% $CaCO_3$ on average. To more rigorously address this issue, Mn enrichment factors and 'excess' Mn are also calculated (Table 2). Manganese enrichment and 'excess' Mn (defined as the non-clay fraction) are largely associated with carbonates and not oxides (Table 2). This is shown by determining the percentage of 'excess' Mn in each respective fraction (oxide verses carbonate) where the clay fraction has already been removed (via calculation of 'excess' Mn). The following expression:

$$((untreated_{excess} - carbonate_{excess}/untreated_{excess})*100)$$

yields only 6·8% of Mn as potentially associated with oxides or oxyhydroxides, based on the average of all five layers (Table 2). Hence, 'excess' Mn from within the

Mn-layers is dominantly associated with carbonates (over 90% on average).

Six untreated samples were also analysed using XRD (PEL-07-layer1, PEL-07-layer2, PEL-07-layer3, PEL-07-layer4, PEL-07-layer5 and PEL-04 at 5 cm depth; Fig. 5). All six XRD diffraction results show the existence of the following phases but in different proportions: quartz, calcite, feldspar, dolomite, rhodochrosite and clay minerals. Manganese is present as carbonate or Mn-Fe carbonate, which may form a complete solid solution (Fig. 5). Manganese carbonate in most samples has a d_{104} peak at *ca* 2·82 Å, except sample PEL-07-layer 4 that has a value of 2·827 Å. The d_{104} peak for the pure Mn end member (rhodochrosite) is 2·84 Å, and that for the pure Fe end member (siderite) is 2·79 Å. The d_{104} of the samples indicates a Mn-rich (>50% Mn) carbonate. One acid treated sample taken from PEL-07 (Mn-layer3; Fig. 5B) yielded no carbonates in the diffraction pattern; this result indicates that the dissolution procedure used here and by Chun *et al.* (2010) effectively removes Mn-carbonate minerals.

To summarize, the X-Ray diffraction and reductive dissolution/acidification results reveal that Mn within the

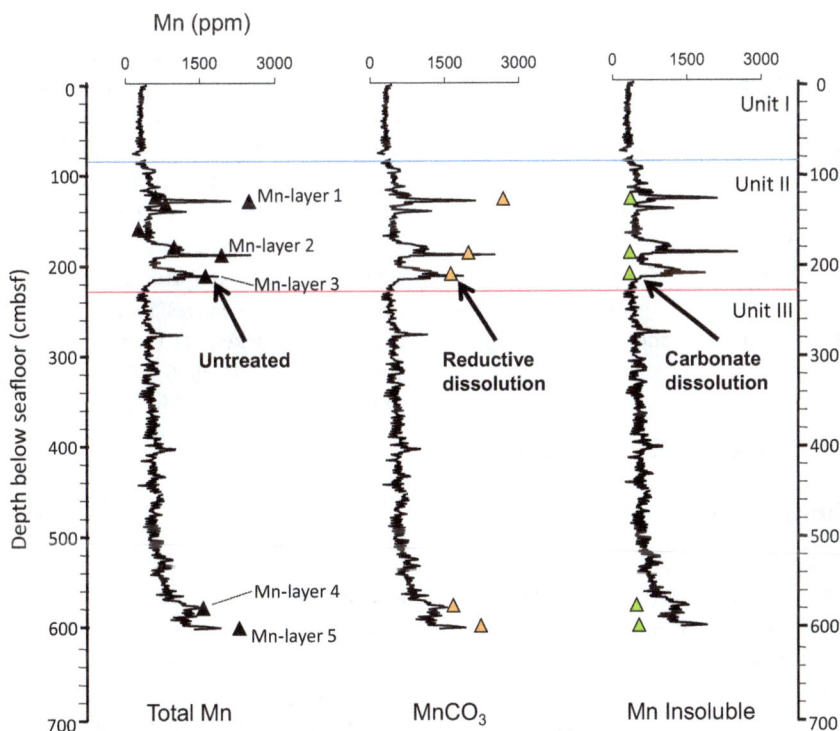

Fig. 4. Calibrated XRF-Mn profile (blue curve) for Core PEL-07 and Mn concentrations (triangles, same scale) following dissolution procedures to operationally determine the likely mineral phases associated with the Mn-layers. From left to right, the red triangles are from untreated sediment samples, orange triangles indicate samples following the reductive dissolution step, and the green triangles represent insoluble (clay) residue following both dissolution procedures. Reductive dissolution removes oxides; hence the remaining concentration comprises carbonate minerals plus insoluble (clay) residue. The following acid-dissolution step removes carbonates, thus only insoluble (clay) residue remains after both steps.

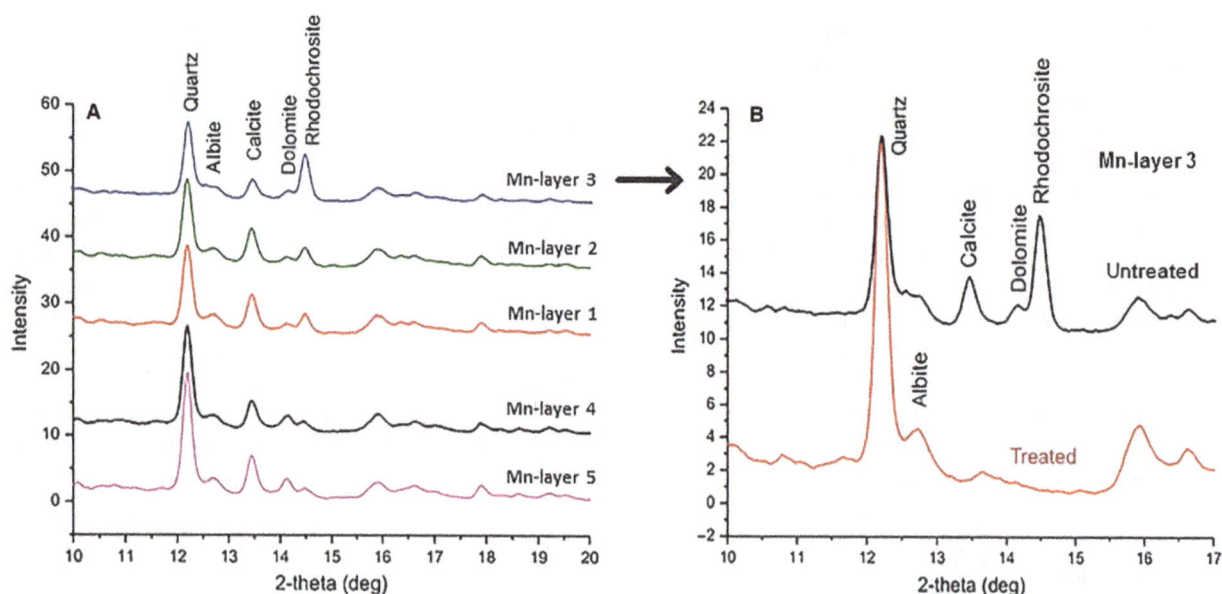

Fig. 5. (A) X-ray diffraction spectra for the five selected Mn-layers in Core PEL-07 (for locations see the numbers on Fig. 4). The mineral diffraction peaks labelled from left to right are quartz, albite, calcite, dolomite and rhodochrosite. The rhodochrosite diffraction peak is present in all samples (Mn-layers), and is largest (most concentrated) in Mn-layer 3. (B) X-ray diffraction spectra for Mn-layer 3, before acid dissolution (black curve), and after acid dissolution (red curve). The other diffraction peaks appear larger, as the carbonate fraction was removed by the acid dissolution, thereby concentrating siliciclastic minerals in the treated sample.

Mn-layers is largely contained in the carbonate phase. Given sample heterogeneity, preferred orientation and different proportions of mineral phases, intensities of the rhodochrosite d_{104} peak is not expected to exactly correlate with the XRF/ICP-AES derived Mn concentration – for example, Mn-layer3 is not the most concentrated based on the XRF profile (Figs 4 and 5). However, Mn-layers 1, 2 and 3 exhibit larger rhodochrosite peaks than Mn-layers 4 and 5 (Fig. 5), which is generally consistent with Mn concentrations from XRF and ICP-AES (Fig. 4; Table 2).

INTERPRETATION AND DISCUSSION

Manganese carbonate layers at MC118

A detailed deep-sea sedimentary record is documented at MC118 in an effort to better understand accumulation of highly enriched Mn-layers and possible mechanisms for their formation. Foremost, it is clear that Mn-layers occur in sediments of different age and generally do not follow lithostratigraphic or chronostratigraphic trends. Hence, their characterization as authigenic deposits, which formed *in situ* through redox processes (Figs 2 and 3), representing diagenetic palaeo-redox horizons. Multi-step chemical dissolution procedures combined with XRD analyses confirms that the Mn-layers are primarily carbonates.

The documentation of such a highly detailed spatio-temporal record of recurring authigenic Mn carbonate layers as observed at MC118 is somewhat unique considering water depth and proximity to the seabed with nearby cold seeps (Canfield *et al.*, 1993; Chun *et al.*, 2010). However, Mn enrichments are often observed in ancient marine sediments or mudstones, along with other elements associated with authigenic deposits, and have been linked to changes in past climate or oceanographic conditions (Dickens & Owen, 1994; Schenau *et al.*, 2002; Tribovillard *et al.*, 2006). The MC118 results indicate that substantial variability in Mn enrichment is present in Recent (Late Pleistocene-Holocene) deep-sea sediments over a relatively small area of the sea floor (*ca* 2 km^2: Fig. 1). This has important implications for interpreting chemostratigraphic records from similar deep-sea depositional environments, as it implies that modern and ancient sedimentary Mn records may reflect localized conditions from depositional heterogeneity, rather than more widespread regional or 'global' events. Thus, subtle variation in depositional or redox environments in deep-marine continental slope sediments may drive noticeable differences in the accumulation of authigenic minerals. The MC118 record shows that authigenic Mn can vary considerably within continental slope sediments over a small area of the sea floor, and suggest linkages to the local gas-hydrate and cold-seep field. This includes sea floor warping, which is well-documented in salt-dominated margins (Jackson, 1995).

Biogeochemistry of the Mn-carbonate layers

Nearly all of the Mn-layers occur at depths well below expected oxygenated pore waters for this depositional setting, consistent with a Mn-carbonate phase preserved in post-oxic sediments. Previous studies have demonstrated that oxygen depletion occurs within 5 cm of the sea floor for similar settings such as the western Gulf of Mexico Shelf (Hu *et al.*, 2011) and the Mississippi Canyon region (Diaz & Trefry, 2006). More recent studies have reported Mn-enrichments (oxides) within 10 cm of the sea floor at similar water depths and close to the MC118 site (Brooks *et al.*, 2015; Hastings *et al.*, 2016). In contrast, most Mn layers at MC118 are found more than a metre below the sea floor, thus they are considered much too deep for active formation of Mn-oxides, which develop near the oxygen depletion depth (Burdige & Gieskes, 1983; Kalhorn & Emerson, 1984; Heggie *et al.*, 1986; Aller, 1990, 1994; Shaw *et al.*, 1990; Reimers *et al.*, 1992). However, it should also be noted that Mn carbonates can develop a protective crust around oxides formed earlier, thereby diminishing their dissolution within post-oxic pore waters (Burdige, 1993). This protective crust may account for the small fraction (under 10%) of Mn-oxide observed in some layers (Table 2).

Repeated Mn cycling (the 'Mn pump'; Sageman & Lyons, 2003) can concentrate dissolved Mn within alkaline pore waters to form Mn-rich carbonates. Thus, Mn enrichment may form along palaeo-redox boundaries, related to the carbonate-Mn pump, where dissolved Mn concentrated in pore waters leads to precipitation of Mn-carbonate minerals under sufficiently alkaline conditions. Mn-oxide dissolution may occur contemporaneously with formation of Mn-carbonates just beneath an active redox front, followed by subsequent burial and preservation of the palaeo-redox boundary (Pedersen & Price, 1982; Burdige, 2006). The rapid burial of highly concentrated Mn-oxides into deeper dysoxic/anoxic and alkaline pore waters, due to a (pulsed) increase in sedimentation rate, would also promote the preservation of Mn-carbonate layers. More generally, sedimentation can influence redox processes through erosion and/or deposition.

Other factors that drive the movement of the Mn-redox boundary include, but are not limited to, (1) changes in organic carbon delivery and reactivity, (2) bottom water oxygen content, or (3) intensity of bioturbation/bioirrigation. All of the above factors influence preservation of 'relict' Mn-peaks (Burdige, 2006). Most of these factors are not explicitly constrained in this study, although previous efforts quantify sedimentation rates,

and organic matter accumulation/composition (Ingram *et al.*, 2013). There is a shift towards more reactive organic matter (Type II) within the shallow Holocene sediments, which is expected to drive the redox boundary upwards, yet this is also balanced by slower sedimentation rates (Ingram *et al.*, 2010). Regardless, there are too many discrete Mn-layers to explain with one stepwise change in sedimentation. Furthermore, while organic matter accumulation is variable with time (Ingram *et al.*, 2013), it should be more uniform spatially in a deep-sea setting – unless it is also influenced by the presence of the seeps. Thus, dynamic sedimentation associated with the mound is hypothesized as a more viable driver to explain spatio-temporal trends of the Mn-peaks across the network of MC118 cores, which suggest the field's influence.

Mapping Mn-carbonate layers: a tool to delineate palaeo-redox conditions

The analysis presented here suggests that the high-resolution Mn record at MC118 primarily reflects the duration of steady-state conditions and the frequency with which they are interrupted, through changes in sedimentation that are driven by local effects from the presence of the elevated bathymetric mound. Relatively more stable sedimentation is expected for core sites farther away from the area of active cold seepage and gas hydrate formation (centred over the bathymetric mound). Such distal sites should be characterized by prolonged periods of steady-state conditions, and hence a more stable palaeo-redox boundary along which more concentrated Mn-layers can form. Conversely, frequent interruption of steady-state conditions is expected in closer proximity to or downslope from the field yielding less-concentrated and more-numerous Mn-layers (Fig. 2). The distribution of Mn-layers at the site is consistent with this hypothesis, as only one or two very-concentrated authigenic Mn-layers (Cores HAT-01, HAT-05 and PEL-02) are observed at locations distal from the field, while a greater abundance of less-concentrated layers are observed closer to the field (Core HAT-03 and HAT-04, Fig. 3).

Authigenic Mn enrichments are well-documented in marine sediments (Burdige, 1993, 2006), however, it is the remarkable detail of the MC118 record in both time and space that makes this study unique, and reveals past movement of the palaeo-redox boundary at various core sites. This record is made possible through the application of high-resolution XRF scanning techniques. The integration of such spatio-temporal data with quantitative biogeochemical models for Mn-cycling should provide a powerful new tool for evaluating redox cycling and/or sedimentation events over a wide range of time scales, for example, quantification of the duration of steady state conditions required to generate a Mn-layer of given concentration (Froelich *et al.*, 1979; Burdige & Gieskes, 1983; Finney *et al.*, 1988; Burdige, 1993; Price, 1998), and deconvolution of local versus regional/global influences on the observed Mn record.

CONCLUSIONS

The results presented here demonstrate that multiple discrete Mn-layers observed at MC118 are authigenic deposits, preserved as carbonates and formed within shallow sediments along a transient palaeo-redox boundary. The Mn-layers occur independently from established lithostratigraphic and chronostratigraphic horizons at the site, and results from sequential extraction procedures, along with XRD analysis, reveal that layers are mostly rhodochrosite ($MnCO_3$). It is not known if the Mn-layers formed initially in shallow sediments as oxides/oxyhydroxides or as carbonates, however, they are presently carbonate minerals preserved within post-oxic conditions.

The spatio-temporal heterogeneity of this Mn record is linked to the MC118 gas-hydrate and cold-seep field, including the related salt diapirism beneath the mound. This sea floor feature and associated release of hydrocarbons in turn influenced palaeo-redox cycling and/or sedimentation processes that impacted the duration of steady-state conditions in the past. Thus, the concentrations and frequency of Mn-layers across the site is a fingerprint of the palaeo-redox boundary and its spatial expression with time. While numerous factors drive changes in redox conditions, it is suggested that variability in sedimentation rate, caused by the presence of the sea floor mound, is a viable mechanism to explain the complexity of the Mn record.

ACKNOWLEDGEMENTS

Financial support for this research was provided by university and departmental funds to Dr. Stephen Meyers (UNC-Chapel Hill and UW-Madison). Additional support was provided by the Gulf of Mexico Gas Hydrate Research Consortium (HRC grants 300212198E (UM 07-01-071) and 300212260E (UM 08-11-047) to C.S. Martens), and funding for ship time on the R/V Pelican was provided by Minerals Management Services (now Department of Ocean Energy) and NOAA's National Institute for Undersea Science and Technology. Funding for the R/V Hatteras Cruise, ship time and coring operations was provided by the Duke/UNC Oceanographic Consortium as part of a joint proposal with other UNC investigators, Kai Ziervogel, Drew Steen and Carol Arnosti. The crew, joint investigators listed above, Sherif Ghobrial, Carol Lutken and Ken Sleeper were all instrumental in the shipboard core collection process that provided the material used for this study.

References

Aller, R.C. (1990) Bioturbation and manganese cycling at the sediment-water interface. *Philos. Trans. R. Soc. London*, **331**, 51–68.

Aller, R.C. (1994) The sedimentary cycle in Long Island Sound: its role as intermediate oxidant and the influence of bioturbation, O_2, and C_{org} flux on diagenetic reaction balances. *J. Mar. Resour.*, **52**, 259–295.

Brooks, G.R., Larson, R.A., Schwing, P.T., Romero, I., Moore, C., Reichart, G.-J., Jilbert, T., Chanton, J.P., Hastings, D.W., Overholt, W.A., Marks, K.P., Kostka, J.E., Holmes, C.W. and Hollander, D. (2015) Sedimentation pulse in the NE Gulf of Mexico following the 2010 DWH blowout. *PLoS ONE*, **10**, 1–24, e0132341.

Brunner, C.A. (2007) *Stratigraphy and Palaeoenvironment of Shallow Sediments from MC118*. Proceedings of the Annual Meeting of the Gulf of Mexico Hydrate Research Consortium, October 10–11, 2007, Oxford, Mississippi.

Burdige, D.J. (1993) The biogeochemistry of manganese and iron reduction in marine sediments. *Earth Sci. Rev.*, **35**, 249–284.

Burdige, D.J. (2006) *Geochemistry of Marine Sediments*. Princeton University Press, Princeton, NJ, 609 pp.

Burdige, D.J. and Gieskes, J.M. (1983) A pore water/solid phase diagenetic model for manganese in marine sediment. *Am. J. Sci.*, **283**, 29–47.

Canfield, D.E., Thamdrup, B. and Hansen, J.W. (1993) The anaerobic degradation of organic matter in Danish coastal sediments: Fe reduction, Mn reduction, and sulfate reduction. *Geochim. Cosmochim. Acta*, **57**, 3867–3883.

Castellini, D.G., Dickens, G.D., Snyder, G.T. and Ruppel, C.D. (2006) Barium cycling in shallow sediments above active mud volcanoes in the Gulf of Mexico. *Chem. Geol.*, **226**, 1–30.

Chun, C.O.J., Delaney, M.L. and Zachos, J.C. (2010) Palaeoredox changes across the Palaeocene-Eocene thermal maximum, Walvis Ridge (ODP Sites 1262, 1263, and 1266): evidence from Mn and U enrichment factors. *Paleoceanography*, **25**, 1–13.

Diaz, R.J. and Trefry, J.H. (2006) Comparison of sediment profile image data with profiles of oxygen and Eh from sediment cores. *J. Mar. Syst.*, **62**, 164–172.

Dickens, G.R. (2001) The potential volume of oceanic methane hydrates with variable external conditions. *Org. Geochem.*, **32**, 1179–1193.

Dickens, G.R. (2003) Rethinking the global carbon cycle with a large dynamic and microbially mediated gas hydrate capacitor. *Earth Planet. Sci. Lett.*, **213**, 169–183.

Dickens, G.R. and Owen, R.M. (1994) Late Miocene-Early Pliocene manganese redirection in the central Indian Ocean:

expansion of the intermediate water oxygen minimum zone. *Paleoceanography*, **9**, 169–181. doi:10.1029/93PA02699.

Dickens, G.R., O'Neil, J.R., Rea, D.K. and Owen, R.M. (1995) Dissociation of oceanic methane hydrate as a cause of the carbon isotope excursion at the end of the Palaeocene. *Paleoceanography*, **10**, 965–971.

Diegel, F.A., Karlo, J.F., Schuster, D.C., Shoup, R.C. and Tauvers, R.C. (1995) Cenozoic structural evolution and tectono-stratigraphic framework of the northern Gulf Coast continental margin. In: *Salt Tectonics: A Global Perspective* (Eds M.P.A. Jackson, D.G. Roberts and S. Snelson), *Am. Assoc. Petrol. Geol. Mem.*, **65**, 109–151.

Feng, D., Birgel, D., Peckmann, J., Roberts, H.H., Joye, S.B., Sassen, R., Liu, X.-L., Hinricks, K.-U. and Chen, D. (2014) Time integrated variation of sources of fluids and seepage dynamics archived in authigenic carbonates from Gulf of Mexico Gas Hydrate Seafloor Observatory. *Chem. Geol.*, **385**, 129–139.

Finney, B.P., Lyle, M.W. and Heath, G.R. (1988) Sedimentation at MANOP site H (eastern Equatorial Pacific) over the past 400,000 years: climatically induced redox variations and their effects on transient metal cycling. *Paleoceanography*, **3**, 169–189.

Froelich, P.N., Klinkhammer, G.P., Bender, M.L., Luedtke, N., Heath, G.R., Cullen, D., Dauphin, P., Hammond, D., Hrtman, B. and Maynard, V. (1979) Early oxidation of organic matter in pelagic sediments of the eastern equatorial Atlantic: suboxic diagenesis. *Geochim. Cosmochim. Acta*, **43**, 1075–1090.

Galloway, W.E., Ganey-Curry, P.E., Li, X. and Buffler, R.T. (2000) Cenozoic depositional history of the Gulf of Mexico basin. *Am. Assoc. Petrol. Geol. Bull.*, **84**, 1743–1774.

Ginsburg, G.D. (1998) Gas hydrate accumulation in deep-water marine sediments. In: *Gas Hydrates: Relevance to World Margin Stability and Climate Change* (Eds J.P. Henriet and J. Mienert), *Geol. Soc. London. Spec. Publ.*, **137**, 51–62.

Haq, B.U. (1998) Natural gas hydrates: searching for the long-term climatic and slope-stability records. In: *Gas Hydrates: Relevance to World Margin Stability and Climate Change* (Eds J.P. Henriet and J. Mienert), *Geol. Soc. London. Spec. Publ.*, **137**, 303–318.

Hastings, D.W., Schwing, P.T., Brooks, G.R., Larson, R.A., Morford, J.L., Roeder, T., Quinn, K.A., Bartlett, T., Romero, I.C. and Hollander, D.J. (2016) Changes in sediment redox conditions following the BP DWH blowout event. *Deep-Sea Res. II*, **129**, 167–178.

Heggie, D.T., Kahn, D. and Fischer, K. (1986) Trace metals in metalliferous sediments, MANOP site M: interfacial pore water profiles. *Earth Planet. Sci. Lett.*, **80**, 106–116.

Hu, X., Cai, W.J., Wang, Y., Guo, X. and Lou, S. (2011) Geochemical environments of continental shelf-upper slope sediments in the northern Gulf of Mexico. *Palaeogeogr. Palaeoclimatol. Palaeoecol.*, **312**, 265–277.

Ingram, W.C., Meyers, S.R., Brunner, C.B. and Martens, C.S. (2010) Late Pleistocene-Holocene sedimentation surrounding an active seafloor gas-hydrate and cold-seep field on the Northern Gulf of Mexico Slope. *Mar. Geol.*, **278**, 43–53.

Ingram, W.C., Meyers, S.R. and Martens, C.S. (2013) Controls on sedimentary Geochemistry and organic carbon burial at a large gas-hydrate and cold-seep field on the northern Gulf of Mexico Slope. *Mar. Pet. Geol.*, **46**, 190–200.

Jackson, M.P.A. (1995) Retrospective Salt Tectonics. In: *Salt Tectonics: A Global Perspective* (Eds M.P.A. Jackson, D.G. Roberts and S. Snelson), *Am. Assoc. Petrol. Geol. Mem.*, **65**, 1–28.

Kalhorn, S. and Emerson, S. (1984) The oxidation state of manganese in surface sediments of the deep sea. *Geochem. Cosmochim. Acta*, **48**, 897–902.

Kvenvolden, K.A. (1988) Methane Hydrate – a major reservoir of carbon in the shallow geosphere? *Chem. Geol.*, **71**, 41–51.

Lapham, L.L., Chanton, J.P., Martens, C.S., Sleeper, K. and Woolsey, J.R. (2008) Microbial activity in surficial sediments overlying acoustic wipeout zones at a Gulf of Mexico cold seep. *Geochem. Geophys. Geosyst.*, **9**, 1–17.

Lapham, L.L., Chanton, J.P., Chapman, R. and Martens, C.S. (2010) Methane under-saturated fluids in deep-sea sediments: implications for gas hydrate stability and rates of dissolution. *Earth Planet. Sci. Lett.*, **298**, 275–285. doi:10.1016/j.epsl.2010.07.016.

Lutken, C.B., Brunner, C.A., Lapham, L.L., Chanton, J.P., Rogers, R., Sassen, R., Dearman, J., Lynch, L., Kuykendall, J. and Lowrie, A. (2006) *Analyses of Core Samples from the Mississippi Canyon 118, Paper OTC 18208.* Offshore Technology Conference, American Association of Petroleum Geologist, May 1–4, 2006, Houston, TX.

Macelloni, L., Simonetti, A., Knapp, J.H., Knap, C.C., Lutken, C.B. and Lapham, L.L. (2012) Multiple resolution seismic imaging of a shallow hydrocarbon plumbing system, Woolsey Mound, Northern Gulf of Mexico. *Mar. Pet. Geol.*, **38**, 128–142.

Macelloni, L., Brunner, C.A., Caruso, S., Lutken, C.B., D'Emidio, M. and Lapham, L.L. (2013) Spatial distribution of seafloor bio-geological and geochemical proxies of fluid regime and evolution of a carbonate/hydrates mound, northern Gulf of Mexico. *Deep Sea Res. Part I*, **74**, 25–38.

Marchitto, T.M. and Wei, K.Y. (1995) History of the Laurentide meltwater flow to the Gulf of Mexico during the last deglaciation, as revealed by reworked calcareous nannofossils. *Geology*, **23**, 779–782.

Martens, C.S., Medlovitz, H.P., Seim, H., Lapham, L. and D'Emidio, M. (2016) Sustained in situ measurements of dissolved oxygen, methane and water transport processes in the benthic boundary layer at MC118 northern Gulf of Mexico. *Deep-Sea Res. II*, **129**, 41–52.

Maslin, M., Mikkelsen, N., Vilela, C. and Haq, B. (1998) Sea-level and gas-hydrate-controlled catastrophic sediment failures of the Amazon Fan. *Geology*, 26, 1107–1110.

Maslin, M., Owen, M., Day, S. and Long, D. (2004) Linking continental-slope failures and climate change: testing the clathrate gun hypothesis. *Geology*, 32, 53–56.

McGee, T. (2006) A seafloor observatory to monitor gas hydrates in the Gulf of Mexico. *Lead. Edge*, 25, 644–647.

McGee, T., Lutken, C., Rogers, R., Brunner, C., Dearman, J., Lynch, F. and Woolsey, R. (2009) Can fractures in soft sediments host significant quantities of gas hydrates? *Geol. Soc. London. Spec. Publ.*, 319, 29–49.

Mienert, J., Vanneste, M., Bunz, S., Andreassen, K., Haflidson, H. and Sejrup, H.P. (2005) Ocean warming and gas hydrate stability on the mid-Norwegian Margin at the Storegga Slide. *Mar. Pet. Geol.*, 22, 233–244.

Milkov, A.V. (2004) Global estimates of hydrate-bound gas in marine sediments: how much is really out there? *Earth Sci. Rev.*, 66, 183–197.

Nisbet, E.G. (1990) The end of the ice-age. *Can. J. Earth Sci.*, 27, 148–157.

Paull, C.K., Buelow, W.J., Ussler, W., III and Borowski, W.S. (1996) Increased continental-margin slumping frequency during sea-level lowstands above gas-hydrate-bearing sediments. *Geology*, 24, 143–146.

Paull, C.K., Brewer, P.G., Ussler, W., III, Peltzer, E.T., Rehder, G. and Clague, D. (2003) An experiment demonstrating that marine slumping is a mechanism to transfer methane from gas-hydrate deposits into the upper ocean and atmosphere. *Geo-Mar. Lett.*, 22, 198–203.

Paull, C.K., Ussler, W., Lorenson, T.D., Winters, W. and Dougherty, J.A. (2005) Geochemical constraints on the distribution of gas hydrates in the Gulf of Mexico. *Geo-Mar. Lett.*, 25, 273–280.

Pedersen, T.F. and Price, N.B. (1982) The geochemistry of manganese carbonate in Panama Basin sediments. *Geochim. Cosmochim. Acta*, 46, 59–68.

Price, B.A. (1998) *Equatorial Pacific Sediments: A Chemical Approach to Ocean History*. PhD Diss. Scripps Institute of Oceanography, UCSD, San Diego, CA, 364 pp.

Reimers, C.E., Jahnke, R.A. and McCorkle, D.C. (1992) Carbon fluxes and burial rates over the continental slope and rise off central California with implications for the global carbon cycle. *Global Biogeochem. Cycles*, 6, 199–224.

Richter, T.O., Van Der Gaast, S., Koster, B., Vaars, A., Gieles, R., De Stigter, H.C., De Haas, H. and Van Weering, T.C.E. (2006) The Avaatech XRF Core Scanner: technical description and applications to NE Atlantic sediments. In: *New ways of Looking at Sediment Core and Core Data* (Ed. R.G. Rothwell), pp. 39–50. Geological Society Special Publication, London.

Rudnick, R.L. and Gao, S. (2003) Composition of the continental crust. In: *Treatise on Geochemistry* (Eds D.H. Heinrich and K.T. Karl), pp. 1–64. Pergamon, Oxford, UK.

Sageman, B.B. and Lyons, T.W. (2003) Geochemistry of fine-grained sediments and sedimentary rocks. In: *Treatise on Geochemistry*, Vol. 7 (Ed. F. MacKenzie), pp. 115–158. Elsevier, New York, NY.

Sassen, R., Roberts, H.H., Carney, R., Milkov, A.V., DeFreitas, D.A., Lanoil, B. and Zhang, C.L. (2004) Free hydrocarbon gas, gas hydrate, and authigenic minerals in chemosynthetic communities of the northern Gulf of Mexico continental slope: relation to microbial processes. *Chem. Geol.*, 205, 195–217.

Sassen, R., Roberts, H.H., Jung, W., Lutken, C.B., DeFreitas, D.A., Sweet, S.T. and Guinasso, N.L. Jr (2006) *The Mississippi Canyon 118 Gas Hydrate Site: A Complex Natural System Paper OTC 18132*. Offshore Technology Conference, May 1–4, Houston, TX.

Schenau, S.J., Reichart, G.J. and De Lange, G.J. (2002) Oxygen minimum zone controlled Mn redistribution in Arabian Sea sediments during the late Quaternary. *Paleoceanography*, 17, 1058. doi:10.1029/2000PA000621.

Shaw, T.J., Gieskes, J.M. and Jahnke, R.A. (1990) Early diagenesis in differing depositional environments: the response of transitional metals in pore water. *Geochem. Cosmochim. Acta*, 54, 1233–1246.

Simonetti, A., Knapp, J.H., Sleeper, K., Lutken, C.B., Macelloni, L. and Knapp, C.C. (2013) Spatial distribution of gas hydrates from high-resolution seismic and core data, Woolsey Mound, Northern Gulf of Mexico. *Mar. Pet. Geol.*, 44, 21–33.

Sleeper, K.A. and Lutken, C. (2008) *Activities Report for Cruise GOM1-08-MC118 Aboard the R/V Pelican Sampling and Deployment Cruise Mississippi Canyon Federal Lease Block 118 Northern Gulf of Mexico April 22–28, 2008*. The Center for Marine Resources and Environmental Technology and the Seabed Technology Research Center, University of Mississippi. Available at: http://www.olemiss.edu/depts/mmri/programs/ppt_list.html.

Sleeper, K.A., Lowrie, A., Bosman, A., Macelloni, L. and Swann, C.T. (2006) *Bathymetric Mapping and High Resolution Seismic Profiling by AUV in MC 118 (Gulf of Mexico)*. Paper OTC 18133, Offshore Technology Conference, Houston, TX.

Tribovillard, N., Algeo, T.J., Lyons, T. and Armelle, R. (2006) Trace metals as palaeoredox and palaeoproductivity proxies: an update. *Chem. Geol.*, 232, 12–32.

7

Archimetrics: a quantitative tool to predict three-dimensional meander belt sandbody heterogeneity

WIETSE I. VAN DE LAGEWEG*, WOUT M. VAN DIJK†, DARREN BOX‡ and MAARTEN G. KLEINHANS*

*Faculty of Geosciences, Universiteit Utrecht, P.O. Box 80115, 3508 TC, Utrecht, The Netherlands
†Department of Geography, Durham University, South Road, Durham, DH1 3LE, UK
‡ExxonMobil Russia, 31 Novinsky Boulevard, 123242, Moscow, Russia (Email: wietse.vandelageweg@gmail.com)

Keywords

Flume experiment, fluvial architecture, meander belt, meandering river, morphodynamic modelling, preservation, sandbody heterogeneity, stratigraphy.

ABSTRACT

Fluvial meander belt sediments form some of the most architecturally complex reservoirs in hydrocarbon fields due to multiple scales of heterogeneity inherent in their deposition. Currently, characterization of meander belt bodies largely relies on idealized vertical profiles and a limited number of analogue models that naively infer architecture from active river dimensions. Three-dimensional architectural data are needed to quantify scales of grain-size heterogeneity, spatial patterns of sedimentation and bar preservation in a direct relationship with the relevant length scales of active river channels. In this study, three large flume experiments and a numerical model were used to characterize and construct the architecture (referred to as 'archimetrics') and sedimentology of meander belt deposits, while taking reworking and partial preservation into account. Meander belt sandbody width-to-thickness ratios between 100 and 200 were observed, which are consistent with reported values of natural meander belts. For the first time, the relief of the base of a meander belt is quantified, enabling improved estimates of connectedness of amalgamated meander belts. A key observation is that the slope and number of lateral-accretion packages within natural point bar deposits can be well predicted from fairly basic observables, a finding subsequently tested on several natural systems. Probability curves of preserved architectural characteristics for three dimensions were quantified allowing estimates of bar dimensions, baffle and barrier spacing distributions and container dimensions. Based on this, a set of rules were identified for combining reservoir parameters with the identified probability curves on sandbody dimensions and character, to help create more realistic geomodels for estimating exploration success on the basis of seismic and core data.

INTRODUCTION

Characterization and prediction of the three-dimensional architecture and fluid-flow behaviour of fluvial hydrocarbon reservoirs and drinking water aquifers is challenging because of the various scales of sediment heterogeneity between and within fluvial deposits (Miall, 1988; Jordan & Pryor, 1992; Pranter et al., 2007; Willis & Tang, 2010). In essence, fluvial deposits are composed of a number of architectural elements (Miall, 1985; Holbrook, 2001) spanning a wide range of spatial and temporal scales (Fig. 1). The heterogeneity of fluvial deposits at the scale of a hydrocarbon field (i.e. heterogeneity levels

1 and 2 of Jordan & Pryor, 1992) primarily depends on channel belt sandbody (i.e. Fig. 1, sixth-order contacts) stacking patterns and the associated connectivity between individual channel belt sand bodies (Allen, 1978; Leeder, 1978; Bridge & Leeder, 1979; Mackey & Bridge, 1995). Heterogeneity within individual channel belt sand bodies also impacts performance and efficiency of reservoirs due to an internal architecture with a complex arrangement of contact surfaces, where contrasting grain size, sorting and other lithologic characteristics make for baffles and barriers to flow (Tyler & Finley, 1991; Pranter et al., 2007; Donselaar & Overeem, 2008; Willis & Tang, 2010).

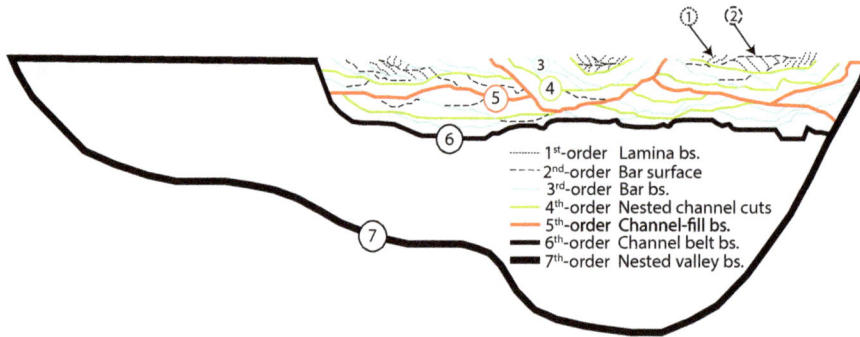

Fig. 1. The hierarchy of fluvial architectural elements. In this study, the focus is on third-order to sixth-order bounding contacts: bar boundaries, nested channel cuts, channel fill and lateral-accretion element boundaries and the external geometry of meander belt sand bodies. Modified from Miall (1985), Holbrook (2001) and Van de Lageweg et al. (2015).

Efforts to quantify fluvial meander belt architecture until now focused on vertical core data (Paola & Borgman, 1991), ignoring the horizontal dimensions. In practice this has the consequence that three-dimensional geomodels developed for flow calculations (i.e. heterogeneity levels 1 and 2 of Jordan & Pryor, 1992) have had to assume either geostatistical or geometrical body dimensions derived from natural rivers (Pranter et al., 2007). However, this approach largely neglects internal heterogeneity and the highly fragmented nature of fluvial three-dimensional bodies (Parker et al., 2013; Van de Lageweg et al., 2013a,b). At the other end of the hierarchical spectrum, quantitative descriptions of preserved laminae, ripples and dunes (Bridge & Best, 1997; Storms et al., 1999; Leclair, 2002) could be used to populate low-order (i.e. Fig. 1, first-order to third-order) architectural bodies. However, populating flow models based on individual order statistics would largely ignore the inherent hierarchical relationships and connectivity between the different architectural levels of fluvial deposits.

Reservoir geocellullar modelling requires quantitative information on the geometries and material properties of reservoir bodies at all hierarchical levels (Fig. 1) to build realistic geological models with predictive flow simulation. However, the large-scale sandbody architecture and stratigraphic and sedimentary heterogeneity within channel belt sand bodies is commonly difficult to model from subsurface data with a typical well spacing between 200 m and 400 m in developed fields (Pranter et al., 2007). This generally results in a lack of architecturally detailed deterministic information. As a consequence, geomodellers often turn to modern analogues (Yang & Sun, 1988; Bhattacharya & Tye, 2004) or outcrops (White & Barton, 1999; Willis & White, 2000) in an effort to obtain quantitative stratigraphic information required to accurately populate the geological model.

The dimensions and distributions of sand bodies and shales are critical parameters in designing a meaningful reservoir model. Sand bodies are generally described by width-to-thickness ratio, net-to-gross (sandstone to shale) ratio, and connectedness ratio (Gibling, 2006; Gouw & Autin, 2008). Point bar deposits are among the largest and architecturally most complex reservoir bodies formed by fluvial meandering systems. In particular, point bar lateral-accretion elements and the associated shale that drape them are potential baffles and barriers to fluid flow and may therefore act to compartmentalize reservoirs (Willis & White, 2000; Pranter et al., 2007; Donselaar & Overeem, 2008; Willis & Tang, 2010). In addition, internal facies trends, for example, the commonly described vertical change from cross-bedded coarse and medium-grained sands to silt laminated ripples in point bar deposits (Nanson, 1980; Bridge et al., 1995), generate a fining-upward succession that potentially can affect the porosity and permeability of reservoirs and bodies.

To improve our ability to build geologically sound models of reservoirs formed by fluvial meandering systems, we must improve our ability to predict the external geometry of meander belt sand bodies, their internal architecture, composed of channel fills and point bar deposits, and the distribution, thickness and orientation of muds. In addition, an explicit link between the causal process of river meandering and the resultant architecture will help elucidate the effect that changing discharge, channel depth and sediment has on the formation and preservation of meander belt deposits, allowing for a better understanding and predictive ability of spatial patterns in sedimentation and preservation. Here, the dimensionless size ratios between the formative meandering channel (i.e. depth, width and meander length of the channel) and the resultant architectural elements (i.e. thickness, width and length of point bar deposits) are developed to make comparisons between systems of different size possible. These dimensionless size ratios between morphological form and depositional product are referred to as archimetrics, inspired by empirical morphometric relationships derived from modern meandering rivers (Leopold & Wolman, 1957; Zeller, 1967; Kleinhans & Van den Berg, 2011). The idea behind the analysis is that we use known morphodynamic relationships as the relevant length scales for the resulting deposits, while incorporating reworking and partial preservation statistics.

The objective of this study is therefore to develop quantitative three-dimensional relationships that describe the stratigraphical and sedimentological architecture of meandering rivers. This depositional model should aid architectural characterization and geomodel building for fluid-flow calculations in the hydrocarbon development and production phases of reservoirs composed of meander belt deposits. Deposits formed by meandering rivers are emphasized because of their considerable lithological variation and because their reservoir behaviour has proven challenging to predict to date (Willis & White, 2000; Pranter et al., 2007; Donselaar & Overeem, 2008; Willis & Tang, 2010). The depositional model is developed using three flume experiments and one numerical model. The specific objectives of this study are to:

1 Quantify the external geometry of meander belts: width, thickness and the relief variation of the base of meander belts.

2 Quantify the three-dimensional internal stratigraphic architecture of meander belts using internal bounding surfaces that are formed by channel migration and bend cut-off.

3 Quantify relationships between dimensions of non-truncated morphological elements (e.g. channel depth and meander length) and dimensions of truncated, partially preserved architectural elements.

4 Evaluate the processes that control the distribution and dimension of fines within point bar deposits, in particular lateral-accretion surfaces.

5 Compare observed lateral slopes of accretion surfaces to predictions by a semi-empirical relationship that predicts these surfaces as a function of observable channel and sediment characteristics.

6 Identify a set of rules for building realistic three-dimensional reservoir models of deposits formed by meandering rivers.

DATA AND METHODOLOGY

Three flume experiments and one numerical model simulation were used in this study. The complementary nature of models and experiments and the benefits of using both have been discussed elsewhere (Kleinhans et al., 2014). Experiment A and the numerical model focused on the formation of channel bounding surfaces. The dimensionless archimetric size ratios from experiment A were quantitatively compared to the numerical model to evaluate possible scale dependence and effects of different styles of bend cut-off. By including flume experiments as well as a numerical model it is intended to move beyond effects of initial setups and specific boundary conditions. When commonalities in, for example, the meander belt width-to-thickness ratios arise for two completely different approaches and scales, it is likely that they are not related to initial and boundary conditions and therefore have a wider validity. Experiments B and C focused on the deposition of fine sediment in point bar deposits, which cannot presently be modelled yet with the detail that experiments deliver.

First, the setups and procedures of experiments A, B and C are described. Second, the numerical model settings are given. Third, the data processing work flow for the flume experiments and the numerical model are explained.

Experimental setups

Experiments A and B have been published elsewhere (Van Dijk et al., 2012, 2013b; Van de Lageweg et al., 2013a,b, 2014) while the present analyses and Experiment C are novel. All experiments were conducted in a flume 6 m wide and 10 m long. The design conditions were not derived from direct scaling of a specific river but are based on a minimization of scaling issues, that is, low sediment mobility, scour hole formation and sediment cohesion (see full review in Kleinhans et al., 2010, 2014). Key dimensionless variables were therefore kept within a range of values to ensure process similarity with natural meandering coarse-grained rivers. For flow similarity, Froude numbers (Fr) must be subcritical ($Fr < 1$) and flow must be turbulent ($Re > 2000$). Rough flow conditions were ascertained by the use of a poorly sorted sand (D_{10}, D_{50} and D_{90} are 0·25, 0·51 and 1·35 mm, respectively). Furthermore, the upstream water and sediment supply point was dynamic and moved transversally in both directions. Pilot experiments (not reported here) showed that meander dynamics were not sustained in the absence of this upstream perturbation. In agreement with (Lanzoni & Seminara, 2006), Van Dijk et al. (2012) showed that the direct effect of the transversal perturbation on downstream bend development was limited to the development of the first bend; yet the presence of the perturbation ensured dynamic meandering with repeated chute cut-offs. Based on observed bend migration rates during the pilot experiments, a constant rate of movement of 1 cm/hour was applied for all flume experiments.

The general settings of experiments A, B and C were similar but a number of specific conditions differed (Table 1). Experiment A had an initial straight channel that was 0·3 m wide, 0·015 m deep and set at a gradient of 0·0055 m/m. Sediment and water were supplied at constant rates of 0·75 l/hr and 1 l/s, respectively. The sediment feed consisted of a mixture of poorly sorted sand and 20 volume per cent of silt-sized silica flour (D_{10}, D_{50} and D_{90} are 3·7 µm, 32 µm and 97 µm, respectively). The silt-sized silica flour formed weakly cohesive deposits and limited channel widening, which would otherwise ultimately have resulted in a braided channel pattern (Peakall et al., 2007; Van Dijk et al., 2012). The duration of experiment A was 260 hours.

Experiment B (Van Dijk et al., 2013b; Van de Lageweg et al., 2014) had an initial straight channel that was 0·15 m wide, 0·01 m deep and set at a gradient of 0·01 m/m (Table 1). A simple stepped shape hydrograph was used with a long (2·5 hours) duration low discharge of 0·25 l/s and a short (0·5 hours) duration high discharge of 0·5 l/s. The stepped shape hydrograph was not intended to represent the flood hydrograph of a specific river but was introduced to stimulate overbank flow and overbank deposition of fines. To this end, an additional 0·5 l of silt-sized silica flour was added during the high discharge stages of the hydrograph. The low discharge sediment feed rate was 0·2 l/hr. This sediment feed had an identical composition for flume experiments A and B. The duration of experiment B was 120 hours.

The design conditions of experiment C were identical to those of experiment B (Table 1). The only difference was that experiment C used walnut shell fragments instead of silt to simulate fines. The walnut shell fragments were well-sorted (between 1·3 mm and 1·7 mm) and behaved as suspended load. It is hypothesized that the silt and walnut shells may capture different depositional behaviour in natural systems. Specifically, the applied walnut shells may represent very fine, highly mobile and mostly suspended materials such as clay and fine silts, whereas the applied silts may be a proxy for coarser, less mobile and bedload to suspended load

materials such as coarse silt and fine sand. During high discharge 1 l of walnut fragments was added. The long duration (multiple weeks) of these large-scale experiments limited the applicability of walnut fragments because they started to decompose after a few weeks. A minor amount of an oily substance was released during this walnut decomposition, which caused cohesive behaviour such that the river planform ossified. We used the initial point bar deposits that were formed prior to ossification to compare the spatial distribution of walnut shells in experiment C to that of the silt in experiment B, because the two different experimental methods to simulate fine-sediment deposits gave complementary results.

Surface elevation data were collected every 8 hours for experiment A and every 3 hours for experiments B and C. A line-laser with a vertical resolution of 0·2 mm recorded the elevation of the fluvial morphology at longitudinal increments of 2 mm. The point cloud from the line-laser was median filtered on to a 10-mm grid for experiment A and on to a 4-mm grid for experiments B and C to produce Digital Elevation Models (DEMs). See Van Dijk et al. (2012) and Van de Lageweg et al. (2013a) for a more detailed description of data reduction from the line laser.

In addition, a high-resolution (0·25 mm ground resolution) camera photographed the fluvial surface in experiments B and C. In experiment B, these high-resolution images were used to derive silt surface concentrations (see also Van Dijk et al., 2013b). The high-resolution camera with RGB-band gives values for green, red and blue, which were transformed to a LAB colour space. In this colour space, L corresponds to luminosity (low = black and high = white), A reflects the position between red/magenta (high values) and green (low values), and B is the position between yellow (high values) and blue (low values). The luminosity was used here to make distribution maps of the highly reflective silt-sized silica flour (Fig. 2). For every survey, 36 individual images (Fig. 2A) were collected and tied to the DEM by the mutual robot coordinate system.

Table 1. Design conditions of flume experiments and numerical model run

Parameter Used for	Flume experiment A Bounding contacts	Flume experiments B and C Fines	Numerical model Bounding contacts
Length	10 m	10 m	10 000 m
Width	6 m	3 m	3000 m
Initial channel width	0·3 m	0·15 m	200 m
Initial channel depth	0·015 m	0·01 m	8 m
Width-to-depth ratio	20	15	25
Discharge	1×10^{-3} m^3/s	0·25 and $0·5 \times 10^{-3}$ m^3/s	2500 m^3/s
Basin slope	$5·5 \times 10^{-3}$	1×10^{-2}	2×10^{-4}

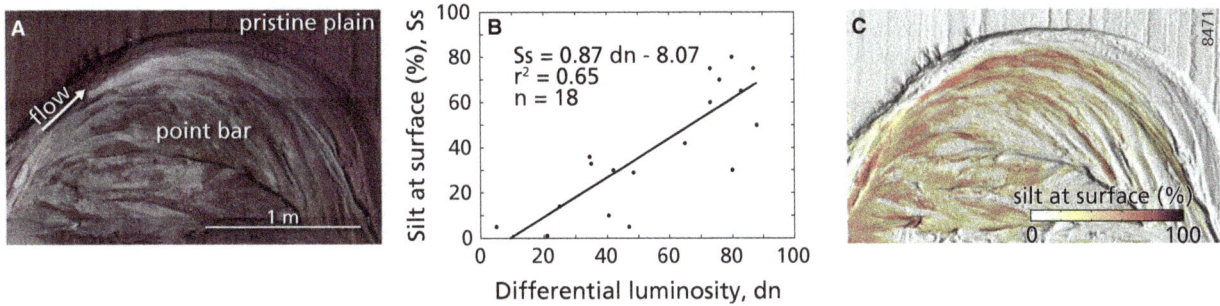

Fig. 2. Illustration of the method used to convert image whiteness to silt concentration. (A) Rectified image of experiment B with silt deposition (white) on top of point bar. (B) Least-squares linear relationship between differential luminosity (quantified by the image digital number *dn*) and measured silt surface concentrations (*Ss*). (C) Silt surface concentrations of image shown in (A). Note that the whiter areas in (A) have a higher silt surface concentration. Modified from Van de Lageweg *et al.* (2014).

Post-experiment, 18 silt samples were collected that were related to the luminosity difference between each survey image and the initial image of the bed for that location (Fig. 2B). Similar relationships were found in other tests and experiments supporting the validity of this relationship. This relationship was then used to convert the luminosity maps to silt surface distribution maps (Fig. 2C). In experiment C, the high-resolution images were used to examine the sedimentation pattern of walnut shell fragments. Experiment C served to compare the spatial distribution of walnut shells to that of silt and therefore only the last time step prior to ossification is presented.

Numerical model setup

A highly sinuous meandering river and resultant sandbody architecture was simulated by the two-dimensional fluid dynamics and morphodynamics code NAYS2D (Asahi *et al.*, 2013; Schuurman *et al.*, 2015; Van de Lageweg *et al.*, 2015). This process-based model is the first to produce river meandering without presuming a fixed relationship between bank erosion and bank accretion, contrary to one-dimensional meander simulation models (Willis & Tang, 2010). This key feature of NAYS2D to simulate inner bend bar accretion and outer bend bank erosion independently allows for channel width variations, which are known to affect hydrodynamics (Zolezzi *et al.*, 2012; Frascati & Lanzoni, 2013) and sediment deposition along meander bends (Eke *et al.*, 2014; Van de Lageweg *et al.*, 2014).

The NAYS2D model parameters were chosen such that a dynamic and sustained freely meandering river was simulated. The full model domain was 3 km wide and 10 km long. Runs started from a straight 200 m wide and 8 m deep channel set at a gradient of 2×10^{-4} m/m (Table 1). To maintain model stability, channels were not allowed to erode beyond 8 m below the initial floodplain level although bars could freely aggrade. Sediment transport rate was computed by the Engelund & Hansen (1967) formulation. A uniform grain size of 2 mm was applied and discharge was maintained constant at 2500 m³/s following similar simplified approaches in Schuurman *et al.* (2013, 2015). In this approach, constant discharge is assumed to be the dominant or effective discharge integrating all the morphodynamic work done over a yearly hydrograph. A constant uniform bed roughness (Nikuradse $k_s = 0.15$ m) was used to parameterize bed form roughness. Similar to the experimental setups, the upstream water and sediment inflow point moved transversely. The rate of movement was maintained constant at 0.65 m/yr and had a maximum amplitude of 300 m. A detailed analysis of the effect of upstream perturbations on the evolution and morphological characteristics of the meandering channel in NAYS2D and other numerical models is provided in and Schuurman *et al.* (2015).

The NAYS2D computational grid was initially straight and consisted of grid cells with a width and length of 20 m. During the run, bank erosion and bank accretion adjusted the bank lines. Curvilinear re-gridding was performed to keep the grid boundary fitted and to maintain the transverse grid lines perpendicular to the channel centre line. The dynamic coordinate system enabled channel migration and neck cut-offs but introduced some instability over the course of the run, which limited planform development to a maximum of 668 modelled time steps. Channel neck cut-offs were modelled as an instantaneous change in planform, which could occur at any time step when two migrating banks met. At the end of the model run, the dynamic coordinates of all recorded time steps were re-sampled to fixed Cartesian ones to generate DEM surfaces and three-dimensional virtual meander belt architectures.

Quantifying three-dimensional meander belt sandbody architecture

Sequential highly detailed DEMs of the meander morphology in the flume experiments and numerical model were used to generate three-dimensional synthetic sandbody architectures that record channel history and the formation of bounding contact surfaces (Fig. 3). The ability to track the morphological channel history while building three-dimensional synthetic meander belt architectures allowed the preserved sandbody architecture to be explicitly related to original meander morphology at the time of sedimentation. From this three-dimensional architectural output, virtual cores, transects and maps were constructed and analysed.

Meander belt sandbody architecture from the flume experiments and numerical model was analysed in three dimensions. To this end, we extracted virtual cores in all three directions through the depositional block to calculate statistics of thicknesses, widths and lengths between bounding surfaces (Fig. 4). Set thickness is defined as a depositional body enclosed by two successive bounding surfaces in a vertical direction (z), set length as a depositional body enclosed by two successive bounding surfaces in a longitudinal direction (x), and set width as a depositional body enclosed by two successive bounding surfaces in a lateral direction (y) (Fig. 4C, see also Van de Lageweg *et al.*, 2013a). The number of grid points and therefore the number of virtual cores was large with, for example, $2 \times 3 \cdot 10^5$ vertical cores for experiment A, $1 \times 2 \cdot 10^6$ vertical cores for experiment B and $3 \times 8 \cdot 10^4$ vertical cores for the numerical model run. It is important to note that the identified surfaces and units merely served as a quantitative framework for the meander belt architecture and do not necessarily relate to lithostratigraphic sand bodies that would define reservoir heterogeneities or to specific scales of depositional bed or bedform. Although heterogeneity patterns can follow depositional bed surface geometry, this is not always the case and has also not been assessed in this study.

In order to generalize results, the extracted set thicknesses, set widths and set lengths were compared with morphometric parameters that are commonly used to describe meander morphology (Fig. 4). For this purpose, the mean channel depth (h_{mean}), mean meander bend length (L_{bend}), mean channel width (w_{mean}) and mean meander bend amplitude (A_{bend}) were used to summarize the meander morphology in the flume experiments as well as the numerical model. For simplicity and easier field application, the A_{bend} was approximated by $0.5 \times$ meander belt width. Archimetrics were then calculated as the dimensionless size ratios between the set statistics characterizing meander belt sandbody architecture and the morphological statistics characterizing the original meander form: set thickness was related to h_{mean}, set length to L_{bend} and set width to A_{bend} and w_{mean}.

The detection of bounding surfaces and contacts in flume experiment A was defined by the vertical resolution

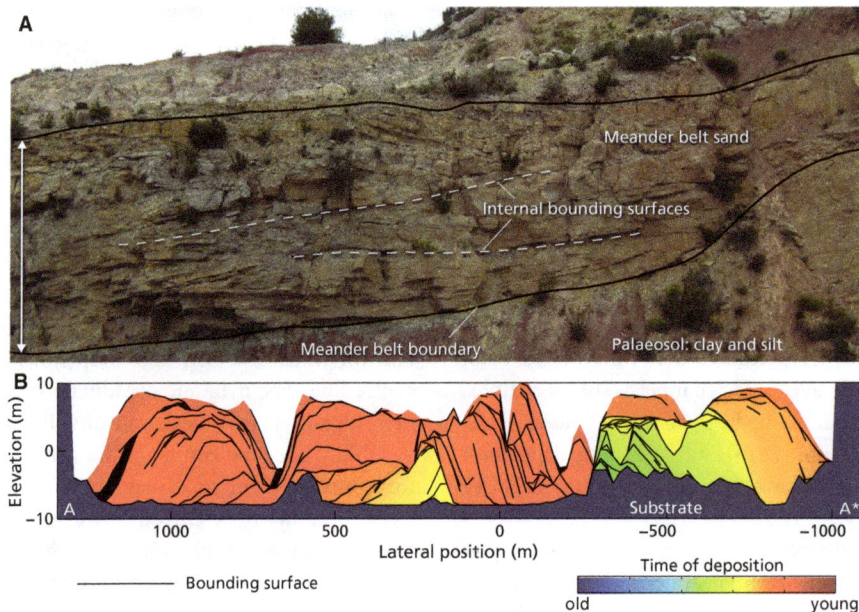

Fig. 3. Meander belt architecture formed by (A) nature and (B) the numerical model NAYS2D. The meander belt architecture resulting from the numerical model and flume experiments is referred to as 'synthetic stratigraphy' and is used to infer time of deposition, channel scour surfaces and typical thicknesses, widths and lengths between channel and scour bounding surfaces. The position of the synthetic slice A-A* is indicated in Fig. 5C.

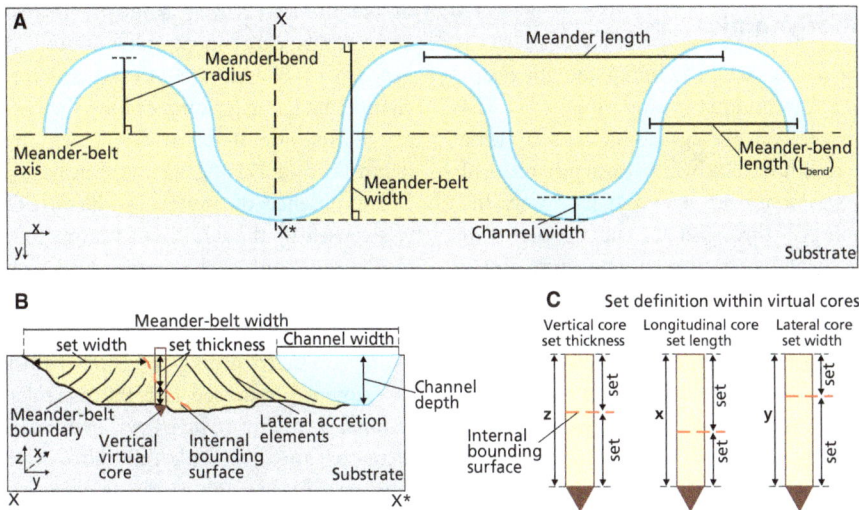

Fig. 4. Schematic drawing of morphometric and archimetric parameters used to quantify meander morphology and architecture. (A) Map view of meander belt with definition of key morphometric parameters. (B) Cross-section of meander belt with definition of key morphometric and archimetric parameters. (C) Set definition within virtual cores. Cores were made in vertical (z), longitudinal (x) and lateral (y) direction. We define set thickness as a depositional body enclosed by two successive bounding surfaces in vertical direction, set length as a depositional body enclosed by two successive bounding surfaces in longitudinal direction, and set width as a depositional body enclosed by two successive bounding surfaces in lateral direction.

of the line laser. All features smaller than 0·2 mm could not be detected and elevation differences between sequential DEMs smaller than or equal to 0·2 mm were therefore removed. In the numerical model, bounding surfaces formed by smaller scale features such as dunes and ripples were not explicitly modelled. Their height was estimated from the prescribed bed roughness as $h_{dune} = 2 \cdot k_s$ (Van Rijn, 1984). This resulted in a typical h_{dune} of 0·3 m. Bounding surfaces resulting from an erosion event with a magnitude of 0·3 m or less were removed, implicitly ensuring that the channel and bar scouring and channel cut-offs were the only formative processes feeding into the synthetic stratigraphy.

Lateral-accretion surface slopes in point bar deposits correspond to the transverse bed slopes of meander inner bends. The transverse bed slope of the meander inner bend in flume experiment B was measured and compared the slope to a transverse bed slope predictor derived from physics. The analytical prediction for the transverse bed slope ($\tan \frac{\delta z}{\delta n}$) was solved at the meander bend apex (Struiksma, 1985; Talmon et al., 1995):

$$\tan \frac{\delta z}{\delta n} = 9 \cdot \left(\frac{D_{50}}{h}\right)^{0.3} \cdot \sqrt{\theta} \cdot \frac{2}{\kappa^2} \cdot \left(1 - \frac{\sqrt{g}}{\kappa \cdot C}\right) \cdot \frac{h}{R} \quad (1)$$

where h is water depth (m), g is gravitational acceleration (m²/s), θ is Shields sediment mobility number, C is the Chézy number (here calculated as $18 \cdot \log\left(\frac{12h}{D_{90}}\right)$), κ is the Von Kármán's constant (0·4) and R is the radius of

curvature of the streamlines. This relationship has been tested successfully in experiments (Van Dijk et al., 2012, 2013a) and in the field (Kleinhans et al., 2012) and in many engineering applications). Following Struiksma (1985), the transverse bed slope was determined by selecting the observations at 0·1w (here 2 cm) from the deepest measurement on the profile and at 0·1w from the water line. The transverse bed slope, h was assumed constant at 0·011 m, while R changed.

RESULTS

The meander morphodynamics are initially described here and typical morphometrics for flume experiment A, flume experiment B and the numerical model are identified. Then, the dimensions of architectural elements of meander belt deposits formed in the flume experiments and the numerical model are quantified. Following earlier work on fluvial architectural elements and contact surfaces (Miall, 1985; Holbrook, 2001), a hierarchical approach is adopted to describe and quantify the architectural elements: starting with the largest element corresponding to the external geometry of meander belt sand bodies (Fig. 1, sixth-order contacts), then focusing on the internal sandbody architecture of a meander belt (Fig. 1, fifth- to third-order contacts) and finally zooming in to the point bar lateral-accretion elements (Fig. 1, third order) and associated shale drapes.

Meander morphodynamics

In flume experiment A (Fig. 5A), the initial straight channel developed into a low-sinuous meandering river. The meandering channel was on average about 0·25 m wide and 0·015 m deep, resulting in a width-to-depth ratio of 17. Typically, L_{bend} was about 3·5 m. Point bars were the dominant morphology throughout the experiment although the initial point bar bodies became increasingly dissected as the experiment progressed.

Initially, a series of point bars raised channel sinuosity, calculated as the ratio between the length along the channel and the straight-line distance between the ends of the channel, to a maximum of about 1·3 in flume experiment A. With this higher sinuosity, water started to flow across the point bar surface and eventually a chute cut-off took place. Former active channels transformed into floodplain lakes, which remained as topographic lows in this experiment because of the limited amount of fine-grained silica flour available as fill. The floodplain depressions were thus easily re-occupied and re-activated by the channel through chute cut-offs. In total, four chute cut-offs were observed during experiment A. The morphology at the end of the experiment consisted of a large number of point bars that were mostly cross-cut by chute cut-offs. A qualitative comparison shows that this morphological configuration has similarities with the 'coarse-grained meandering stream': model 5 of Miall (1985), which agrees with the analysis presented here that the experimental meandering rivers represent gravel-bed rivers best (Kleinhans et al., 2014).

In flume experiment B (Fig. 5B), the meandering river was characterized by sustained lateral migration of the channel and the absence of cut-offs in the middle section

of the flume. Channel sinuosity increased to 1·4 at the end of this experiment. Due to the lower discharge in experiment B compared to experiment A (Table 1), the channel had smaller dimensions: the channel was on average 0·2 m wide, 0·011 m deep, with a resultant width-to-depth ratio of 17, and had a typical L_{bend} of about 2·2 m.

The absence of chute cut-offs in flume experiment B was caused by the addition of more cohesive fine material to the sediment feed compared to experiment A. These cohesive fines were deposited on the floodplain when the water level exceeded bankfull conditions. Initially, this mainly occurred during the brief high-flow stages. As the experiment progressed and the meander bend sharpened, channel gradient reduced so that water levels generally increased and overbank flow also occurred during low-flow stages. Although overbank deposition of the fine silt only resulted in thin sheet splays and levees, depressions on the point bar surface could not be re-activated and chute cut-offs did not occur.

In the numerical model (Fig. 5C), the initial straight channel developed into a highly sinuous ($S \sim 1·5$–2) meandering channel. The sinuous single-thread channel was on average 8 m deep but locally reached depths of up to about 20 m, measured from the top of aggraded bars to the channel bottom. The deepest parts corresponded to outer bends while inner bends were notably shallower. Channel width also varied greatly with an average of 250 m, resulting in a typical width-to-depth ratio of 31, but locally it was only 180 m wide while the maximum observed channel width was 280 m. Meanders had a typical L_{bend} of 1000 m. The channel was locally straightened by neck cut-offs from about 350 time steps onwards. In total, we observed four neck cut-offs. Qualitatively, the channel width-to-depth ratio and channel

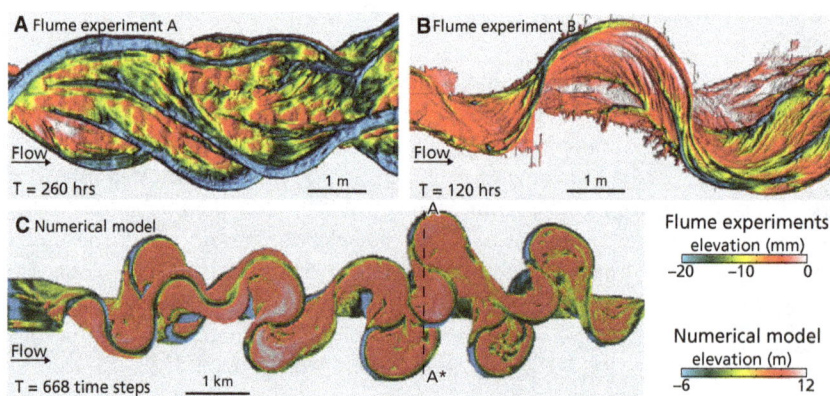

Fig. 5. Digital Elevation Models (DEMs) of meander belts that were formed in the flume experiments and numerical model. (A) DEM of flume experiment A in which a meander belt formed during constant bankfull discharge (Van de Lageweg et al., 2013b; Van Dijk et al., 2012). (B) DEM of flume experiment B in which a meander belt formed as the result of a variable discharge (Table 1) and addition of fine cohesive silt during the high-flow stages (Van Dijk et al., 2013a; Van de Lageweg et al., 2014). (C) DEM of meander belt simulated with the numerical model NAYS2D (Van de Lageweg et al., 2015). Slice A-A* is presented in Fig. 3.

planform generated with the numerical model have similarities with the 'classic meandering stream': model 6 of Miall (1985).

Meander belt architecture

Meander belt external sandbody geometry

In the numerical model, the meander belt sandbody had an average width of 1230 m at the end of the simulation (Fig. 6A). This is equal to about 5 w_{mean}. The meander belt sandbody had an average thickness of 8·4 m, which is in close agreement with the h_{mean} of 8 m (Fig. 6C). The relief at the base of the meander belt sandbody generally fell within h_{mean} (Fig. 6E). The base of the sandbody was deepest in the axial zone of the meander belt and persistently shallowed towards the margins of the sandbody, as evidenced by the Z_{10} distribution.

Longitudinal variations in elevation relief of the sandbody base were laterally homogeneous, as seen from the difference between the bed level Z_{10} and Z_{90} percentile profiles (Fig. 6E).

In experiment A, the meander belt sandbody had an average width of 2·1 m at the end of the experiment (Fig. 6B). This is equal to about 8 w_{mean}. The thickness of the meander belt was on average 1·2 cm, which is close to the h_{mean} of 1·5 cm (Fig. 6D). The relief at the base of the meander belt sandbody was generally captured within h_{mean} (Fig. 6F). Re-occupation of older unfilled channels and resultant repetitive scouring and deepening of these channels led to two substantial dips in base elevation at the margins of the sandbody, most clearly seen in the Z_{10} percentile profile (Fig. 6F). Longitudinal variations in elevation relief of the sandbody base were laterally homogeneous, except for the margins of the sandbody base where a general deepening was seen on both sides of the

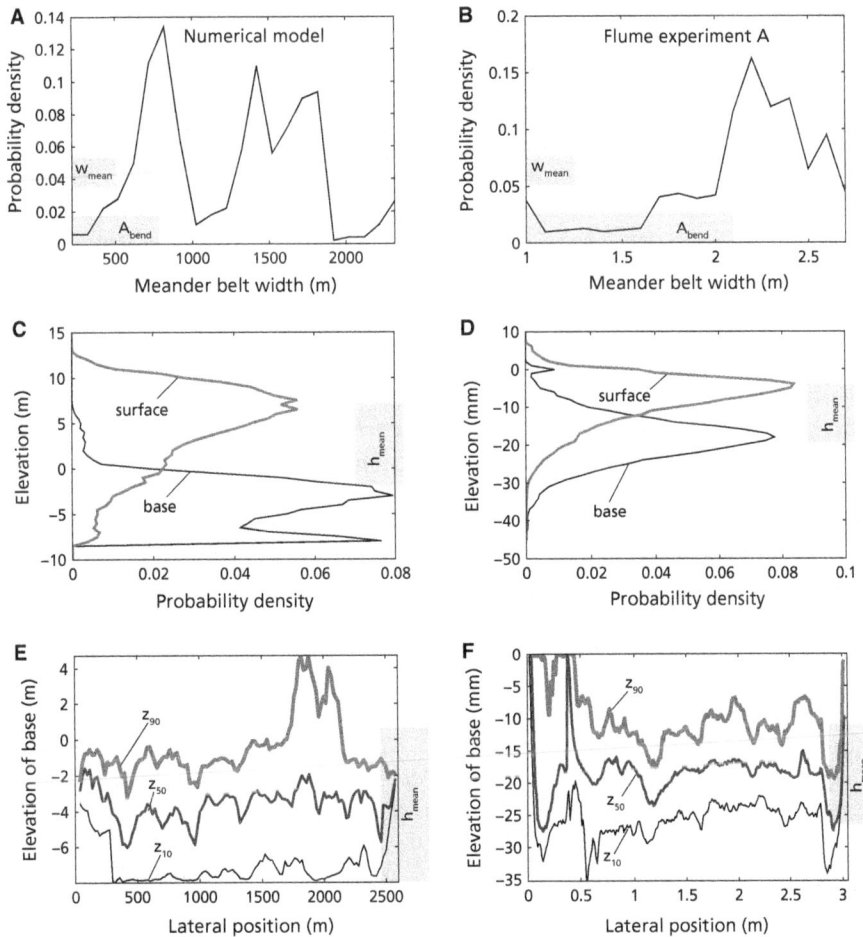

Fig. 6. Quantification of the external geometry of the meander belt sandbody. Probability density distributions of the width of the sandbody in (A) the numerical model and (B) flume experiment A. Probability density distributions of the relief in elevation at the surface (i.e. topography) and at the base of the meander belt sandbody in (C) the numerical model and (D) flume experiment A. Percentiles of the base relief as a function of the lateral position within the meander belt in (E) the numerical model and (F) flume experiment A.

sandbody, most evident from the Z_{50} percentile profile (Fig. 6F). Nevertheless, the downstream half of the left-hand margin of the sandbody base was much shallower, as indicated by the Z_{90} percentile profile. The relief in elevation of the base of the sandbody was generally captured by h_{mean} for the numerical model as well as flume experiment A (Fig. 6E and F). Base relief was primarily caused by in-channel depth variations. Persistence of the meandering channel to remain in the same location, deep outer bends, cut-off channels and confluence scours all resulted in a deepening of the base of the sandbody. In the numerical model, the relief was homogeneous and the base shallowed towards the margins while in flume experiment A the relief of the base was more irregular and a deepening of the base towards the margins of the sandbody was present. These different base expressions may either be the result of the different styles of meandering in flume experiment A and the numerical model, or an artefact of the numerical model's limitations in representing the deepest channel scours.

In flume experiment A, the base relief of the sandbody was similar to the topographical relief at the surface of the sandbody (Fig. 6D). In the numerical model, the deepest channel scours were not well reproduced (Fig. 6C). Consequently, the relief in elevation at the base of the sandbody was smaller than the topographical relief.

Combining the width and thickness statistics into dimensionless sandbody geometrical ratios resulted in similar outcomes for flume experiment A and the numerical model (Fig. 7A). Flume experiment A had a sandbody width-to-thickness ratio of 175 and the numerical model had a ratio of 150 at the end of the simulation. In flume experiment A, however, the spread in thickness around the average sandbody thickness of the belt was larger than in the numerical model (Fig. 7A). In flume experiment A, sandbody thicknesses of up to 4·5 cm (ca 3 h_{mean}) were observed while the maximum sandbody thickness in the numerical model was 20 m (ca 2·5 h_{mean}). In contrast, the spread in sandbody width was larger in the numerical model than in experiment A (Fig. 7A). In the numerical model, the width of the sandbody varied between 500 m (ca 2 w_{mean}) and 2000 m (ca 8 w_{mean}) while in flume experiment A the sandbody width was more uniform and ranged between 2 m (ca 8 w_{mean}) and 2·5 m (ca 10 w_{mean}).

Internal sandbody architecture

The dimensions of the meandering channel (depth, meander length, meander amplitude) as well as the succession of migrating and scouring channels determined the formation and preservation of bounding surfaces and contacts within the sandbody. Vertically, the deepest channels

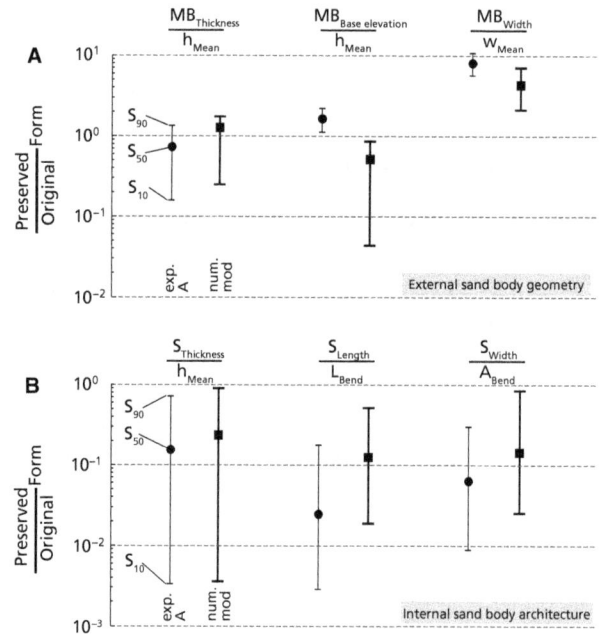

Fig. 7. Relation between preserved truncated and original non-truncated channel form. (A) Characterization of the external sandbody geometry of a meander belt (MB) by percentiles of the ratio between $MB_{thickness}$ and h_{mean}, between $MB_{baseelevation}$ and h_{mean}, and between MB_{width} and w_{mean}. (B) Characterization of the internal sandbody architecture of a meander belt by percentiles of the ratio between $S_{thickness}$ and h_{mean} between S_{length} and L_{bend} and between S_{width} and A_{bend}.

left behind the thickest channel and bar deposits. These were then reworked and dissected and decreased in thickness by shallower channels as the flume experiment and numerical simulation progressed (Fig. 3). Horizontally, the meander bends with the largest length and amplitude formed the longest and widest deposits, which were eventually cross-cut and reworked by cut-offs at a later stage. This conceptual framework highlights that the succession of the meandering channels migrating across the fluvial landscape ultimately determined where the internal sandbody architecture consisted of thick, long and wide deposits and also where these deposits were reworked and therefore thinned, shortened and narrowed again.

Deposits were generally more fragmented in flume experiment A than in the numerical model (Fig. 8B). Experiment A produced a more mature channel belt in the sense that more cut-offs were recorded that reworked previously deposited bar deposits. In the numerical model, the deposits were on average 77% shorter, 67% narrower and 61% thinner than the original (L_{bend}, A_{bend} and h_{mean}) morphological form. In flume experiment A, the deposits were on average 87% shorter, 88% narrower and 71% thinner than the original morphological form. Reworking cross-cuts older deposits and thus thins,

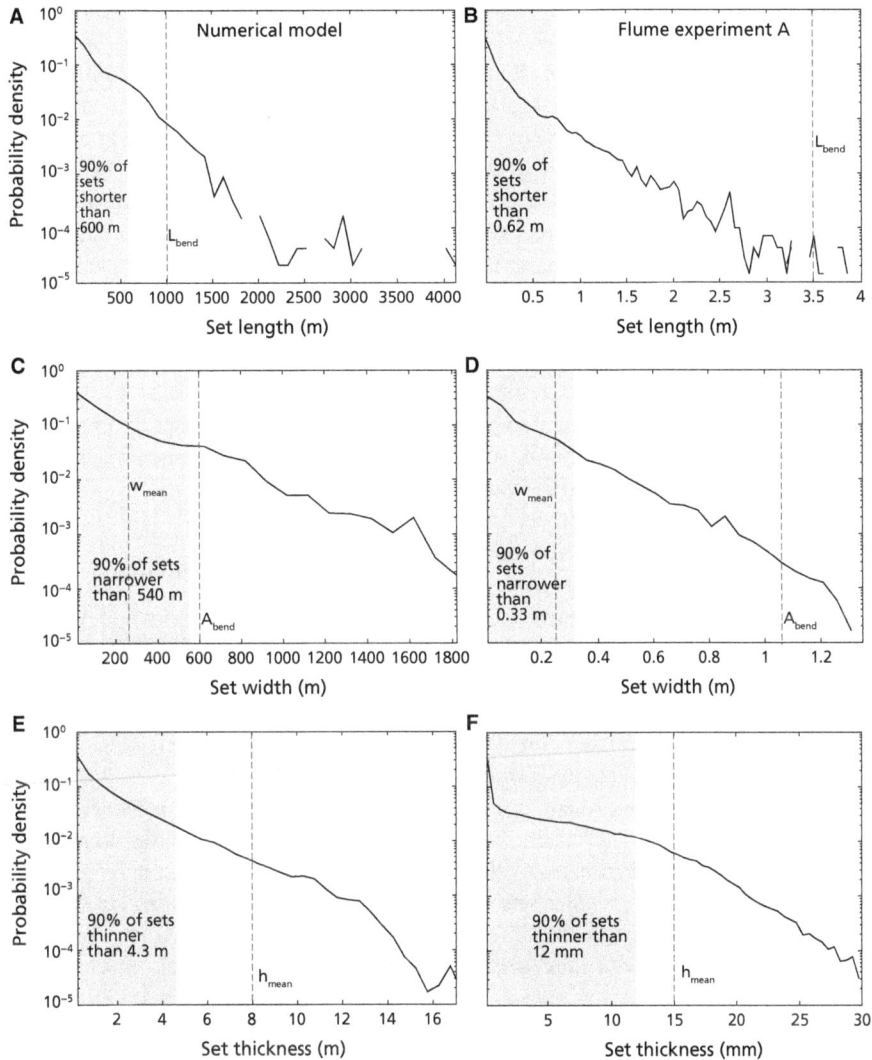

Fig. 8. Descriptive statistics of the internal sandbody architecture of meander belts. Probability density distributions of set length for (A) the numerical model and (B) flume experiment A. Probability density distributions of set width for (C) the numerical model and (D) flume experiment A. Probability density distributions of set thickness for (E) the numerical model and (F) flume experiment A. Note that most of the internal sandbody architecture is fragmented and notably shorter, narrower and thinner than the original (L_{bend}, A_{bend} and h_{mean}) channel form.

shortens and narrows them. The number of stacked sets in a vertical virtual core was used as a measure to quantify and compare the degree of reworking in flume experiment A and in the numerical model. In flume experiment A, we observed an average of 2·61 stacked sets per core and in the numerical model an average of 1·86 stacked sets per core. This indicates that the deposits in the numerical model were less frequently reworked, which explains the overall lower degree of fragmentation of the internal sandbody architecture in all three dimensions.

The higher degree of fragmentation in flume experiment A was also seen in the dimensionless ratios between set length, set width and set thickness. In the numerical model, the set length-to-width ratio was 1·2, the set

width-to-thickness ratio was 64 and the set length-to-thickness ratio was 74. In flume experiment A, the set length-to-width ratio was 1·1, the set width-to-thickness ratio was 30 and the set length-to-thickness ratio was 33. This shows that sets were notably narrower and shorter relative to their thickness in flume experiment A compared to the numerical model, which can have been due to the larger number of bend cut-offs recorded in flume experiment A.

Systematic lateral differences in scouring of the meandering channel and resultant reworking of older point bar deposits were observed in flume experiment A as well as the numerical model. In flume experiment A and in the numerical model, most of the morphological activity (i.e.

point bar formation and cut-offs) took place in a narrow zone surrounding the meander belt axis (Fig. 9). For example, four major cut-offs were observed in flume experiment A and the numerical model which took place within a corridor corresponding to 50% and 20% of the meander belt width, respectively. In this axial zone of the meander belt, point bar deposits that were formed early in the experiment and numerical simulation were likely reworked and cross-cut later. Closer to the margins of the meander belt, point bar deposits were less frequently and less substantially reworked.

Logically, the lateral differences in scouring and reworking frequency of the meandering channel resulted in lateral differences within the sandbody architecture (Fig. 10). For an arbitrary division into a central (i.e. middle third of total width) and marginal zone (i.e. summation of outer thirds of total width) of the meander belt width (Table 2), the sets were about 20% and 30% thinner in the central zone compared to the marginal zone in experiment A and the numerical model, respectively. In contrast to set thickness, no clear lateral trends in width and length of the sets were observed for flume experiment A and the numerical model (Table 2). The set distributions showed a tendency towards longer sets for the marginal zones in the numerical model (Fig. 10A) while instead shorter sets were more likely to be present in the marginal zones in flume experiment A (Fig. 10B). For both the numerical model and flume experiment A, the set distributions showed a tendency towards wider sets in the central zone of the meander belt although the differences between axial and marginal zones were small (Fig. 10C and D).

The style of channel cut-offs had a large impact on set width. In the chute-cut-off-dominated flume experiment

A, sets were on average 88% narrower than the original meander bend amplitude. In the neck-cut-off-dominated numerical model, sets were on average 67% narrower than the original meander bend amplitude. The reason for this difference in sandbody architecture was that the neck cut-offs in the numerical model occurred in a narrow zone (<20% of the meander belt width) close to the meander belt axis (Fig. 9). In contrast, the chute cut-offs in flume experiment A took place in a much wider zone, much farther from the meander belt axis. Consequently, reworking in the numerical model took place in a narrow axial zone over and over again, leaving marginal deposits relatively undisturbed while in flume experiment A even the margins of the sandbody were reworked, resulting in more, but on average narrower, sets across the sandbody in the lateral direction.

Lateral-accretion elements

Flume experiment B showed a variety of processes that controlled the distribution and dimension of fines within point bar deposits. In experiment B, the fine silt-sized silica flour was dominantly deposited in the upstream half of the experimental reach (Fig. 11). This indicates that most of the silt behaved as bedload and that only the finest fraction was able to spill overbank. In these overbank areas, flow velocities rapidly decreased. The fine silt that spilled overbank was therefore generally deposited close to the main channel forming small levees and splays.

Silt was also deposited at the point bar surface for which two depositional mechanisms were observed (Figs 11 and 12A). The first mechanism proceeded from water spilling onto the point bar surface thus depositing a silt drape on the point bar surface. This occurred throughout the experiment but intensified as the meander bend sharpened during the later phases of the experiment. The second mechanism was related to the variable discharge used in this experiment. During the high-flow stages, sediment was often eroded upstream of the point bar and was deposited again along the point bar forming a new scroll bar. During the low-flow stages, flow velocities generally decreased but the fine silt was still mobile and could settle on this newly formed scroll bar.

A more detailed analysis of fine silt deposition at the point bar surface (Fig. 12A) shows that the fines were mainly deposited at the point bar head (Fig. 12B) and in topographically lower point bar swales (Fig. 12C). During initial bar formation, high-flow velocities limited the settling of the fine-grained silt on the bar. Lateral accretion increased the bar amplitude, which raised water levels above bankfull level and resulted in overbank flow onto the point bar head, which explains the increase in silt in these regions. Since flow velocities were low on the point

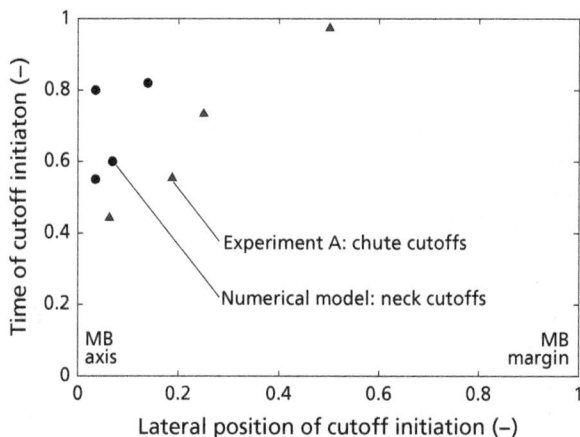

Fig. 9. Lateral position of channel cut-off initiation within the meander belt sand bodies formed in flume experiment A and the numerical model.

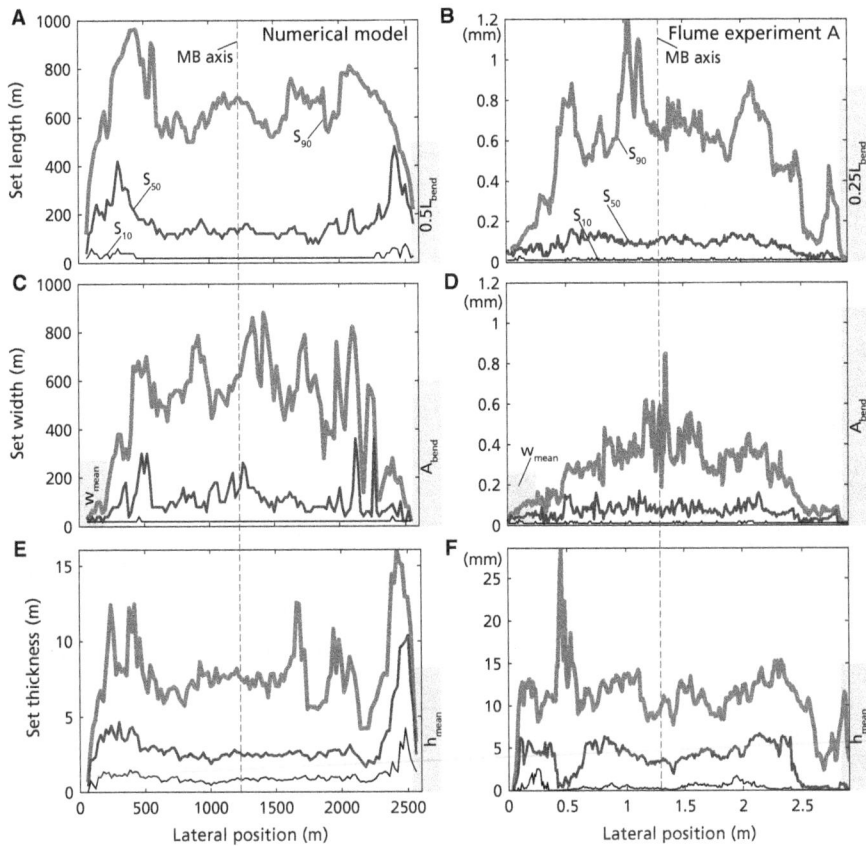

Fig. 10. Lateral differences in the internal sandbody architecture of meander belts. Percentiles of set length as a function of the lateral position within the meander belt for (A) the numerical model and (B) flume experiment A. Percentiles of set width as a function of the lateral position within the meander belt for (C) the numerical model and (D) flume experiment A. Percentiles of set thickness as a function of the lateral position within the meander belt for (E) the numerical model and (F) flume experiment A.

Table 2. Lateral differences in internal sandbody architecture of meander belts. The meander belt was arbitrarily divided into three equally wide zones with the middle zone defined as the centre of the meander belt. The set statistics of the two outer zones were averaged and here defined as the margin of the meander belt. Set thickness was normalized by h_{mean}, set length by L_{bend} and set width by A_{bend}

	Numerical model		Flume experiment A	
	Margin	Centre	Margin	Centre
Set thickness	0·44	0·31	0·30	0·25
Set length	0·21	0·20	0·09	0·08
Set width	0·23	0·26	0·11	0·13

bar surface, fines were predominantly deposited forming a thin silt sheet at this head area (Fig. 12B). Interestingly, the silt did not spread uniformly across the point bar surface (Fig. 12C). Five individual silt deposition peaks were identified, of which those at around 25 cm and 40 cm from the origin were related to increased silt deposition at the point bar head. The other three peaks were part of

a gradual rise in silt deposition from 60 cm onwards due to point bar lateral extension and meander bend sharpening. The three peaks in silt intensity as shown in Fig. 12C corresponded to topographically lower swales in which the silt was predominantly deposited. The result of this silt intensity pattern was an alternating topographical and associated grain-size pattern: high ridges that consisted of coarse sediment and low swales that were partly filled with fine silt.

Notably, the alternating surface grain-size pattern with concentrated peaks of silt was also present in the subsurface point bar deposits (Fig. 13). In the synthetic (Fig. 13A) and real (Fig. 13B) sediment peels of the point bar deposit in flume experiment B, laterally inclined surfaces were recognized, parts of which had fine silt draped on them.

A number of trends in silt deposition within the point bar deposit were identified. First, the thickest silt deposits were found at the point bar surface in the low-lying swales. Tracking these silt deposits from top to bottom revealed that they gradually thinned down the point bar

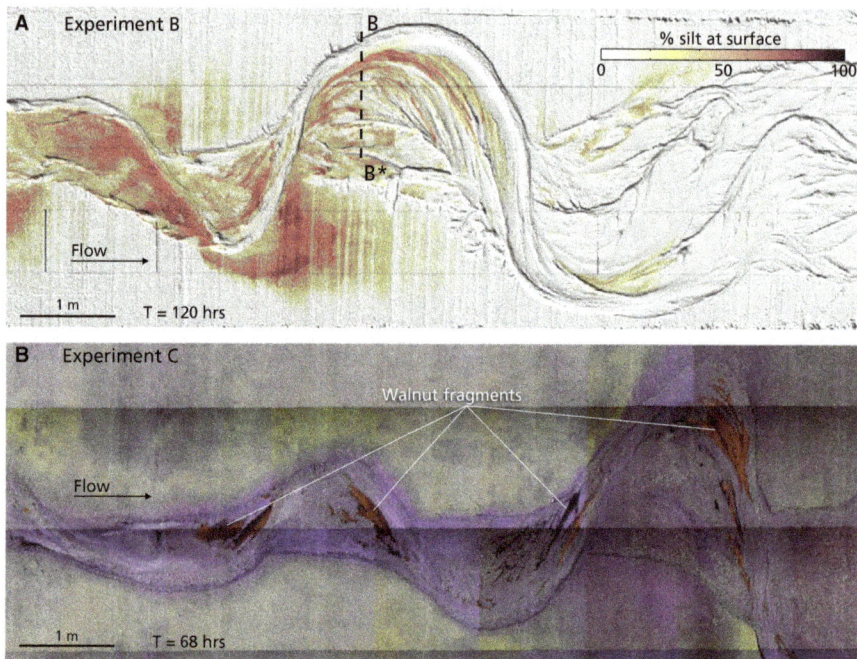

Fig. 11. Deposition of silt and walnut fragments. (A) Silt deposition in flume experiment B. Most of the silt is deposited in the upstream half of the experimental reach. Silt is predominantly deposited overbank forming splays, small levees and thin silt drapes on the point bar surface. Slice B-B* is presented in Fig. 13. (B) Deposition of walnut shell fragments in flume experiment C. Walnut shell fragments are mainly deposited in the downstream half of the point bar.

deposit along the lateral-accretion surfaces. The inclined silt drapes were mostly restricted to the upper 75% of the point bar deposit. Second, the outer point bar deposits, corresponding to later times of deposition (Fig. 13A), received more silt than the inner point bar. Third, silt was nearly absent in the downstream half of the point bar.

The slopes of the silt-draped lateral-accretion surfaces in flume experiment B corresponded well to the predicted transverse bed slope from Eq. 1 (Fig. 14). The slope of the lateral-accretion elements was typically about 5°. However, the variation between individual DEMs was considerable and the transverse bed slopes ranged from about 2° to 10° during the course of the experiment. Both the measured and predicted transverse bed slopes increased for a decreasing bend radius.

DISCUSSION

External geometry of meander belt sand bodies

The observed meander belt width-to-thickness ratios of experiment A and the numerical model are consistent with those reported for a range of ancient natural systems in Gibling (2006). The width-to-thickness ratios also

compare well with those reported for a variety of Holocene meander belts within the Rhine-Meuse delta and Lower Mississippi Valley (Gouw & Berendsen, 2007; Gouw & Autin, 2008). This indicates that a lower bound width-to-thickness ratio of 100 and an upper bound width-to-thickness ratio of 200 provide a robust estimate of the range of natural geometric variability of meander belt sand bodies.

Meander belt width-to-thickness ratios of 100 to 200 indicate wide and thin sand bodies. Importantly, this thin shape of meander belts suggests that most of the variation in sandbody geometry, facies and lithology is likely to be found laterally and not vertically. It is important to note that this is true for meander belt sand bodies formed in aggrading settings too. Although precise sedimentation conditions are unknown for the meander belts reported in Gibling (2006), it is likely that some of them were formed in aggrading settings. The much larger lateral than vertical extent of the meander belt sandbody, even for highly aggrading meander belts, was confirmed in a morphodynamic modelling study (Van de Lageweg *et al.*, 2015), which also related the volume of the sandbody to that of the floodplain in an effort to predict net-to-gross (i.e. the sand percentage) ratio.

A key result of this study is the quantification of the base relief of the meander belt sandbody. Similar to Best

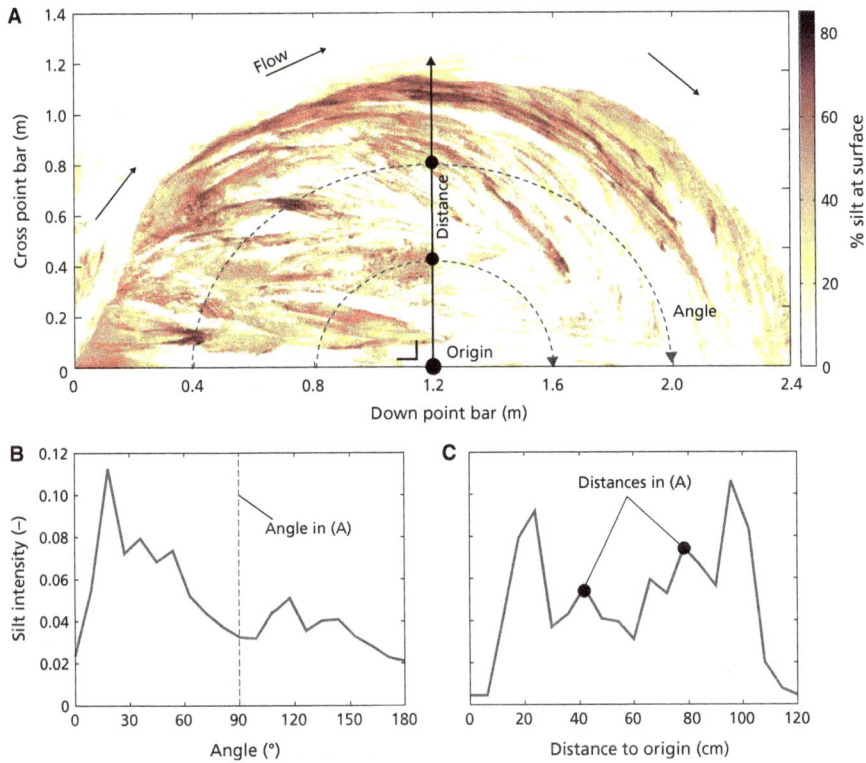

Fig. 12. Trends in silt intensity at the point bar surface that formed in flume experiment B. (A) Map view of silt concentration at point bar surface with definitions of origin, sector angle and distance. (B) Silt intensity as a function of the angle to the origin. (C) Silt intensity as a function of the distance to the origin. Silt concentrations are integrated for all analysed angles and distances to the origin and divided by the total silt concentration for all these points to calculate a silt intensity. Note that the silt intensity is high at the upstream side of the point bar and also increases with distance to the origin.

& Ashworth (1997) who reported on some channel junction scours in modern systems with 4–5 times greater incision than mean channel depth, a mean channel depth to deepest scour ratio of 3·5 was observed in this study. In turn, these topographical variations were reflected in the basal relief of the belts. Comprehensive characterization of the belt basal relief allows for better estimates of sand quantities, stacking patterns and connectedness of amalgamated meander belt sandstone bodies (Allen, 1978; Leeder, 1978; Bridge & Leeder, 1979; Bridge & Mackey, 1993; Bryant et al., 1995; Heller & Paola, 1996; Ashworth et al., 1999; Sheets et al., 2002; Straub et al., 2009; Hajek et al., 2010). Well-exposed outcrops could theoretically be used to partially observe and quantify the relief at the base of natural ancient meandering systems (Holbrook, 2001) but this renders a far from complete three-dimensional image. Flume experiments and numerical models can help to reveal and map this hidden component of meander belt sand bodies and to that end, this study provides the first statistical distributions of the sandbody base relief.

In this study, two statistics were distilled that will be of use to characterize the base relief for natural meander belt systems. First, most of the base relief falls is captured within h_{mean} since the majority of the topography is formed by features within and smaller than the main channel and deep scours are a relatively rare feature within the riverscape An estimate of h_{mean} is therefore required to characterize the base relief of modern and ancient meander belts, which can perhaps be obtained from preserved channel fills at the margins of the meander belt. Second, flume experiment A showed that the probability density distributions of the meander belt topographical surface and sandbody base have a similar shape. This indicates that for modern meander belts an estimate of the base relief can be made based on the probability density of the meander belt topographical surface elevation. In turn, this also implies that modern analogues of ancient meander belts may provide some indication of the base relief of ancient systems.

Internal architecture of meander belt sand bodies

The internal architectures of the meander belt sand bodies generated in the numerical model and flume experiment A were quantitatively similar. Despite the differences in

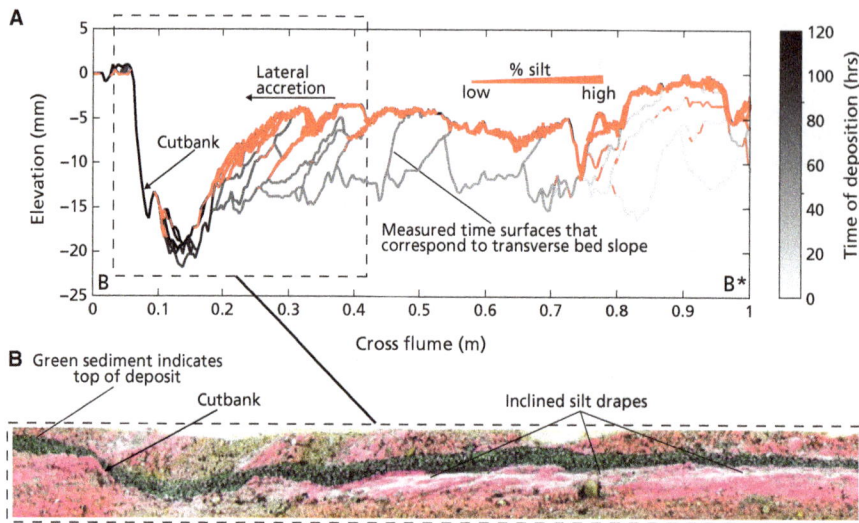

Fig. 13. Synthetic and real sediment peel of the point bar deposit in flume experiment B. The position of the slice B-B* is indicated in Fig. 11. (A) Synthetic slice of point bar deposit showing time of deposition and the amount of silt. Point bar deposits have lateral-accretion surfaces corresponding to the migrating transverse bed slope of the river channel. The measured time surfaces correspond to this transverse bed slope. (B) Sediment peel showing a detail of the point bar deposit in (A). This peel provides information on sedimentary structures and particle size trends. The silt-sized silica flour is white, the coloured green sediment indicates the top of the deposit and the pink colour is caused by the dye that was used to colour the water. Modified from Van de Lageweg et al. (2014).

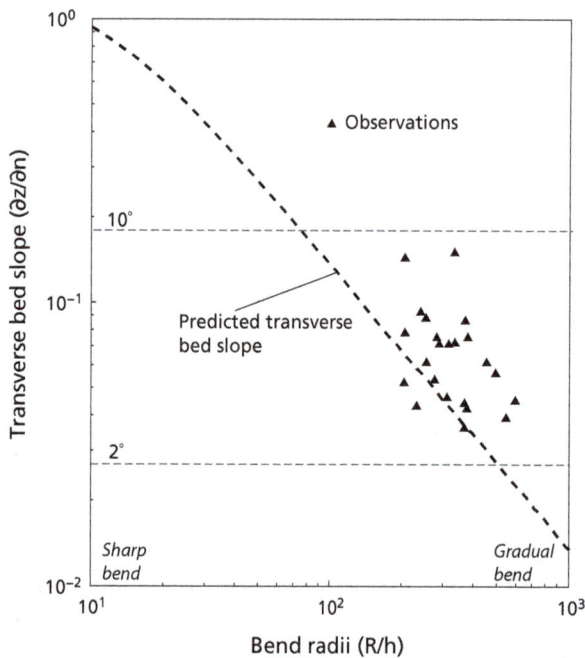

Fig. 14. Comparison of measured and predicted (Eq. 1) transverse bed slopes in flume experiment B. The transverse bed slope is measured at the meander bend apex for all time steps.

scale between the experimental and numerical rivers and resultant sand bodies, preservation ratios were consistent. For example, the ratio between the average preserved bar

deposit thickness and average depth of the meandering channel was 40% in the numerical model and 30% in flume experiment A. Such vertical (i.e. thickness) preservation ratios are consistent with those predicted for natural braided river systems (Paola & Borgman, 1991) but lower than those observed in the sandy braided South Saskatchewan River, Canada, pre-flood and post-flood (Parker et al., 2013). Interestingly, similar vertical preservation ratios between the preserved deposit and original morphological form have been observed for a range of depositional environments. The ratio of mean cross-set thickness to mean dune height is 30% (Leclair, 2002; Jerolmack & Mohrig, 2005) and a similar preservation ratio exists for ripples (Storms et al., 1999). This quantitative evidence highlights that in addition to meander belt sand bodies many other preserved architectures are also built up from truncated deposits, which are challenging to relate to the original non-truncated morphological form.

The ability to observe the formative processes is pivotal to understanding and explaining spatial differences in internal sandbody architectures between flume experiment A and the numerical model. The moderately sinuous chute-cut-off-dominated meandering river in flume experiment A and the highly sinuous neck-cut-off dominated meandering river in the numerical model both resulted in multiple, thin, stacked deposits in the axial zone of the sandbody with a gradual thickening of the deposits towards the margins of the sand bodies. Similar lateral architectural trends have been reported for Quaternary meandering river systems (Lewin & Macklin, 2003), highlighting that fully preserved channel fills are most likely recovered close to the margins of a meander belt.

The difference in meandering style of chute cut-offs in flume experiment A and neck cut-offs in the numerical model affected the internal sandbody architectures. In this specific case of cut-off styles, the difference in architecture was most clearly seen in the width of the resultant deposits, with on average wider deposits produced by the neck-cut-off meandering system in which the cut-offs took place in a narrow zone. It is, however, likely that this specific case of chute-cut-off meandering style versus neck-cut-off meandering style can be extended to the majority of meandering processes leaving a specific imprint on the internal architecture of meander belt sand bodies, because the style of cut-off is a function of the degree of floodplain formation in freely meandering rivers (Kleinhans & Van den Berg, 2011).

In addition to internal meandering dynamics, external conditions are also likely to impact the internal sandbody architecture of meander belts. For example, the degree of valley confinement is known to affect the style of meandering and thus the formation of the sandbody architecture. Ielpi & Ghinassi (2014) provide an example of how meandering style may affect sandbody architectures in their study on expansional and downstream-migrating point bar deposits in an exhumed Jurassic meander plain. They relate expansional point bars to meander belts in unconfined settings and find largely undisturbed deposits due to the limited reworking for this meandering style. In contrast, downstream-migrating point bars are often associated with a high degree of spatial confinement resulting in a high degree of reworking, low preservation of bar deposits, and a complex internal sandbody architecture (Alexander et al., 1994; Ielpi & Ghinassi, 2014).

Lateral-accretion elements within point bar deposits

General grain-size trends derived from real and virtual sediment peels in flume experiment B are consistent with grain-size trends observed in natural coarse-grained point bar deposits (McGowen & Garner, 1970; Jackson, 1976; Bridge et al., 1995). More important in the design of reservoir models, however, is a characterization of the dimensions and distributions of shales within point bar deposits. Primarily because the shales cause pronounced anisotropy in permeability (Novakovic et al., 2002; Pranter et al., 2007; Donselaar & Overeem, 2008; Willis & Tang, 2010), sometimes up to the point of unconnected chambers separated by barriers where disregarding the

shales would have led to the suggestion that the entire point bar is the container.

Flume experiments B and C showed a number of trends in the spatial distribution of fines that may help to refine depositional models of point bars (Figs 11 and 12). In flume experiment C, walnut shells were used instead of silt to simulate flow-limiting fines, while all other parameters were identical to those in experiment B. In both experiments, the fine-simulating sediments were predominantly deposited in the upper outer point bar deposits (Fig. 13). Inclined well-defined silt and walnut drapes were observed in the upper outer point bar bodies, which were interpreted to translate to inclined shales in natural point bars. The silt and walnut drapes were continuous in the upper half of the point bar, then became thinner and discontinuous when tracking them downward, with generally no fines present in the lower 25% of the point bar deposit. Importantly, these results from two different experiments independently indicate that shale drapes are most likely encountered in the upper outer point bar deposits, suggesting that in those areas anisotropy in permeability and compartmentalization may be largest.

Flume experiments B and C also differed in fine-sediment deposition in a number of ways (Fig. 11), highlighting our incomplete understanding of and limited ability to replicate these complex point bar deposits. First, the walnut fragments were predominantly deposited in the downstream half of the point bar, rather than the upstream half when using silt. The walnuts were deposited on the lee side of the point bar body where flow velocities were sufficiently decreased and the walnut fragments could settle. This shows that the difference in mobility between the walnut fragments and the silt had a large impact on the spatial distribution of fines within the point bar deposits. Second, walnut drapes were generally thicker and therefore laterally more continuous than the silt drapes. This may be a result of the specific material properties but the different behaviour hints at a potential relationship between fine-sediment properties and continuity of flow-limiting layers.

Dedicated flume experiments and numerical simulations are needed to further investigate relationship between fine-sediment properties and continuity of flow-limiting layers and to better characterize the spatial distribution and continuity of fines in point bar deposits. Walnut fragments and silt are only two means to represent fines in flume experiments and already show different depositional behaviour, potentially capturing different depositional behaviour in natural systems. The complexity of natural point bar deposits suggests that a combination of silt and walnuts and perhaps other materials is needed to comprehensively represent natural depositional behaviour and the spatial distribution and continuity of fines in experimental point bar deposits. Combined with numerical simulations and detailed sedimentological studies across a range of natural point bars (McGowen & Garner, 1970; Jackson, 1976; Bridge *et al.*, 1995; Pranter *et al.*, 2007; Donselaar & Overeem, 2008), incorporating fine-grained and coarse-grained systems, this provides a powerful strategy to further improve our depositional models of these complex systems.

The dip of point bar lateral-accretion elements and the associated shale drapes is the second key parameter in the design of a reservoir model, in addition to the afore-discussed spatial distribution of shales. Flume experiment B provides quantification for this key parameter too. First, the dip of the point bar lateral-accretion elements in flume experiment B agrees well with theory (Eq. 1). Second, the dip of the silt drapes in flume experiment B is consistent with reported values for natural coarse-grained meandering systems (McGowen & Garner, 1970; Jackson, 1976; Bridge *et al.*, 1995). The dip of point bar lateral-accretion elements, that is the transverse bed slope, is a function of channel geometry and sediment composition. The large natural variation in these two parameters explains the large range of reported lateral-accretion dip slopes in natural point bars. Generally, sandy and sand-silt-mud rivers have a low width-to-depth ratio and can reach dip slopes up to 15° (Puigdefabregas & Van Vliet, 1978; Nanson, 1980; Mossop & Flach, 1983; Smith, 1987; Crerar & Arnott, 2007; Musial *et al.*, 2012). Gravelly and gravel-sand rivers typically have higher width-to-depth ratios resulting in lower (*ca* 5°) lateral-accretion dip slopes (McGowen & Garner, 1970; Jackson, 1976; Bridge *et al.*, 1995).

It is important to realize that the dip slope of potentially flow-limiting lateral-accretion elements of natural point bar deposits can be predicted from a few fairly basic morphological and sedimentological parameters (Table 3 and Fig. 15). This is illustrated for two modern and two ancient, and for two coarse-grained and two finer-grained meandering systems by making a comparison between the reported and predicted dip slopes. For this comparison, we selected the modern coarse-grained river Rhine, Germany (Erkens *et al.*, 2009), the modern finer-grained Daule river, Ecuador (Smith, 1987), the ancient coarse-grained Castisent Formation, Spain (Nijman & Puigdefabregas, 1978) and the ancient finer-grained McMurray Formation, Canada (Mossop & Flach, 1983; Crerar & Arnott, 2007; Musial *et al.*, 2012). Although a number of assumptions were made regarding h_{mean} for the ancient systems, the predicted dip slopes generally agreed well with the reported dip (Table 3) or observed dip by the authors (Fig. 15B). This demonstrates that robust predictions of the dip slopes of point bar lateral-accretion elements and the associated shale drapes can be made from readily available parameters.

Table 3. Application of transverse bed slope predictor (Eq. 1) to a number of modern and ancient fluvial meandering systems. Assumed for all systems was a valley slope of 10^{-5} m/m, a Chézy number of 25 $m^{0.5}$/s, a κ of 0.4 and a g of 9.81 m/s^2. The first component of Eq. 1 ($9 \cdot \left(\frac{D}{h}\right)^{0.3}$) is often calibrated and generally between 0.5 and 1.5 (Talmon *et al.*, 1995). A value of 1.0 was assumed for this component for all fluvial systems. Data for the river Rhine, Upper Rhine Graben (Germany) were derived from Erkens *et al.* (2009); for the Daule river (Ecuador) from Smith (1987); for the Eocene Castisent Formation (Spain) from Nijman & Puigdefabregas (1978); and for the Cretaceous McMurray Fm (Canada) from Mossop & Flach (1983), Crerar & Arnott (2007) and Musial *et al.* (2012)

Parameter	Rhine	Daule	Castisent	McMurray
h_{mean}	8 m	12 m	8 m	25 m
Sediment	Gravel to sand	Sand-silt-mud	Gravel to sand	Sand-silt-mud
Typical grain size	0.5 mm	0.125 mm	0.25 mm	0.125 mm
$\tan\left(\frac{\delta z}{\delta n}\right)$	$0.37\frac{R}{h}$	$0.15\frac{R}{h}$	$0.26\frac{R}{h}$	$0.11\frac{R}{h}$
Typical $\frac{R}{h}$	125	40	100	40
Predicted dip	1–2°	8–9°	2°	13°
Observed dip	4–7°	12°	5°	8–12°

Application of archimetrics for reservoir characterization

The identified probability distributions on key archimetric parameters provide powerful new tools for geomodellers in characterizing reservoirs. Probability distributions allow geomodellers to obtain probabilities that reservoirs of certain dimensions or character may exist (Capen, 1992; Wood, 2004). This enables geomodellers to make quantitative judgements on the likelihood that their expert choices of sandbody dimensions and connectivity actually occur in the field – allowing for objective, probability-based, estimates of exploration success.

To assist geomodellers with making quantitative judgements on sandbody channel belt dimensions, probability density curves of the external geometry (Fig. 6) and internal architecture (Fig. 7) of meander belt sand bodies were constructed in this manuscript. Transforming these probability density curves into cumulative probability density curves allows for a probability-based assessment of the occurrence of sandbody dimensions. For example, the width of the meander belt sandbody generated with the numerical model has a P_{50} of 1320 m, with 80% of the sandbody width observations ranging between a P_{10} of 660 m and a P_{90} of 1820 m. Similarly, the thickness of the meander belt sandbody generated with the numerical model has a P_{50} of 6.4 m, with a less than 10% probability of finding a thickness greater than 9.7 m ($P_{90} = 9.7$ m) or less than 1.9 m ($P_{10} = 1.9$ m). Such simulated width and thickness dimensions are consistent

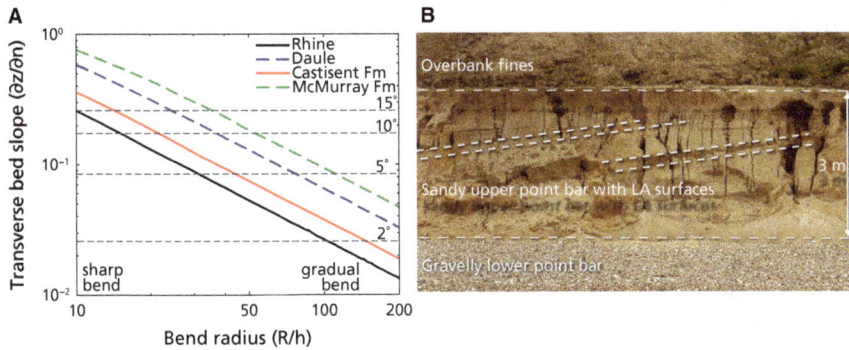

Fig. 15. Prediction of transverse bed slope for a number of modern and ancient fluvial meandering systems. (A) Transverse bed slope as a function of bend radius. (B) Section of a point bar deposit of the river Rhine in a quarry in Rheinberg, Germany (courtesy G. Erkens). The observed dip of the lateral-accretion surfaces varies between 4° and 7°, and is slightly higher than predicted from Eq. 1. The parameters used for the prediction of the transverse bed slopes are given in Table 3.

with the common ranges observed for natural preserved channel belt bodies of meandering rivers in the Holocene Rhine-Meuse delta (Gouw, 2008), the Holocene Lower Mississippi Valley (Aslan & Autin, 1999; Gouw & Berendsen, 2007; Gouw & Autin, 2008), and many other ancient fluvial meandering systems (Gibling, 2006; Rygel & Gibling, 2006). Notably, the width as well as the thickness of meander belt sand bodies are relatively easy to measure in outcrops and from three-dimensional seismic allowing knowledge of sandbody channel belt dimensions to be extrapolated from natural meandering systems to better inform geomodelling applications.

To generalize the findings of this study, normalized dimensionless sandbody dimensions and probability curves were derived for the flume experiment as well as the numerical model (Fig. 7). Dimensionless body dimensions and probability curves enable the results of this study to be used by geomodellers in characterizing reservoirs, if a number of key morphometrics such as typical channel depth and typical meander length of the target system are known. For example, the dimensionless ratio between the thickness of bar deposits and the depth of the meandering channel (i.e. the ratio between the preserved and original form) in the flume experiment has a P_{50} of 0·14, with 80% of the ratio observations ranging between a P_{10} of 0·003 and a P_{90} of 0·72. This means that for a natural target system with a typical channel depth of 10 m, 50% of the bar deposits would have a thickness of less than 1·4 m and 10% would have a thickness of more than 7·2 m, based on the probability curve derived from the flume experiment. Such a highly fragmented internal organization of meander belts is qualitatively substantiated by the identification of many architectural elements and hierarchy of bounding surfaces (Miall, 1985, 1988; Jordan & Pryor, 1992; Holbrook, 2001), emphasizing the heterogeneous nature of these deposits. Quantitative

support from natural systems for the highly fragmented internal organization of meander belts observed in this study arises from a study by Parker et al. (2013) in which repeated surface and subsurface surveys were used to establish how much of the barforms became truncated in response to a large flood. They found that unit bar deposits were truncated primarily in width (60%), in height (20%), and in length (32%) as compared with the formative bedform. With such a high degree of fragmentation of bar deposits in three dimensions following a single flood, the internal organization of a meander belt, which integrates the scouring and reworking effects of many floods, is likely to be composed of highly truncated deposits with fully preserved bar deposits and channel fills recorded although representing rare cases, as also quantified by the identified probability curves in this study (Fig. 7).

Archimetrics allow subsurface information to be combined with probability and statistics on reservoir dimensions and character, while taking reworking and partial preservation into account. Typically, a number of reservoir parameters (e.g. width and thickness of a sandstone body) are known in the development and production phases of meander belts serving as reservoirs. For the sake of simplicity and lacking information, geomodellers generally assume full preservation of geometrical body dimensions (e.g. point bar) derived from natural rivers to populate their geological models (Pranter et al., 2007). Combining the reservoir parameters with the identified probability curves and statistics on reworked and partially preserved sandbody dimensions and character in this study provides a set of rules to refine the geological models. In particular, the probability curves provide geomodellers with a powerful quantitative tool to characterize a range of geometrical bodies and spatial trends in preservation of these bodies, in addition to the identified information on the spatial distribution, dip angle, and continuity of flow-limiting shale drapes within point bar deposits. As a result, such refined geological models are anticipated to lead to more sophisticated lithology and porosity models, which ultimately cascades into better and more realistic flow simulations. The following scheme is proposed to maximize the applicability of the identified probability-based archimetrics:

1 Define the meander belt sandbody geometry (Fig. 1, sixth-order contacts). Two observables need to be estimated: average width and thickness of the sandbody. Typically, the width of the sandbody and its variation can be determined from seismic surveys. If not, the width should be estimated based on the thickness from well data and reported width-to-thickness ratios ranging from 100 to 200.

Well data are useful to quantify the meander belt thickness. This study indicates that most of the thickness variation is found laterally. Therefore, an effective strategy to quantify thickness variations, among other important stratigraphic and sedimentary parameters, is to drill wells along lateral transects.

A third parameter needed to quantify the sandbody geometry is the base relief. For the first time, this study provides a quantification of the base relief and shows that the relief is typically smaller than h_{mean}. h_{mean} can be estimated in two ways: from a fully preserved channel fill, which is most likely recovered close to the margins of a meander belt, or by estimating h_{mean} from the thickness of the meander belt sandbody. Observations in this study indicate that h_{mean} is equal to about one-third of the maximum sandbody thickness of the meander belt. This ratio was, however, obtained for a zero-net-depositional system and will be lower for net-depositional systems. Because the sedimentation conditions for most natural systems are unknown, the first approach to estimate h_{mean} is preferred.

2 Define the dimensions of the original meandering river. This is a crucial step because it allows one to connect the external geometry of the sandbody (Fig. 1, sixth-order contacts) to its internal architecture (Fig. 1, lower order contacts with a focus on third-order to fifth-order). Key morphometric parameters are h_{mean}, w_{mean}, L_{bend} and A_{bend} (Fig. 4). As stated in the previous step, h_{mean} is best estimated from a fully preserved channel fill. If such a channel fill is recovered, an estimate of w_{mean} can also be made. L_{bend} can be approximated in two ways. First, derive a typical L_{bend} from the undulating margins of meander belts for reservoirs with high-quality seismic data. Second, empirical relationships between L_{bend} and w_{mean} may be used for reservoirs with lower quality or no seismic data. Empirical relationships from modern rivers indicate that L_{bend}

is often equal to about 2·5–5 w_{mean} (Leopold & Wolman, 1957; Zeller, 1967). For easier field application, we approximated A_{bend} by 0·5 meander belt width. Dimensionless probability-based ratios between these key morphometric parameters and the identified archimetrics can now be used to further populate the reservoir (Fig. 7).

3 Use the archimetrics to build the internal sandbody architecture of meander belts (Fig. 1, third-order to fifth-order). As a first-order approximation of the dimensions of individual channel and bar architectural elements, a third of the original dimensions is reasonable. This indicates that a depositional body on average has a length of 0·3 L_{bend}, a width of 0·3 A_{bend} and a thickness of 0·3 h_{mean}.

Lateral architectural trends exist within the meander belt sandbody, with a general thickening of the deposits towards the margins of the meander belt. Therefore, further refinement of the internal sandbody architecture may be achieved by employing the entire dimensionless probability curves, although the number of available wells will determine the feasibility and confidence one can have in further refining the internal architecture. If a sufficient number of wells are available, it may be possible to estimate the degree of fragmentation from the number of stacked bar deposits. A higher number of stacked deposits is indicative of a higher degree of fragmentation, which logically results in shorter, narrower and thinner deposits.

4 Insert lateral-accretion surfaces within the individual channel and bar architectural elements. This implies that for every depositional body that is identified in the previous step an estimate of the presence, spatial distribution and dip of these surfaces needs to be made. Again, the number of wells defines to what level of detail this is possible. Currently, a robust and accurate depositional model predicting the spatial distribution of shale drapes acting as baffles or barriers to flow is lacking although our results suggest an increased likelihood of anisotropy and compartmentalization in upper outer point bar deposits, where shale drapes are best developed.

The dip angle and number of potentially flow-limiting lateral-accretion packages within point bar deposits can, however, be predicted robustly. A description of the parameters required to estimate the dip angle of the packages is given in Table 3. With this dip angle, the number of lateral-accretion packages within a point bar body can then be estimated from elementary geometrical considerations (Fig. 16). For example, for the McMurray Formation with a h_{mean} of 25 m, an average dip angle of about 10° and a typical R of 1000 m (Table 3), the average horizontal length of a lateral-accretion package is expected to be about 140 m. With a typical R of 1000 m, the point

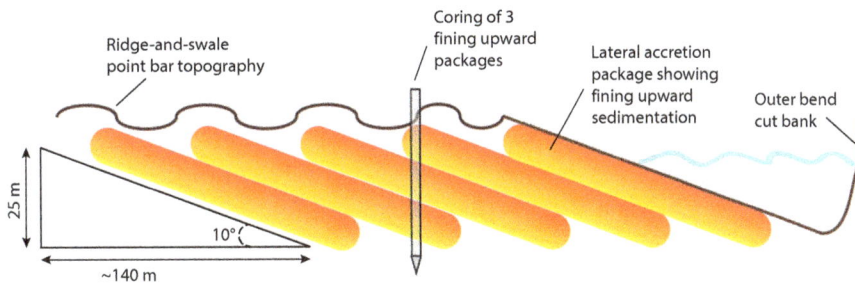

Fig. 16. Geometrical calculation of the number of potentially flow-limiting lateral-accretion packages within a point bar body. This example is based on the parameters derived for point bar deposits in the McMurray Formation with a typical channel depth of 25 m and a typical dip angle of the lateral-accretion packages of 10° (see Fig. 15 and Table 3). Based on these two parameters, a geometrical upper and lower bound prediction can be made on the number of flow-limiting packages a point bar body can host laterally and vertically. Note the vertical exaggeration of 2×.

bar body is expected to host a minimum of seven lateral-accretion packages and a maximum of about 15, depending on the degree of overlap applied between consecutive lateral-accretion packages. As a result of this lateral arrangement of the packages, two to four vertically stacked lateral-accretion packages are expected within this 25 m thick point bar body, where the number of vertically stacked packages depends on the position within the body and the lateral arrangement of the packages. These predictions are consistent with observations from photographs of point bar deposits of the McMurray Fm as seen in Musial *et al.* (2012, figures 8, 9 and 10) and supported by field observations made by author DB. Such straightforward geometrical considerations may therefore provide a simple and first-order estimate of the number of potentially flow-limiting lateral-accretion packages to be expected for a typical point bar in the field of interest, before gathering additional information to obtain a more detailed picture of the dimensions and distribution of these packages.

The presented archimetrics provide quantitative information on the external geometry and internal architecture of meander belt sand bodies. This is a first step towards a more quantitative understanding and characterization of meander belt sandbody architecture. A logical next step includes identification and characterization of truly three-dimensional architectural elements rather than providing probability curves for *x*, *y* and *z* independently. The latter approach lumps the architectural information on individual bodies such as channel fills and point bars into a single probability curve, while the former provides three-dimensional statistics describing individual architectural bodies. Also, Nicholas *et al.* (2016) present a simplified rule-based case to identify different populations associated with specific deposits such as dunes and unit bars within braided river deposits, though no attempt is made to quantify the different populations and their three-dimensional expression.

Object-based analysis (Deutsch & Tran, 2002; Novakovic *et al.*, 2002; Blaschke, 2010) can perhaps be extended to three dimensions to identify and discriminate between characteristics of channel fills, point bar sands, and point bar shales based on geometrical and sedimentary properties changing through time in experiments and models. Although computationally more demanding, this would provide for a more a sophisticated analysis and a further refinement of the quantitative characterization of meander belt sandbody architecture feeding into reservoir models.

CONCLUSIONS

Three flume experiments and a numerical model were used to quantitatively describe the three-dimensional sandbody architecture of meander belts. The direct linkage between causal meander morphodynamics and resultant sandbody architecture allows the preserved and truncated deposits to be related to the original non-truncated meander form in three dimensions. The dimensionless results are general and applicable in detailed reservoir architecture studies. A hierarchical approach of bounding contacts was adopted to analyse and characterize the external geometry and internal sandbody architecture of meander belts, referred to as archimetrics. Results demonstrate that:

The external sandbody geometry of a meander belt is typically described by width-to-thickness ratios between 100 and 200. The relief at the base of meander belt sand bodies is smaller than h_{mean}. These ratios facilitate improved estimates of connectivity between amalgamated meander belt deposits.

The internal sandbody architecture consists of depositional bodies with on average a thickness of about 30% of the mean channel depth, 30% of the mean length of a meander and 30% of the mean amplitude of a meander. Although fully preserved channel fills and meander bends are recovered, most of the internal architecture of meander belts consists of fragmented deposits.

The structure of the internal sandbody architecture is closely related to the meandering style. A general thickening of the deposits towards the margins of meander belts is seen although this trend is affected by processes such as, for example, neck-cut-off versus chute-cut-off and expansional-migrating versus downstream-migrating point bar style.

The transverse bed slope of the experimental lateral-accretion surfaces agrees well with a semi-empirical relationship. The application of this relationship with a number of natural modern and ancient meandering systems demonstrates that it predicts the slope and number of lateral-accretion packages within natural point bar deposits.

The identified probability distributions on key archimetric parameters provide powerful tools for geomodellers in characterizing reservoirs. A set of rules is outlined for building realistic three-dimensional reservoir models enabling geomodellers to make quantitative, probability-driven, judgements on the likelihood that their choices of sandbody dimensions occur in the field of interest.

ACKNOWLEDGEMENTS

We would like to thank two anonymous reviewers for comments that significantly improved the manuscript. WIvdL was supported by ExxonMobil Upstream Research Company (grant EM01734 to MGK and G. Postma). MGK and WMvD were supported by the Netherlands Organisation for Scientific Research (NWO) (grant ALW-VIDI-864.08.007 to MGK). We are grateful to G. Erkens and K.M. Cohen for discussion. The authors contributed in the following proportions to conception and study design, data collection, analysis and conclusions, and manuscript preparation: WIvdL (40, 60, 60, 90%), WMvD (10, 40, 10, 0%), DB (10, 0, 10, 0%), MGK (40, 0, 20, 10%).

References

Alexander, J., Bridge, J.S., Leeder, M.R., Collier, R.E.L.L. and Gawthorpe, R.L. (1994) Holocene meanderbelt evolution in an active extensional basin, southwestern Montana. *J. Sed. Res.*, **64**, 542–559.

Allen, J.R.L. (1978) Studies in fluviatile sedimentation: an exploratory quantitative model for the architecture of avulsion-controlled alluvial suites. *Sed. Geol.*, **21**, 129–147.

Asahi, K., Shimizu, Y., Nelson, J. and Parker, G. (2013) Numerical simulation of river meandering with self-evolving banks. *J. Geophys. Res. Earth Surf.*, **118**, 2208–2229. doi:10.1002/jgrf.20150.

Ashworth, P.J., Best, J.L., Peakall, J. and Lorsong, J.A. (1999) Influence of aggradation rate on braided alluvial architecture: field Study and physical scale-modelling of the Ashburton river gravels, Canterbury plains, New Zealand. In: *Fluvial Sedimentology VI* (Eds N.D. Smith and J. Rogers), pp. 333–346. Blackwell Publishing Ltd., Oxford, UK. doi: 10.1002/9781444304213.ch24.

Aslan, A. and Autin, W.J. (1999) Evolution of the Holocene Mississippi River Floodplain, Ferriday, Louisiana: insights on the origin of fine-grained floodplains. *J. Sed. Res.*, **69**, 800–815.

Best, J.L. and Ashworth, P.J. (1997) Scour in large braided rivers and the recognition of sequence stratigraphic boundaries. *Nature*, **387**, 275–277. doi:10.1038/387275a0.

Bhattacharya, J.P. and Tye, R.S. (2004) Searching for modern Ferron analogs and application to subsurface interpretation. In: *Analog for Fluvial–Deltaic Reservoir Modeling: Ferron Sandstone of Utah* (Eds T.C. Chidsey Jr., R.D. Adams and T.H. Morrison), *Am. Assoc. Petrol. Geol. Stud. Geol.*, **50**, 39–58.

Blaschke, T. (2010) Object based image analysis for remote sensing. *ISPRS J. Photogramm. Remote Sens.*, **65**, 2–16. doi:10.1016/j.isprsjprs.2009.06.004.

Bridge, J.S. and Best, J.L. (1997) Preservation of planar laminae due to migration of low-relief bed waves over aggrading upper-stage plane beds: comparison of experimental data with theory. *Sedimentology*, **44**, 253–262. doi:10.1111/j.1365-3091.1997.tb01523.x.

Bridge, J.S. and Leeder, M.R. (1979) A simulation model of alluvial stratigraphy. *Sedimentology*, **26**, 617–644.

Bridge, J.S. and Mackey, S.D. (1993) A revised alluvial stratigraphy model. In: *Alluvial Sedimentation* (Eds **M. Marzo, C.Puigdefabregas.**), *Int. Assoc. Sed. Spec. Publ.*, **17**, 319–336.

Bridge, J.S., Alexander, J., Collier, R.E.L., Gawthorpe, R.L. and Jarvis, J. (1995) Ground-penetrating radar and coring used to study the large-scale structure of point-bar deposits in three dimensions. *Sedimentology*, **42**, 839–852.

Bryant, M., Falk, P. and Paola, C. (1995) Experimental-study of avulsion frequency and rate of deposition. *Geology*, **23**, 365–368.

Capen, E. (1992) *The Business of Petroleum Exploration. AAPG Treatise of Petroleum Geology, Tulsa, United States, Dealing with Exploration Uncertainties*, pp. 29–62.

Crerar, E.E. and Arnott, R.W.C. (2007) Facies distribution and stratigraphic architecture of the Lower Cretaceous McMurray Formation, Lewis Property, northeastern Alberta. *Bull. Can. Pet. Geol.*, **55**, 99–124.

Deutsch, C.V. and Tran, T.T. (2002) FLUVSIM: a program for object-based stochastic modeling of fluvial depositional systems. *Comput. Geosci.*, **28**, 525–535. doi:10.1016/S0098-3004(01), 00075-9.

Donselaar, M.E. and Overeem, I. (2008) Connectivity of fluvial point-bar deposits: an example from the Miocene

Huesca fluvial fan, Ebro Basin, Spain. *AAPG Bull.*, **92**, 1109–1129. doi:10.1306/04180807079.

Eke, E., Parker, G. and Shimizu, Y. (2014) Numerical modeling of erosional and depositional bank processes in migrating river bends with self-formed width: morphodynamics of bar push and bank pull. *J. Geophys. Res. Earth Surf.*, **119**, 1455–1483. doi:10.1002/2013JF003020.

Engelund, F. and Hansen, E. (1967) *A Monograph on Sediment Transport in Alluvial Streams*. Teknisk Forlag, Kobenhavn, Denmark.

Erkens, G., Dambeck, R., Volleberg, K.P., Bouman, M.T.I.J., Bos, J.A.A., Cohen, K.M., Wallinga, J. and Hoek, W.Z. (2009) Fluvial terrace formation in the northern Upper Rhine Graben during the last 20000 years as a result of allogenic controls and autogenic evolution. *Geomorphology*, **103**, 476–495. doi:10.1016/j.geomorph.2008.07.021.

Frascati, A. and Lanzoni, S. (2013) A mathematical model for meandering rivers with varying width. *J. Geophys. Res. Earth Surf.*, **118**, 1641–1657. doi:10.1002/jgrf.20084.

Gibling, M.R. (2006) Width and thickness of fluvial channel bodies and valley fills in the geological record: a literature compilation and classification. *J. Sed. Res.*, **76**, 731–770. doi:10.2110/jsr.2006.060.

Gouw, M.J.P. (2008) Alluvial architecture of the Holocene Rhine-Meuse delta (The Netherlands). *Sedimentology*, **55**, 1487–1516. doi:10.1111/j.1365-3091.2008.00954.x.

Gouw, M.J.P. and Autin, W.J. (2008) Alluvial architecture of the Holocene Lower Mississippi Valley (U.S.A.) and a comparison with the Rhine-Meuse delta (The Netherlands). *Sed. Geol.*, **204**, 106–121. doi:10.1016/j.sedgeo.2008.01.003.

Gouw, M.J. and Berendsen, H.J.A. (2007) Variability of channel-belt dimensions and the consequences for alluvial architecture: observations from the Holocene Rhine-Meuse delta (The Netherlands) and Lower Mississippi Valley (USA). *J. Sed. Res.*, **77**, 124–138.

Hajek, E.A., Heller, P.L. and Sheets, B.A. (2010) Significance of channel-belt clustering in alluvial basins. *Geology*, **38**, 535–538. doi:10.1130/G30783.1.

Heller, P.L. and Paola, C. (1996) Downstream changes in alluvial architecture: an exploration of controls on channel-stacking patterns. *J. Sed. Res.*, **66**, 297–306.

Holbrook, J. (2001) Origin, genetic interrelationships, and stratigraphy over the continuum of fluvial channel-form bounding surfaces: an illustration from middle Cretaceous strata, southeastern Colorado. *Sed. Geol.*, **144**, 179–222.

Ielpi, A. and Ghinassi, M. (2014) Planform architecture, stratigraphic signature and morphodynamics of an exhumed Jurassic Meander plain Scalby Formation, Yorkshire, UK. *Sedimentology*, **61**, 1923–1960. doi:10.1111/sed.12122.

Jackson, R.G. (1976) Depositional model of point bars in the Wabash River. *J. Sed. Petrol.*, **46**, 579–594.

Jerolmack, D.J. and Mohrig, D. (2005) Frozen dynamics of migrating bedforms. *Geology*, **33**, 57–60.

Jordan, D.W. and Pryor, W.A. (1992) Hierarchical levels of heterogeneity in a Mississippi River Meander belt and application to reservoir systems. *AAPG Bull.*, **76**, 1601–1624.

Kleinhans, M.G. and Van den Berg, J.H. (2011) River channel and bar patterns explained and predicted by an empirical and a physics-based method. *Earth Surf. Proc. Land.*, **36**, 721–738. doi:10.1002/esp.2090.

Kleinhans, M.G., van Dijk, W.M., van de Lageweg, W.I., Hoendervoogt, R., Markies, H. and Schuurman, F. (2010) *From Nature to Lab: Scaling Self-Formed Meandering and Braided Rivers*. River.ow 2010, volume 2, edited by Dittrich, Koll, Aberle and Geisenhainer, pp. 1001–1010. Bundesanstalt fur Wasserbau.

Kleinhans, M.G., Haas, T.D., Lavooi, E. and Makaske, B. (2012) Evaluating competing hypotheses for the origin and dynamics of river anastomosis. *Earth Surf. Proc. Land.*, **37**, 1337–1351. doi:10.1002/esp.3282.

Kleinhans, M.G., van Dijk, W.M., van de Lageweg, W.I., Hoyal, D.C.J.D., Markies, H., van Maarseveen, M., Roosendaal, C., van Weesep, W., van Breemen, D., Hoendervoogt, R. and Cheshier, N. (2014) Quantifiable effectiveness of experimental scaling of river- and delta morphodynamics and stratigraphy. *Earth-Sci. Rev.*, **133**, 43–61. doi:10.1016/j.earscirev.2014.03.001.

Lanzoni, S. and Seminara, G. (2006) On the nature of meander instability. *Water Resour. Res.*, **111**, F04006.

Leclair, S.F. (2002) Preservation of cross-strata due to the migration of subaqueous dunes: an experimental investigation. *Sedimentology*, **49**, 1157–1180.

Leeder, M. (1978) A quantitative stratigraphic model for alluvium, with special reference to channel deposit density and interconnectedness. In: *Fluvial Sedimentology* (Ed. A.D. Miall), *Bull. Can. Soc. Petrol. Geol. Mem.*, **5**, 587–596.

Leopold, L.B. and Wolman, M.G. (1957) River channel patterns: braided, meandering and straight. U.S. *Geol. Survey Prof. Paper*, **282-B**, 39–85.

Lewin, J. and Macklin, M.G. (2003) Preservation potential for Late Quaternary river alluvium. *J. Quatern. Sci.*, **18**, 107–120.

Mackey, S.D. and Bridge, J.S. (1995) Three-dimensional model of alluvial stratigraphy: theory and application. *J. Sed. Res.*, **B65**, 7–31.

McGowen, J.H. and Garner, L.E. (1970) Physiographic features and stratification types of coarse-grained point bars: modern and ancient examples. *Sedimentology*, **14**, 77–111.

Miall, A.D. (1985) Architectural-element analysis: a new method of facies analysis applied to fluvial deposits. *Earth Sci. Rev.*, **22**, 261–308.

Miall, A.D. (1988) Reservoir heterogeneities in fluvial sandstones: lessons from outcrop studies. *AAPG Bull.*, **72**, 682–697. doi:10.1306/703C8F01-1707-11D7-8645000102C1865D.

Mossop, G.D. and Flach, P.D. (1983) Deep channel sedimentation in the Lower Cretaceous McMurray Formation, Athabasca oil Sands, Alberta. *Sedimentology*, **30**, 493–509.

Musial, G., Reynaud, J., Gingras, M.K., Fnis, H., Labourdette, R. and Parize, O. (2012) Subsurface and outcrop characterization of large tidally influenced point bars of the Cretaceous McMurray Formation (Alberta, Canada). *Sed. Geol.*, **279**, 156–172. doi:10.1016/j.sedgeo. 2011.04.020.

Nanson, G.C. (1980) Point bar and floodplain formation of the meandering Beatton River, northeastern British Columbia, Canada. *Sedimentology*, **27**, 3–29.

Nicholas, A.P., Sambrook Smith, G.H., Amsler, M.L., Ashworth, P.J., Best, J.L., Hardy, R.J., Lane, S.N., Orfeo, O., Parsons, D.R., Reesink, A.J.H., Sandbach, S.D., Simpson, C.J. and Szupiany, R.N. (2016) The role of discharge variability in determining alluvial stratigraphy. *Geology*, **44**, 3–6. doi:10.1130/G37215.1.

Nijman, W. and Puigdefabregas, C. (1978) *Fluvial Sedimentology*. Canadian Society of Petroleum Geologists Memoir, Coarse-Grained Point Bar Structure in a Molasse-Type Fluvial System, Eocene Castisent Sandstone Formation, South Pyrenean Basin, **5**, 487–510.

Novakovic, D., White, C.D., Corbeanu, R.M., Hammon, W.S., III, Bhattacharya, J.P. and McMechan, G.A. (2002) Hydraulic effects of shales in fluvial-deltaic deposits: ground-penetrating radar, outcrop observations, geostatistics, and three-dimensional flow modeling for the Ferron Sandstone, Utah. *Math. Geol.*, **34**, 857–893. doi:10.1023/A:1020980711937.

Paola, C. and Borgman, L. (1991) Reconstructing random topography from preserved stratification. *Sedimentology*, **38**, 553–565.

Parker, N.O., Sambrook Smith, G.H., Ashworth, P.J., Best, J.L., Lane, S.N., Lunt, I.A., Simpson, C.J. and Thomas, R.E. (2013) Quantification of the relation between surface morphodynamics and subsurface sedimentological product in sandy braided rivers. *Sedimentology*, **60**, 820–839. doi:10.111/j.1365-3091.2012.01364.x.

Peakall, J., Ashworth, P.J. and Best, J.L. (2007) Meander-bend evolution, alluvial architecture, and the role of cohesion in sinuous river channels: a flume study. *J. Sed. Res.*, **77**, 197–212.

Pranter, M.J., Ellison, A.I., Cole, R.D. and Patterson, P.E. (2007) Analysis and modeling of intermediate-scale reservoir heterogeneity based on a fluvial point-bar outcrop analog, Williams Fork Formation, Piceance Basin, Colorado. *AAPG Bull.*, **91**, 1025–1051. doi:10.1306/ 02010706102.

Puigdefabregas, C. and Van Vliet, A. (1978) Meandering stream deposits from the Tertiary of the Southern Pyrenees. In: *Fluvial Sedimentology* (Ed. A.D. Miall), *Bull. Can. Soc. Petrol. Geol. Mem.*, **5**, 543–576.

Rygel, M.C. and Gibling, M.R. (2006) Natural geomorphic variability recorded in a high-accomodation setting: fluvial architecture of the Pennsylvanian Joggins Formation of Atlantic Canada. *J. Sed. Res.*, **76**, 1230–1251.

Schuurman, F., Kleinhans, M.G. and Marra, W.A. (2013) Physics-based modeling of large braided sand-bed rivers: bar pattern formation, dynamics, and sensitivity. *J. Geophys. Res. Earth Surf.*, **118**, 2509–2527. doi:10.1002/ 2013JF002896.

Schuurman, F., Shimizu, Y., Iwasaki, T. and Kleinhans, M. (2015) Dynamic perturbation in response to upstream perturbation and floodplain formation. *Geomorphology*, **253**, 94–109. doi:10.1016/j.geomorph.2015.05.039.

Sheets, B.A., Hickson, T.A. and Paola, C. (2002) Assembling the stratigraphic record: depositional patterns and time-scales in an experimental alluvial basin. *Basin Res.*, **14**, 287–301. doi:10.1046/j.1365-2117.2002.00185.x.

Smith, D.G. (1987) Recent developments in fluvial sedimentology. *Society of Economic Paleontologists and Mineralogists Special Publication*, Tulsa, Meandering River Point Bar Lithofacies Models: Modern and Ancient Examples Compared, **39**, 83–91.

Storms, J.E.A., Dam, R.L.V. and Leclair, S.F. (1999) Preservation of cross-sets due to migration of current ripples over aggrading and non-aggrading beds: comparison of experimental data with theory. *Sedimentology*, **46**, 189–200. doi:10.1046/j.1365-3091.1999.00212.x.

Straub, K.M., Paola, C., Mohrig, D.C., Wolinsky, M.A. and George, T. (2009) Compensational stacking of channelized sedimentary deposits. *J. Sed. Res.*, **79**, 673–688. doi:10.2110/ jsr.2009.070.

Struiksma, N. (1985) Prediction of 2-D bed topography in rivers. *J. Hydraul. Eng.*, **111**, 1169–1182.

Talmon, A.M., Struiksma, N. and van Mierlo, M.C.L.M. (1995) Laboratory measurements of the direction of sediment transport on transverse alluvial-bed slopes. *J. Hydraul. Res.*, **33**, 495–517.

Tyler, N. and Finley, R. (1991) The three-dimensional facies architecture of terrigenous clastic sediments and its implications for hydrocarbon discovery and recovery. *SEPM Concepts in Sedimentology and Paleontology. Architectural Controls on the Recovery of Hydrocarbons from Sandstone Reservoir*, **3**, 1–5.

Van de Lageweg, W.I., Van Dijk, W.M. and Kleinhans, M.G. (2013a) Morphological and stratigraphical signature of floods in a braided gravel-bed river revealed from flume experiments. *J. Sed. Res.*, **83**, 1032–1045. doi:10.2110/ jsr.2013.70.

Van de Lageweg, W.I., Van Dijk, W.M. and Kleinhans, M.G. (2013b) Channel belt architecture formed by a meandering river. *Sedimentology*, **60**, 840–859. doi:10.1111/j.1365-3091.2012.01365.x.

Van de Lageweg, W.I., Van Dijk, W.M., Baar, A.W., Rutten, J. and Kleinhans, M.G. (2014) Bank pull or bar push: what drives scroll-bar formation in meandering rivers? *Geology*, **42**, 319–322. doi:10.1130/G35192.1.

Van de Lageweg, W.I., Schuurman, F., Cohen, K.M., Van Dijk, W.M., Shimizu, Y. and Kleinhans, M.G. (2015) Preservation of meandering river channels in uniformly aggrading channel belts. *Sedimentology*, **63**, 586–608.

Van Dijk, W.M., Van de Lageweg, W.I. and Kleinhans, M.G. (2012) Experimental meandering river with chute cutoffs. *J. Geophys. Res. Earth Surf.*, **117**, F03023. doi:10.1029/2011JF002314.

Van Dijk, W.M., Teske, R., Van de Lageweg, W.I. and Kleinhans, M. (2013a) Effects of vegetation distribution on experimental river channel dynamics. *Water Resour. Res.*, **49**, 7558–7574. doi:10.1002/2013WR013574.

Van Dijk, W.M., Van de Lageweg, W.I. and Kleinhans, M. (2013b) Formation of a cohesive floodplain in a dynamic experimental meandering river. *Earth Surf. Proc. Land.*, **38**, 1550–1565. doi:10.1002/esp.3400.

Van Rijn, L.C. (1984) Sediment transport, part III: bed forms and alluvial roughness. *J. Hydraul. Eng.*, **110**, 1733–1754.

White, C.D. and Barton, M.D. (1999) Translating outcrop data to flow models, with applications to the Ferron sandstone. *SPEREE J.*, **2**, 341–350.

Willis, B.J. and Tang, H. (2010) Three-dimensional connectivity of point-bar deposits. *J. Sed. Res.*, **80**, 440–454.

Willis, B.J. and White, C.D. (2000) Quantitative outcrop data for.ow simulation. *J. Sed. Res.*, **70**, 788–802.

Wood, L.J. (2004) Predicting tidal sand reservoir architecture using data from modern and ancient depositional systems. Integration of outcrop and modern analogs in reservoir modeling. *AAPG Mem.*, **80**, 45–66.

Yang, C.S. and Sun, U. (1988) Tide-influenced sedimentary environments and facies. In: *Tidal Sand Ridges on the East China Sea Shelf* (Ed. D. Boston), pp. 23–38. Reidel Publishing, Dordrecht.

Zeller, J. (1967) Meandering channels in Switzerland. *Int. Assoc. Sci. Hydrol. Publ.*, **75**, 174–186.

Zolezzi, G., Luchi, R. and Tubino, M. (2012) Modeling morphodynamic processes in meandering rivers with spatial width variations. *Rev. Geophys.*, **50**, RG4005. doi:10.1029/2012RG000392.

8

Climatically forced moisture supply, sediment flux and pedogenesis in Miocene mudflat deposits of South-East Kazakhstan, Central Asia

SILKE VOIGT* (iD), YUKI WEBER*,† (iD), KONSTANTIN FRISCH*, ALEXANDER BARTENSTEIN*, ALEXANDRA HELLWIG*, RAINER PETSCHICK*, ANDRÉ BAHR*,‡, JÖRG PROSS*,‡, ANDREAS KOUTSODENDRIS‡ (iD), THOMAS VOIGT§, VERENA VERESTEK¶ and ERWIN APPEL¶ (iD)

*Institute of Geosciences, Goethe University Frankfurt, Altenhöferallee 1, 60438 Frankfurt, Germany (E-mail: s.voigt@em.uni-frankfurt.de)
†Department of Earth and Planetary Sciences, Harvard University, 20 Oxford Street, Cambridge, MA 02138, USA
‡Institute of Earth Sciences, Heidelberg University, Im Neuenheimer Feld 234-236, 69120 Heidelberg, Germany
§Institute of Geosciences, Friedrich Schiller University Jena, Burgweg 11, 07749 Jena, Germany
¶Institute of Geosciences, University Tübingen, Sigwartstrasse 10, Hölderlinstrasse 12, 72074 Tübingen, Germany

Keywords
Central Asia, climate, Miocene, pedogenesis, sediment flux, terrestrial sedimentation.

ABSTRACT

The continental settings of Central Asia witnessed increased desertification during the Cenozoic as a result of mountain uplift and the Paratethys retreat. The interaction of these tectonic-scale processes with orbitally forced climate change and their influence on Asia's atmospheric moisture distribution are poorly constrained. A Miocene succession of continental mudflat deposits, exposed in the Aktau Mountains (Ili Basin, south-east Kazakhstan), has great potential as a terrestrial palaeoclimate archive. About 90 m of the 1700 m thick succession comprise alluvial mudflat deposits and appear as cyclic alternation of coarse sheet floods, mudflat fines and semi-arid hydromorphic soils. In this study, bulk-sediment mineralogy and geochemistry, magnetic susceptibility, sediment colour and palynology are used to reconstruct environmental conditions by determining changes and forcing mechanisms in the intensity of sediment discharge, weathering and pedogenesis. The results presented here indicate four major periods of arid soil formation and one palustrine interval characterized by higher evaporation rates under highly alkaline/saline conditions. A positive correlation between weathering indices and the Mg/Al ratio suggest that these horizons correspond to maximum rates of evapotranspiration and aridity. The formation of mudflat fines is, instead, interpreted as representing higher detrital sediment production by more intense alluvial fan activity during times of higher precipitation. Time series analysis of weathering indices, colour and magnetic susceptibility data yields cycle-to-frequency ratios with the potential to represent Milankovitch cyclicity with short and long eccentricity as dominant periodicities. Periods of pronounced aridity, paced by long eccentricity forcing, reflect changes in moisture availability. On longer tectonic timescales, the persistent appearance of gypsum indicates a shift towards more arid conditions. This trend in climate is considered to result from the closure of the eastern gateway of the Mediterranean to the Indian Ocean that restricted circulation and enhanced salinity within the Eastern Paratethys.

INTRODUCTION

The evolution of Cenozoic climate is characterized by global cooling, increased meridional temperature gradients and the expansion of polar ice sheets in Antarctica and the Northern Hemisphere since ca 35 Ma and ca 15 Ma, respectively (Zachos et al., 2008; De Vleeschouwer et al., 2017). The mechanisms behind this global cooling, its regional differentiation and the feedbacks involved are still a matter of debate. Continental settings of Central Asia witnessed increased desertification and the establishment of a monsoonal climate during the Cenozoic as a result of India's collision with Asia and the Paratethys retreat (Molnar & Tapponnier, 1975; Ramstein et al., 1997). However, the timing of this continent-scale climate shift relative to global climate evolution, the interplay between regional and global factors and the effects of orbital-scale processes are not yet well constrained.

Based on loess deposits in China, the existence of energetic winter monsoon winds and large source areas for aeolian dust in the interior of Asia has been traced back to 22 Ma (Guo et al., 2002). Alternatively, desert areas in inner Asia north of the uplifting Pamir and Tian Shan mountain chains may have been mainly influenced by westerly wind flow since Eocene to Oligocene times (Sun et al., 2010; Caves et al., 2015). A variety of proxy records suggests a temporally differentiated pattern for the onset of desertification in Central Asia, ranging from the Eocene/Oligocene transition in north-east Tibet and south-western Mongolia (Dupont-Nivet et al., 2007; Sun & Windley, 2015) to the mid-to-late Miocene north of Tibet (Dettman et al., 2003; Kent-Corson et al., 2009; Sun et al., 2015), and the mid-Pliocene on the Chinese Loess Plateau (Wang, 2006). Mammal diversity changes in Oligocene–Miocene successions in Mongolia provide evidence for intermittent episodes of increased precipitation (Harzhauser et al., 2016) and the aeolian origin of the Valley of Lakes successions was questioned by results of clay mineralogy (Richoz et al., 2017). The relative intensities of the westerlies and monsoonal wind systems played an important role in transporting moisture into Asia's continental interior (Caves et al., 2015). Climate modelling results suggest reduced moisture transport to inner Asia by weakened westerlies and monsoonal winds after the global shift to cooler climate conditions in the Oligocene (Licht et al., 2014).

The mid-Miocene (17 to 14 Ma) was one of the last warm periods of the Neogene (Zachos et al., 2008; Holbourn et al., 2014, 2015). While the proxy evidence for a warm and relatively humid mid-Miocene world is clear, the mechanisms responsible for this climate state are not. Atmospheric pCO_2 variations are supposed to drive changes in the global carbon reservoirs, implying changing rates of silicate weathering and global carbon sequestration (Holbourn et al., 2015). A factor recently invoked to explain mid-Miocene warmth is a lower continental topography than today promoting a more zonal atmospheric circulation with a westerly flow over lowered mid-latitude plateaus (Henrot et al., 2010). However, available proxy data yield somewhat contradictory climate scenarios for the mid-Miocene of Central Asia. While

records from Mongolia and China indicate increased desertification since Oligocene to early Miocene times (Guo et al., 2002; Sun & Windley, 2015), the regionally widespread formation of lacustrine deposits in eastern/south-eastern Kazakhstan and the Tarim Basin during the Miocene, described as "the great lacustrine stage", suggest increased atmospheric moisture transport to Central Asia (Akhmetyev et al., 2005; Liu et al., 2014). Palynological data of mid-Miocene age indicate warm-temperate conditions for the Junggar Basin and the north-east Tibetan Plateau (Hui et al., 2011; Miao et al., 2011; Tang et al., 2011b) pointing towards a transient episode of increased humidity.

A terrestrial alluvial floodplain succession of mid-Miocene age, exposed in the Aktau Mountains of the Ili Basin, south-eastern Kazakhstan, has the potential to provide insights into the Miocene climate evolution in Central Asia (Fig. 1). In this study, bulk-sediment mineralogy and geochemistry (element geochemistry, $CaCO_3$/$CaSO_4$ content), magnetic susceptibility (MS), sediment colour and palynological data are used to decipher the regional response of sedimentary particle supply, chemical weathering intensity and pedogenesis to changes in regional moisture supply by precipitation and subsurface aquifer recharge. Furthermore, the results provide insights into climate and environmental conditions in the context of atmospheric moisture transport to Central Asia. Time series analysis of chemical weathering indices, (MS) and sediment colour data are used to decipher potential orbital forcing mechanisms.

GEOLOGICAL SETTING

The Ili Basin is a closed (endorheic) basin within the intracontinental Tian Shan mountain system. It is surrounded by continuously uplifting mountain ranges of the Zailiysky and Dzungarian Alatau and became progressively contracted by N-S shortening and fragmented due to the activation of intrabasinal basement uplifts (Kober et al., 2013; Macaulay et al., 2014). The basin fill covers a time span from the late Eocene to present. The Tian Shan mountain ranges grew as a result of India's collision with Asia with current crustal shortening (ca 20 mm y^{-1}) accounting for nearly half of India's convergence with Eurasia (Abdrakhmatov et al., 1996). Exhumation and unroofing ages indicate the initial uplift of the Tian Shan to have occurred in the Oligocene to early Miocene (Hendrix et al., 1994; Sobel et al., 2003, 2006; Macaulay et al., 2014). During these times, the central Ili Basin mainly accommodated distal, low-energy sediments on a regionally extensive peneplain while the basin margins were accompanied by alluvial fans (Kober et al., 2013).

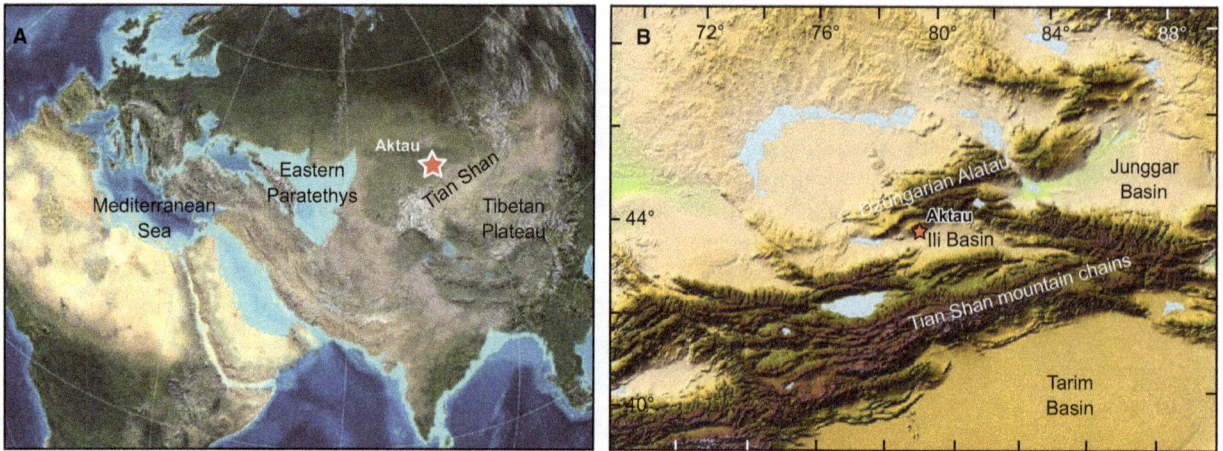

Fig. 1. Map showing the position of the Aktau succession (A) on a palaeogeographic reconstruction of the Miocene (Deep Time Maps™) and (B) on the present-day topography (Amante & Eakins, 2009).

In the Aktau Mountains, a Neogene exposure of Ili Basin sediments south of the Dzungarian Alatau (44°0′9.58″N, 79°14′56.94″E; Fig. 1), a 1700 m thick succession of fine-grained deposits forms an asymmetric anticline with a steeply to vertical dipping southern limb and a gently dipping northern flank (Bazhanov & Kostenko, 1961). Alluvial sediments of Pleistocene age disconformably overlie the succession and are involved in very young folding activity. Aktau means "White Mountains" in the Kazakh language and the succession is characterized by spectacular colour banding of its deposits (Fig. 2). The overall succession and its facies have been documented in a series of earlier studies and expose a quasi-continuous Eocene/Oligocene to Pliocene terrestrial record (Bazhanov & Kostenko, 1961; Bodina, 1961; Lucas et al., 1997; Kordikova & Mavrin, 1996). Previous authors have introduced diverging lithostratigraphic schemes that are summarized by Kober et al. (2013). Here, the formation names of Bodina (1961) are followed together with thicknesses given by Bazhanov & Kostenko (1961) and Kordikova & Mavrin (1996), as both yield the best agreement with field observations (Fig. 3).

The succession is based in red-coloured clays and sandstones of a river system (Arasan Formation, 63 m) that grade into reddish to brown-red mud-dominated deposits, which contain distal alluvial, meandering river deposits and gypsum beds of a saline mudflat with evidence of an ephemeral playa lake (Alakul Fm, 115 m). Higher up the succession, a transition to cross-bedded fluvial sandstones of a meandering river occurs above a significant disconformity (Aidarly Fm, 130 to 140 m) overlain by cyclically bedded, reddish-brown mudflat deposits (Bastau Fm, 90 m). The upper part of the succession consists of greenish grey gypsisols and ephemeral playa lake deposits, as well as perennial lacustrine limestones with freshwater

charophytes, ostracods and gastropods, and intercalated coal seems (Koktal and Kokterek Fms, 460 m). The top of the succession is represented by silty mudflat deposits with intercalations of lacustrine, fossil-rich fresh water limestones and channel sandstones indicative of a permanent river system with an adjacent fresh water lake (Ili Fm, ca 880 m; Fig. 2). Parent rocks for the alluvial plain deposits of the Arasan, Alakul, Bastau and Koktal formations are Permo-Carboniferous volcanics exposed in the Katutau and Dzungarian Alatau mountain ranges at the northern edge of the Ili Basin today. They comprise andesites, rhyolites and trachytes that belong to the Palaeozoic accretionary arc complex of the Central Asian Orogenic belt (Jahn et al., 2000, 2004; Li et al., 2015). The provenance of major parts of the basin fill from these sources is proven by proximal alluvial fan deposits close to the basin margins. In contrast, sandstones of the Arasan, Aidarly and Ili formations were mainly derived from a distant quartz-rich and mica-rich source often mixed with volcanics derived from local sources (Lucas et al., 1997; Kober et al., 2013).

Biostratigraphic ages are available from the fluvial Arasan and Aidarly formations and lacustrine floodplain deposits of the Ili Formation. Dating of the late Eocene (Ergilian) Arasan Formation is based on mammal bones of Brontotheriidae and the hyracodontid Ardynia sp. (Gromova, 1952; Lucas et al., 1997). The lower Aidarly Formation is of late Oligocene age based on occurrences of the giant rhinoceros Indricotherium (Lucas et al., 1997), while the upper Aidarly Formation is placed into the late Burdigalian to Langhian mammal zones MN4 to MN5 based on records of plants and mammals (rodents, carnivores, insectivores, the odd-toed Gomphotherium and early deers such as Stephanocemas and Lagomeryx) (Fig. 3; Lucas et al., 1997; Kordikova, 2000; Kordikova & de Bruijn, 2001). The lower Ili Formation yields charophytes

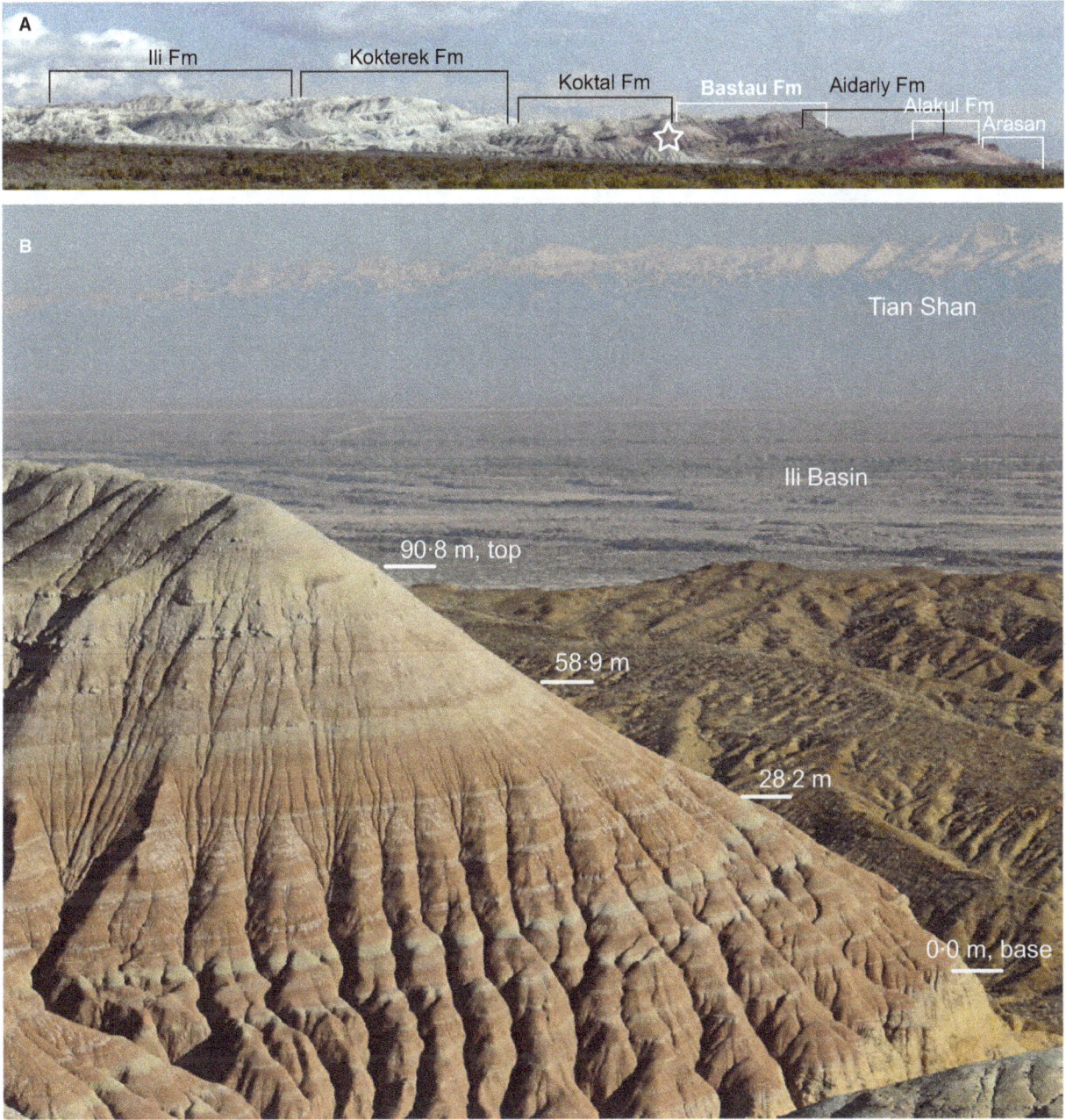

Fig. 2. Outcrop images showing the Aktau succession. (A) View of the western flank of the Aktau Hills with the range of lithostratigraphic formations. Length of exposure is ca 5 km. The hill in (B) is marked by a star. (B) Southward view from the studied Bastau Formation (Fm) with the Ili Basin and Tian Shan Mountains in the background.

typical for the late Miocene to early Pliocene (Dzhaman-garaeva, 1997). Early palaeontological excavations in the Ili Formation provided records of the gomphotere *Anancus avernensis*, an extinct elephant living 7 to 1·8 Ma ago (Bazhanov & Kostenko, 1961). Based on these biostratigraphic data, the age of the here studied Bastau Formation is constrained to the mid-Miocene between 17 and 14 Ma (Fig. 3).

METHODS

During field work in 2011 and 2015, the 91 m thick Bastau Formation was logged in centimetre-to-decimetre-scale resolution. Direct measurements of bed-thickness were adjusted to the total thickness of the sedimentary succession determined with a laser distance and angle meter to correct for erroneous thickness measurements related to variable slope

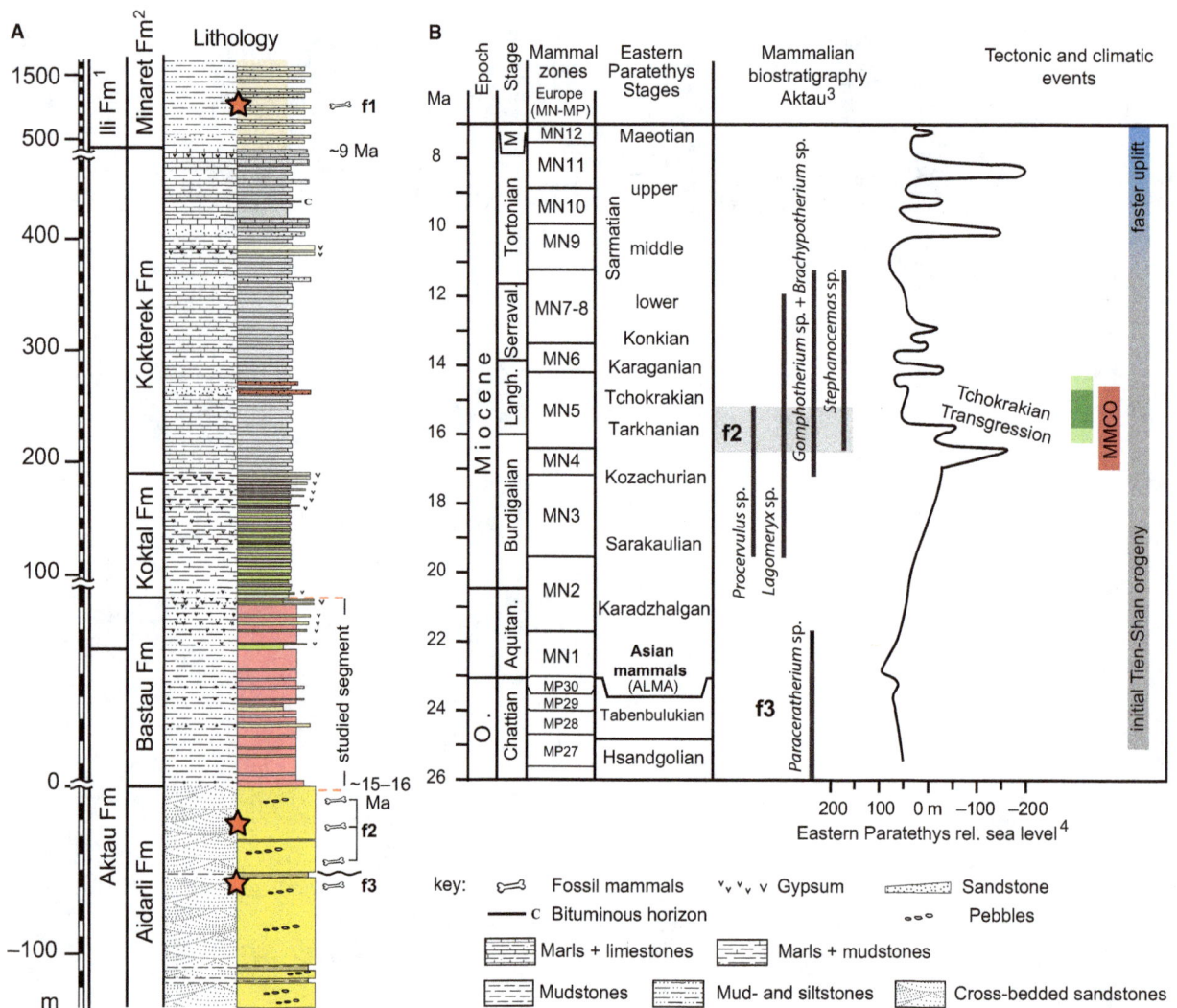

Fig. 3. Stratigraphic overview of parts of the Aktau Mountains succession with (A) the lithology and position of biostratigraphic marker beds (stars) in the Miocene (f2, f3, Lucas *et al.*, 1997) and Late Miocene to Pliocene (f1, Bazhanov & Kostenko, 1961; Dzhamangaraeva, 1997), and (B) an overview of regional divisions, biozonations, tectonic and climate events in Europe and Central Asia and the estimated age for the Bastau Formation (green bar). Ages are after GTS12. Lithostratigraphic formations are from Bazhanov & Kostenko (1961)[1] and Bodina (1961)[2]. Mammalian biostratigraphy[3] is from Bazhanov & Kostenko (1961) and Lucas *et al.* (1997), and the relative Paratethys sea-level curve[4] is after Popov *et al.* (2010). Uplift ages of the Tian Shan are from Sobel *et al.* (2003, 2006) and Macaulay *et al.* (2014). MMCO, Middle Miocene Climatic Optimum.

angles. The succession was sampled in 20 to 25 cm spacing for bulk-sediment geochemistry. Potentially prospective strata were also sampled for palynology. Sediment colour scans were performed in 5 to 8 cm spacing with a Konica Minolta CM-700d spectrophotometer, and volume-specific magnetic susceptibility (MS) was measured every 10 cm with a hand-held SM 30 magnetic susceptibility meter (ZH Instruments, Brno, Czech Republic).

Element geochemistry

To account for the different components of Bastau Formation sediments (silicates, carbonates, sulphates and

salts) and to isolate the silicate fraction, the geochemical composition of powdered sediment samples was analysed in two steps, by separating the acetic acid-leachable fraction from the non-leachable fraction. Following the method of Goldberg & Humayun (2010) 400 to 600 mg of sample powder was treated with 4·5 ml of 50% acetic acid at 50 to 75°C for *ca* 18 h; 250 μl of the leached solvent was then analysed by ICP-OES. The remaining sediment was repeatedly washed to ensure complete removal of dissolved ions from the sediment pore water. The 100 mg of decalcified sample powder was then digested in HF-enhanced agua regia with the microwave Multiwave 3000™ by Anton Paar.

An internal reference material and a blank were included in each session.

Geochemical analyses were carried out by ICP-OES on an iCAP6300 Duo™ by Thermo Scientific. Matrix concentrations were reduced by diluting the primary preparation solutions by a factor of three. Yttrium was added to all samples as an internal reference. Calibration solutions were prepared separately for the leached and bulk-digested samples using certified single and multi-element standards (SPEXCertiPrep). Acetic acid-leached and bulk samples were then analysed in separate sessions. Reproducibility of repeated sample and standard measurements was within 5% (2σ) for most elements.

The weight loss caused by leaching of the rock powders ranged from 3·8% to 75%. As not only calcite was removed from the sediment, the carbonate content (calcite) was calculated by using the Ca concentration in the leachable fraction (Calcite [wt%] = mCa*100·1). Gypsum abundances were assessed using the sulphur content of totally dissolved samples, which show a strong correlation of sulphate (SO_4) and Ca for sulphate concentrations above 1 wt%. The content of gypsum was determined by assuming that all sulphur in the non-leachable fraction is related to gypsum (Gypsum [wt%] = mS (silicate) * 172·14). Some samples experienced a mass loss during the washing process after leaching. Stoichiometric calculations of the Ca content in the leachable fraction and the bulk sediment show the mass loss to be related to dissolution of gypsum. This mass loss has been added to the calculated gypsum content from the non-leachable fraction. Element concentrations of the non-leachable fraction are used to calculate indices of chemical weathering (see below).

Grain size and sediment mineralogy

Grain-size and X-ray diffractometry (XRD) analyses were performed on a subset of samples of the Bastau ($n = 14$) and Alakul formations ($n = 10$) and from stratigraphic equivalent horizons ($n = 18$) from the more proximal Kendyrlisay Valley section (Hellwig et al., 2017). For grain-size analyses, 20 mg of each sample was decalcified with 20% formic acid, and after neutralization and homogenization wet sieved to remove grain sizes >100 μm. The fine fraction (<100 μm) was held in suspension in a Na-polyphosphate solution before 5 ml of the suspension was measured using a HORIBA LA-950 laser particle analyzer.

For XRD analyses, powdered rock samples were mounted on sample holders using the back-loading technique to reveal poor orientation and texturation. The measurements were performed on a PANALYTICAL X'Pert Bragg-Brentano diffractometer, using a copper beam powered by 30 mA and 40 KV generator current, Ni filter, programmable divergence slit, sample spinning and X'Celerator 1D detector. The characteristic diffraction maxima of each identified mineral phase was determined using MacDiff software. Intensities were converted to fixed 1° divergence characteristics and weighted by reference intensity ratios (RIR) to calculate the relative contribution from each mineral phase. Samples with a clay mineral composition containing palygorskite and/or mixed layer structures with expandable layers were treated for 24 h with ethylene glycol to aid in the identification of such phases. Palygorskite needles were also identified by Scanning Electron Microscopy.

The following phases were detected (in brackets: main diffraction maxima positions and RIR value as used for semiquantitative data calculation): mixed layer illite–smectite (around 12 to 13 Å, 0·4), palygorskite (10·35 Å, 0·52), illite/muscovite (10 Å, 0·43), chlorite/clinochlore (14, 7·1, and 3·54 Å, 1·0), quartz (4·26 and 3·34 Å, 3·03), K-feldspar (3·23 to 3·25 Å, 0·6), albite (3·18 to 3·2 Å, 0·64), calcite (3·04 Å, 3·32), ankerite (2·91, 3·15), dolomite (2·9 Å, 2·51), gypsum (7·6 Å, 1·7), halite (2·82 Å, 4·71).

Scanning electron microscopy (SEM)

Single samples from the Bastau Formation were prepared for SEM with a Zeiss Sigma VP. The suspended sediment was mounted on a slice, dried at 40° and afterwards sputtered with platinum. SEM microscopy was performed with a voltage of 10 to 15 kV.

Palynology

Six samples from the Bastau Formation were processed for palynological analysis using standard techniques previously applied to lake sediments from Central Asia including freeze-drying, weighing, HCl (30%) and HF (40%) treatment, and sieving through a 10 μm nylon mesh (Herb et al., 2015). At least 300 pollen grains were counted per sample under 400 × magnification. Identification of taxa and nomenclature followed Hoorn et al. (2012), Han et al. (2016) and Miao et al. (2016).

Time series analysis

Time series of colour data, MS and element geochemistry were used for spectral analysis. In particular, the time series of the Red/Blue colour ratio (700/480 nm), the chemical proxy of alteration (CPA) and the Ti/Al ratio are used because of its high sensitivity to variations in the detrital sediment flux, redox conditions and degree of weathering (Salminen et al., 2005; Buggle et al., 2011). Spectral analysis was performed on each time series following the method of Weedon (2003) in order to identify dominant cycle lengths. Prior to the algorithm, each record was normalized by mean value subtraction and sampled evenly by linear interpolation. The record of MS was plotted on a logarithmic scale to achieve variance stabilization. Redfit power spectra were calculated with the "PAST" software (Hammer *et al.*, 2001) following the algorithm by Schulz & Mudelsee (2002). Dominant cycle frequencies were used for Gaussian band-pass filtering in order to identify potential cycle-frequency ratios typical for orbital forcing in the Miocene Bastau Formation. In addition, average spectral misfit (ASM) calculations and evolutive harmonic analysis (EHA) was performed with the Mg/Al time series using the astrochron software package by Meyers (2014). For ASM analysis, candidate frequencies which reflect possible Milankovitch forcing were identified from the Mg/Al redfit power spectrum at 90% confidence level. Miocene orbital target frequencies and their uncertainties are derived from Laskar *et al.* (2004) following the approach in Meyers *et al.* (2012).

WEATHERING INDICES

Chemical weathering indices rely on the concept that mobile elements are selectively removed from weathering profiles relative to rather immobile elements. A number of element indices have been applied to different terrestrial sediments as palaeoenvironmental indicators. Here, a modified Ca-free version of the chemical index of alteration (CIA; Nesbitt & Young, 1982), the CPA (Buggle *et al.*, 2011), and the molar Mg/Al ratio were chosen as an analogue for magnesium-bearing minerals (Maynard, 1992).

The CIA, derived from the silicate fraction, is a quantitative measure of feldspar weathering by relating Al, enriched in the residues, to Na, Ca and K removed from a soil profile by plagioclase and K-feldspar weathering (CIA = [Al_2O_3/(Al_2O_3 + Na_2O + CaO* + K_2O)]*100; Nesbitt & Young, 1982). Changes in sediment provenance, hydraulic sorting and post-depositional processes lead to K^+ addition, as for instance, diagenetic illitization. Illitization is also reported as pedogenic process in soils forming under arid climates (Singer, 1988) when smectite is altered during repeated wetting and drying cycles in the presence of K^+ (Eberl *et al.*, 1986). The most interfering element for Bastau Formation sediments, however, is Ca, which is commonly present both in detrital plagioclase and pedogenic carbonates and sulphates. Some of the Ca content of the acid insoluble fraction is related to gypsum, therefore we used a Ca-free version CIA-Ca of the CIA.

The CPA, defined as the molar ratio of Al and Na (CPA = [Al_2O_3/(Na_2O + Al_2O_3)]*100), is a weathering index for carbonate-rich shales, siltstones and sandstones

because of the small ionic radius of Na and its interference with non-silicate minerals in non-saline soils (Buggle *et al.*, 2011). The paired elements, Na and Al, minimize biases due to variable mineralogical composition of the parent material.

Climates with low to moderate precipitation reduce the intensity of weathering. Soluble cations such as Mg^{2+} can accumulate in soil pore waters by the limited flux of water through the soil profile which leads to the formation of alkaline and alkaline earth-rich secondary minerals (e.g. smectite and carbonates; Calvo *et al.*, 1999; Sheldon & Tabor, 2009; Torres & Gaines, 2013). In highly alkaline/saline solutions, rich in dissolved silica, Mg^{2+} is incorporated into trioctahedral clay minerals as Mg-smectite and sepiolite (Deocampo, 2004, 2015; Cuadros *et al.*, 2016). Here, the molar Mg/Al ratio from the total dissolved fraction is used as a measure of clay authigenesis in times of elevated rates of evaporation and higher groundwater table.

In addition, the molar ratio of Ti/Al from the non-leachable fraction (Ti*100/Al) is used as a geochemical index for palaeoenvironmental interpretation. The Ti/Al ratio is a classical indicator for sediment provenance, the more Ti present the more mafic the parent rock is (Salminen *et al.*, 2005; Sheldon & Tabor, 2009). Higher Ti/Al ratios indicate higher abundances of heavy minerals such as rutile, anatase, brookite, titanite and/or ilmenite (titanomagnetite) or detrital Ti-rich pyroxenes and amphibols in the catchment area. If the chemical composition of the parent rock in the source area remains unchanged through time, the Ti/Al ratio can be interpreted as an indicator of weathering intensity and sedimentary discharge from the catchment area (Sheldon & Tabor, 2009). Physical weathering readily removes Ti from igneous and metamorphic rocks where it subsequently becomes enriched in the fine fractions of floodplain sediments (Salminen *et al.*, 2005; Taboada *et al.*, 2006; Minyuk *et al.*, 2014). The Ti/Al ratio is used here as a proxy for the intensity of alluvial sediment discharge in times of unchanged provenance.

DATA AND RESULTS

Sedimentary facies

The Bastau Formation consists of reddish-brown mudstones with intercalated greyish-green and reddish sandstones that appear cyclically throughout the succession (Figs 2B and 4A). A typical sedimentary cycle begins with thin (5 to 20 cm) beds of medium-grained to coarse-grained sandstones, composed of several units separated by thin pelitic layers, finally grading into several metres thick mudstones. The base of the single sandstone beds may be slightly erosive or channelized. Although sandstone units can be traced over hundreds of metres, individual layers pinch out over short distances (10 to 50 m). Grain size varies in different layers from well-rounded granules to medium-grained sand of moderate roundness. Especially, thin lobate units have significant matrix content, pointing to hyper-concentrated flows. The topmost centimetres of single sandstone layers often show secondary clay infiltration. The poorly sorted sandstones, rich in unweathered volcanic rock fragments and plagioclase grains, are interpreted as representing distal lobes of

sheet flood deposits of terminal splays and their related feeder channels (Fig. 4C).

The mudstones are homogeneous, structureless rocks on average with less than 1% to 2% sand content. Often, they display a mottled texture or polyedric fracturing. They yield secondary carbonates and salts and the grain-size distribution (<100 μm) is bimodal with modal peaks at 0·2 to 0·3 μm and 9 to 10 μm, respectively (Fig. 5). The small grain-size fraction is mainly represented by authigenic components while the larger modal peak is indicative of detrital silt particles. The grain-size pattern is supported by

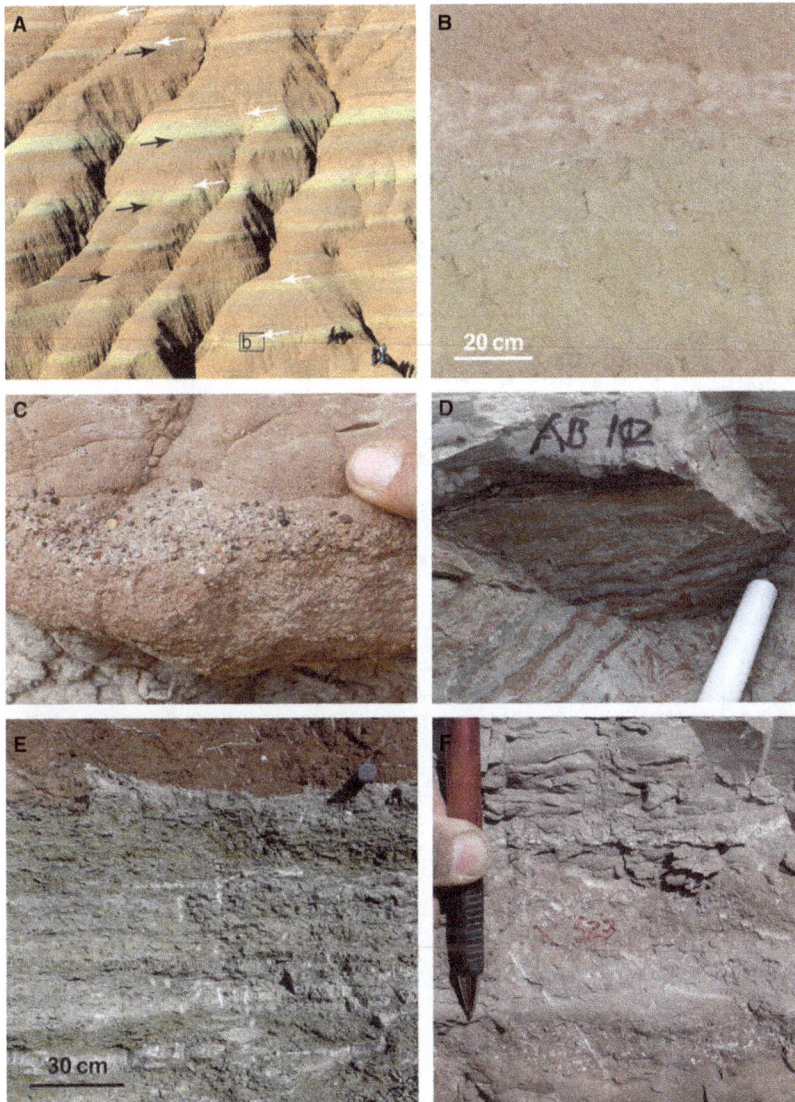

Fig. 4. Outcrop images showing the Bastau Formation facies types: (A) alternation of greyish sandstones and reddish-brown mudstones between 9 and 32 m representing sheet flood deposits (black arrows) and phreatic carbonates (white arrows) in a mudflat, (B) detail of (A) showing nodular phreatic carbonates on top of a bleached sandstone at 9 m, (C) enrichment of unweathered rock fragments in badly sorted sandstones, (D) reddish mottled grey gleysol at 16·2 m, (E) well-bedded calcareous marl with gypsum deposited in a playa lake system overlain by reddish mudstones at 61·4 m, and (F) abundant occurrence of displacive gypsum in mudstone deposits above the first lake (Horizon IV). Note people in bottom right corner of A.

the results of powder XRD analyses (Fig. 5). Relatively high abundances of unweathered minerals (quartz + albite + K-feldspar) sum up to 60% of the mudstone's composition. The mean clay mineral content is 29% and comprises mixed layer illite/smectite (0·7%), palygorskite (7·8%), illite (15·1%) and chlorite (5·0%). The relatively high abundance and needle-like preservation of palygorskite underlines its authigenic formation under arid to semi-arid depositional conditions (Fig. 5). At some horizons, distinct nodular calcareous horizons occur (Fig. 4A

and B). Mostly, they form 5 to 10 cm thick beds above massive sheet flood deposits. The nodular appearance of carbonates demonstrates its phreatic origin from saturated solutions in times of elevated groundwater table. Episodes of elevated groundwater table and pedogenic reworking are also evident from reddish or greyish mottling structures (Fig. 4D). Accordingly, the mudstones are interpreted as dry mudflat deposits, homogenized by plant growth and bioturbation and, in part, overprinted by in situ weathering and authigenesis.

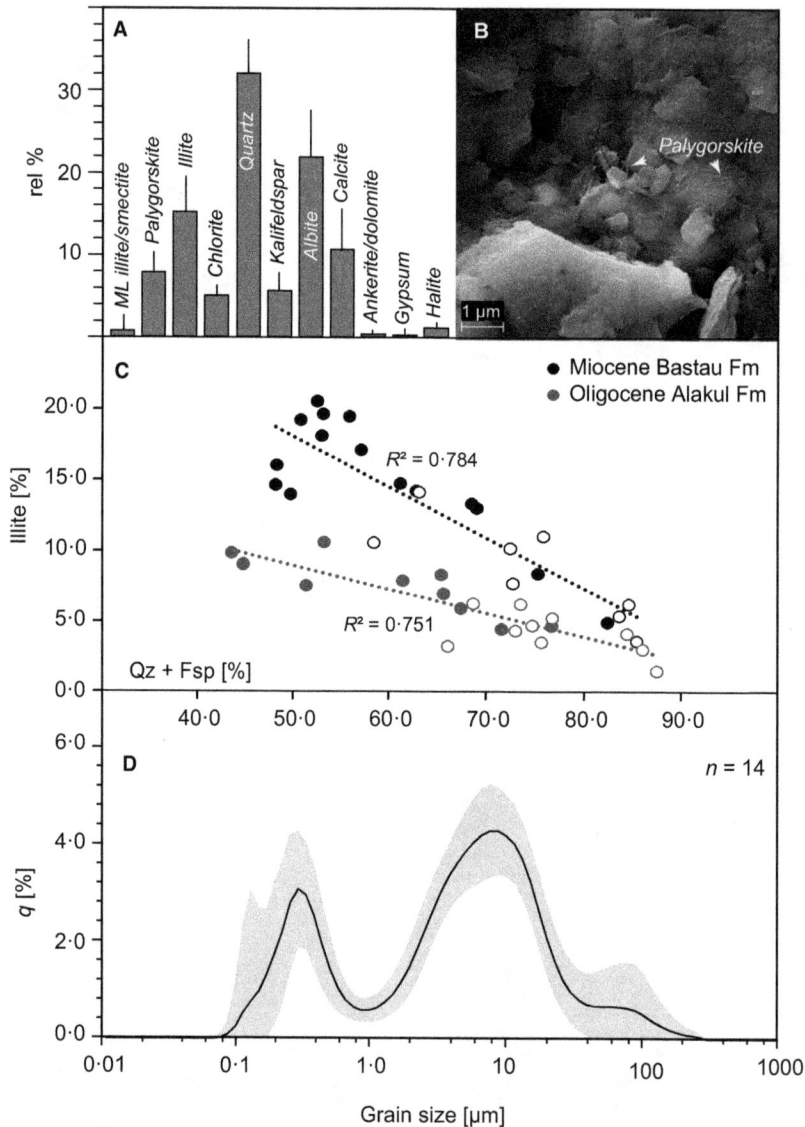

Fig. 5. Overview of the mineralogical composition and grain size of Bastau Formation sediments, (A) mean relative abundance of mineral phases estimated by powder X-ray diffractometry (XRD), (B) Scanning electron microscopy (SEM) image of sample AB 133, (C) relation of relative percentages of illite to quartz and feldspar in the Bastau and Alakul formations (closed circles) in comparison to a more proximal site (open circles, see text), (D) mean grain-size distribution of the <100 μm fraction with the 1 σ standard deviation (grey area). Unweathered source rock minerals sum up to 60% (albite + kalifeldspar + quartz). Abundant occurrence of small palygorskite needles in the mudstones refer to their authigenic formation. The bimodal grain-size distribution shows the grain-size separation of authigenic and detrital components.

Above 59 m, a prominent change in sedimentary facies occurs with a 3 m thick green horizon ("green band", GB) of calcareous mudstones (Fig. 4E). It consists of bedded nodular carbonates in its lower part and increased contents of gypsum towards the top (Fig. 6). This horizon is interpreted as the first occurrence of prolonged palustrine conditions in the Aktau succession. The shift from carbonates to gypsum represents an increase in salinity from almost freshwater to hypersaline conditions indicative of high rates of evaporation. Higher up, the succession consists again of alternations of sandstones and mudstones, however, with abundant gypsum

(Fig. 4F). Gypsum is present as single idiomorphic lenticular crystals within the sediment and forms up to 1 m thick crusts of chicken wire gypsum on top of the Bastau Formation (Fig. 2B). Thin sections show the gypsum also as secondary precipitates in the pore space suggesting a phreatic zone origin. Lenticular gypsum crystals and chicken wire-textured massive gypsum indicate mechanical replacement of soft, water-saturated mud. The abrupt appearance of gypsum above 60 m corresponds to elevated rates of evaporation in a progressively hydrologic restricted basin and indicates a facies shift from dry to saline mudflat and playa lake environments (Fig. 6).

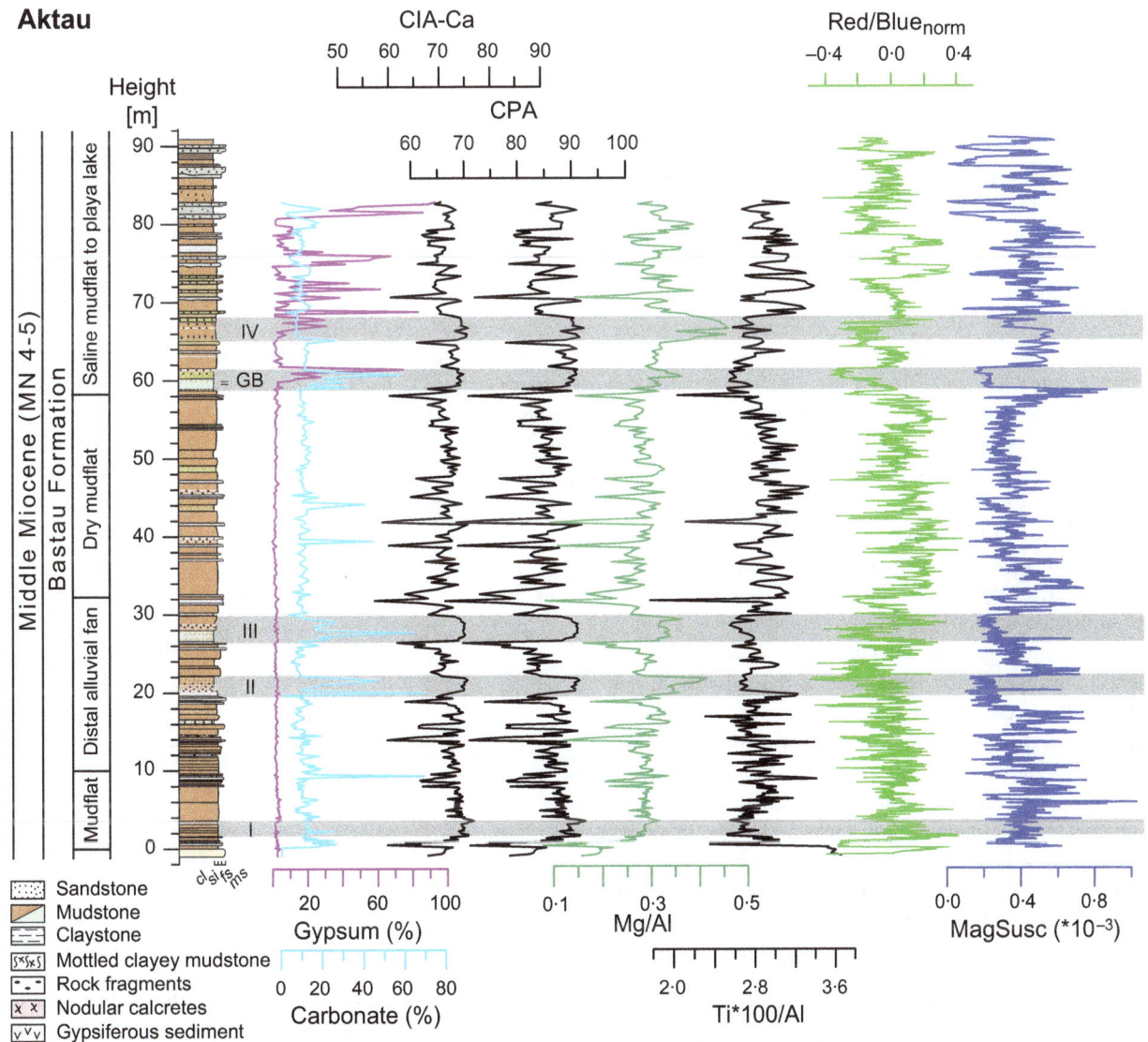

Fig. 6. Lithologic log of the Bastau Formation with carbonate and gypsum content, weathering indices in the acid-leachable fraction (CPA, CIA-Ca, Mg/Al; see text), Ti/Al ratio, sediment colour (normalized Red/Blue ratio) and MS. Numbered grey bars mark horizons of intensified weathering and clay mineral authigenesis. Arrows mark the position of productive palynological samples.

Element geochemistry, sediment colour and magnetic susceptibility

The carbonate content of mudstones in the Bastau Formation is on the order of 10% to 15% (Fig. 6). Elevated values between 20% and 30%, with single maximum values up to 60%, occur in beds with phreatic cementation above sheet flood deposits and in the palustrine horizon between 58·2 m and 61·4 m. The gypsum content is negligibly small below 60 m. A first significant occurrence of 10% to 75% at 60·0 to 61·4 m is associated with the upper part of the lacustrine horizon.

The weathering indices CIA-Ca, CPA and Mg/Al have mean values of 71 ± 3, 86 ± 5 and 0·29 ± 0·05, respectively, typical for sustained chemical weathering conditions. In addition, the three indices show relatively similar variations (Fig. 6). Elevated values occur in four horizons characterized by pedogenic reworking evident from mottling and associated colour changes. Namely these horizons occur at 1 to 3 m (I), 19 to 23 m (II), 27 to 30 m (III) and 65 to 68 m (IV). A fifth horizon is marked by the palustrine horizon GB (58 to 61·4 m). Very low weathering indices are associated with coarse sheet flood deposits rich in unweathered rock fragments. The CPA shows an almost identical pattern of variability as the CIA-Ca but differs from it by having higher amplitude variations. The Mg/Al ratio shows similar relative trends as the CPA and CIA-Ca but displays pronounced maxima which are caused by additional Mg enrichment in horizons of elevated pedogenesis (Figs 6 and 7). Prominent Mg enrichment occurs in Horizon IV.

Median concentrations of TiO_2 are 0·7 ± 0·1% and the Ti*100/Al ratio displays small variations around a mean value of 2·8 to 2·9, typical for a source rock with rather uniform chemistry (Fig. 6). Elevated values at the base of the succession are the only exception and refer to a different provenance for the fluvial Aidarly Formation sediments. Throughout the remainder of the succession, low Ti/Al ratios occur in the horizons of pedogenic reworking, and chemical weathering paces the more elevated values into 20 to 30 m long depositional cycles. A similar pattern exposes the Red/Blue colour ratio with low and high frequency variations indicative of changes in lithology and redox conditions of the sediment. The low frequency variations display 20 to 30 m long cycles separated by lower values in horizons of elevated weathering. Superimposed variations of higher frequency are associated with colour banding.

The MS displays cyclic variations with significant drops in the more weathered horizons. The minima are stratigraphically more expanded and include horizons of mudstone mottling reflecting changes in iron mobility and thus the sediment redox state after deposition.

Elevated values occur mainly in the lower and upper part of the succession (e.g. at 7 to 8 m and 22 to 23 m or beneath the palustrine horizon (GB) associated with enhanced detrital input.

Palynology

Only two of the six analysed samples (KAZ-10 and KAZ-11 at 62·20 m and 62·55 m, respectively) yielded moderately preserved palynological assemblages consisting predominantly of pollen and spores; in addition, the assemblages contain abundant non-pollen palynomorphs, including algal cysts of unknown affinity, rare organic-walled dinoflagellate cysts of presumably freshwater origin and fungal spores (Fig. 8).

The pollen and spore assemblages extracted from samples KAZ-10 and KAZ-11 contain substantial numbers of conifer-derived bisaccate pollen grains (i.e. *Pinuspollenites*, *Piceapollenites* and *Abiespollenites* in order of decreasing abundance). They make up 70·2% and 14·2% of the assemblages, respectively; *Taxodiaceapollenites* reaches 0·6% and 7·3%. With regard to pollen from deciduous trees, the assemblages are dominated by (in order of decreasing abundance) *Ulmipollenites* (9·9% and 55·0%, respectively) and *Pterocaryapollenites* (1·4% and 3·3%); other deciduous tree-pollen taxa occurring in low (i.e. ≤2%) percentages are *Alnuspollenites*, *Carpinuspollenites*, cf. *Caryapollenites*, *Fraxinoipollenites*, cf. *Juglanspollenites*, *Quercoidites*, *Striaticolpites*, *Tiliaepollenites* and *Triporopollenites*. Non-arboreal pollen grains are mainly from *Cyperaceaepollis* (6·8% and 4·8%, respectively) and *Graminidites* (5·1% and 13·0%); other non-arboreal pollen taxa occurring in low (i.e. ≤2%)

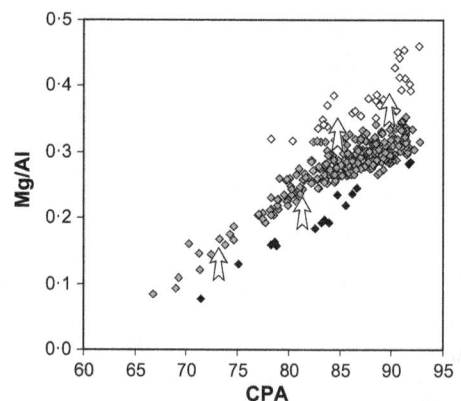

Fig. 7. Cross plot of CPA and Mg/Al ratios for Bastau Fm sediments that demonstrates the process of Mg enrichment during early diagenetic authigenic clay formation (white arrows). Black symbols represent the lowermost 2 m of the succession; white symbols mark horizons of phreatic carbonate precipitation and evaporative enrichment, and grey diamonds all other samples (see text).

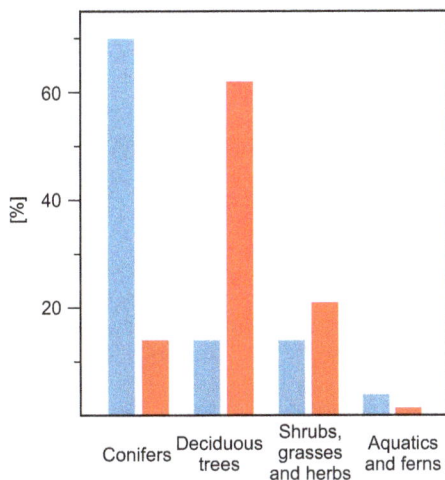

Fig. 8. Percentages of the different pollen and spore groups identified in samples KAZ-10 (blue bars) and KAZ-11 (red bars).

percentages are *Chenopodipollis*, *Compositoipollenites* and *Ephedripites* spp. Fern spores account for 3·7% of the pollen and spore assemblage of sample KAZ-10, and pollen from aquatic plants (i.e. *Sparganiaceaepollenites*) account for 1·5% of the pollen and spore assemblage of sample KAZ-11.

Time series analysis

Average Middle to Late Miocene sedimentation rates of the Aktau succession (bases Bastau to Ili formations) are on the order of 5·0 to 8·5 cm ka^{-1} based on its overall thickness of 530 ± 40 m and biostratigraphic age (8 ± 1 Ma; Bazhanov & Kostenko, 1961; Bodina, 1961). The prevalence of fine-grained sediments indicates rather stable subsidence without tectonically enhanced deepening of the Ili Basin.

Time series analysis was performed on the records of CPA, Ti/Al, Mg/Al, Red/Blue and MS. Spatial resolution of the CPA and Ti/Al time series (25 ± 6 cm, 1σ) is not sufficient for a robust detection of precession periods. Spatial resolutions of MS and the Red/Blue ratio are higher (6 ± 2 cm and 9 ± 5 cm, 1σ), however, both time series are measured directly on rock fragments in the field and can be inaccurate because the lack of plain surfaces may cause signal noise and distortion. Redfit spectra were calculated to identify cycle lengths of dominant periodicities and filter outputs of significant cycles were generated to identify cycle-frequency ratios diagnostic for orbital forcing. ASM was calculated for the Mg/Al time series (Meyers, 2014). The method offers a statistical test to reject the null hypothesis (no orbital signal) at a certain significance level. If the null hypothesis can be rejected, the ASM metric estimates the most probably

sedimentation rate for a stratigraphic interval by comparing candidate frequencies to the fixed target frequencies from the orbital solution for a given range of sedimentation rates (Meyers & Sageman, 2007). In addition, EHA was performed to test the stability of sedimentation rate.

Redfit spectra of the four studied time series display different dominant cycle lengths (Fig. 9). The Ti/Al ratio has only one dominant cycle 27 to 28 m in length, with more than 99% significance. Similar periodicities are also visible in the logMS (22 to 30 m, >95%) and the Red/Blue (20 to 40 m, >99%) time series. While the Ti/Al ratio shows a clear peak, the broad range of cycle lengths of the logMS and Red/Blue time series refers to the amalgamation of different cycles, which cannot be addressed because of the short length of the time series. The CPA, MS and Red/Blue ratio show dominant peaks between 6·4 to 7·3 m (>90 to 95%) and 5·0 to 5·4 m (>95%). Their common occurrence in different time series with different spatial resolution argues for a common forcing. Further significant peaks (>95%) occur at variable cycle lengths between 1·5 m and 2·5 m, with the four time series lacking consistency.

Results from the ASM calculation of the Mg/Al time series show that for a sedimentation rate of 5·1 cm kyr^{-1} the null hypothesis can be rejected with a H$_0$ significance level of 0·18% (Fig. 10A and B). This sedimentation rate would assign the 5·0 to 5·4 m cycle to the periodicity of short eccentricity. Furthermore, the EHA normalized amplitude spectrum shows spectral power for the frequencies of long and short eccentricity and obliquity (Fig. 10C). A shift towards higher sedimentation rates higher occurs above 27 m and explains the increased cycle length of 6·4 to 7·3 m for the short eccentricity signal (Fig. 10C). Short eccentricity is weakly developed between 35 m and 55 m where the dominance of a *ca* 3 m cycle suggests a stronger control by obliquity. Gaussian bandpass filter outputs were generated for the 6·4 m cycle in the CPA, the 27 m, 22 m and 7·3 m cycle in the MS, the 27 m cycle in the Ti/Al, and the 30 m, 20 m, 7 m and 5·1 m cycle in the Red/Blue ratio (Fig. 9). The long 22 to 30 m cycle displays minima in the horizons of elevated weathering intensity. Here, the 6·4 to 7·3 m cycle has its highest amplitudes (grey bars in Fig. 9) while it weakens in the intervals between. The 6·4 to 7·3 m filters of MS and CPA display anti-phase correlated cycles. The filter output of the 5·1 m cycle in the Red/Blue ratio shows the 5·0 to 5·4 m cycle closely related to the 6·4 to 7·3 m cycle representing similar sedimentary cycles in horizons of lower sedimentation rate.

The arrangement of spectral peaks (frequency ratios), such as the 1 : 4 relationship between long (405 ka) and short (*ca* 100 ka) eccentricity is robustly expressed by the 20 to 22 m and 5·0 to 5·4 m cycles in the lower part of

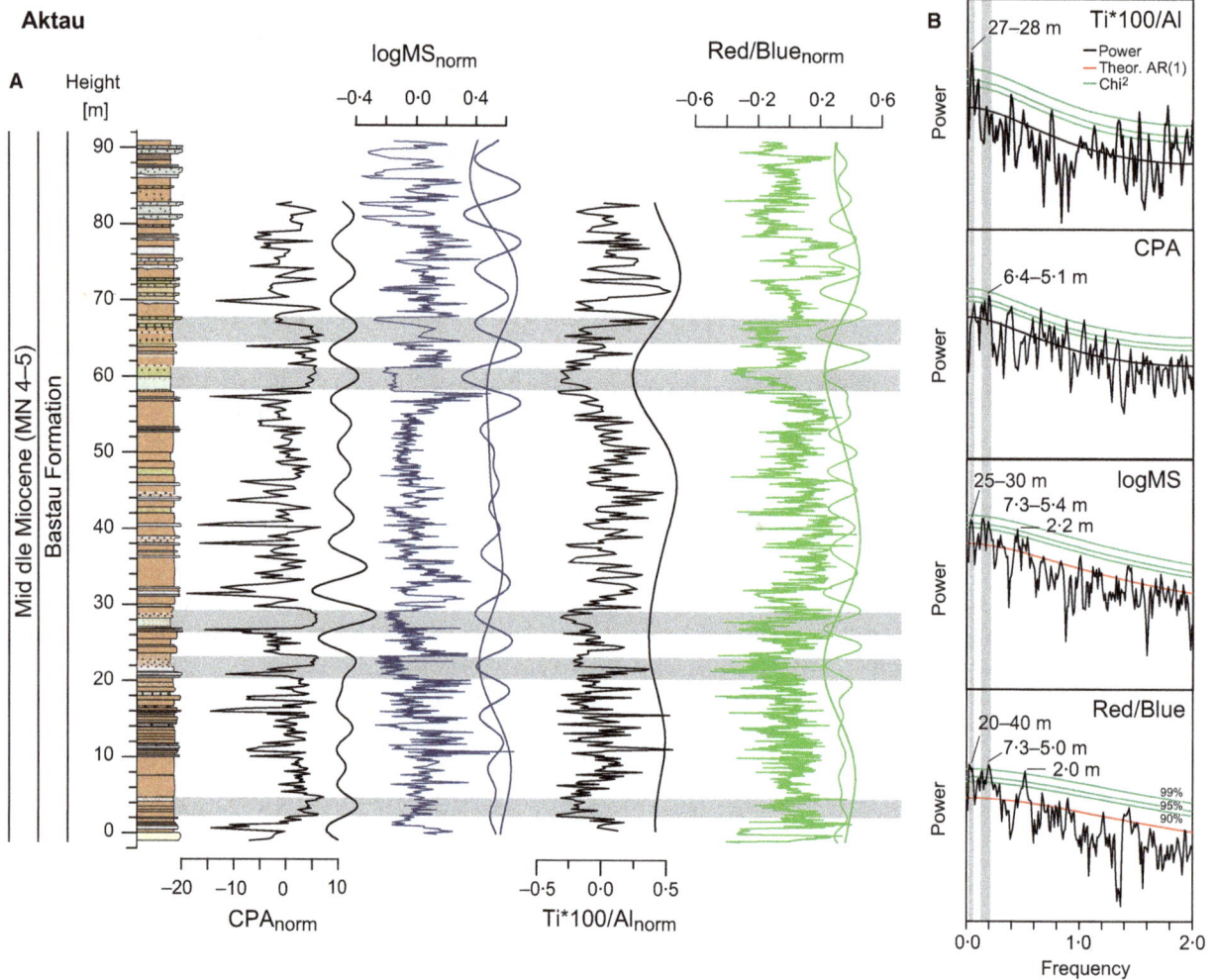

Fig. 9. Gaussian filter outputs (A) and Redfit spectra (B) for dominant frequencies of CPA, logMS, Ti/Al and the Red/Blue colour ratio. Filter frequencies for the time series are 0·037 ± 0·015 cycles m^{-1} for logMS, Ti/Al and Red/Blue, 0·155 ± 0·047 cycles m^{-1} for CPA, 0·137 ± 0·041 cycles m^{-1} for logMS and 0·191 ± 0·057 cycles m^{-1} for Red/Blue. The 6·4 to 7·3 m and 5·0 to 5·4 m cycles, probably related to the 100 ka short eccentricity cycle, are prominently developed in the CPA, MS and Red/Blue time series. The 27 to 30 m cyclicity is significantly developed in the logMS, Ti/Al and Red/Blue data and is interpreted as representing the 405 ka long eccentricity cycle. Further subsidiary cycles occur between 1·5 m and 2·5 m in the CPA, logMS and Red/Blue data.

the succession and the 27 to 30 m and the 6·4 to 7·3 m cycles in the upper part evident in the different studied times series. The resolution of the MS and Red/Blue time series is high enough to detect precession and obliquity periods. However, no clear signal could be identified in the Redfit spectra of both records because of signal noise at higher frequencies. Given the low temporal resolution of biostratigraphy and the limitations of the presented data sets, we caution against explicit development of an astrochronology for this interval. This work motivates the development of improved independent age controls for the entire Aktau succession, such as palaeomagnetic stratigraphy and radiometric dating. There is limited discussion below on the identification of potential orbital

frequencies and the assessment of their effects on climatically forced sedimentation.

DISCUSSION

Weathering and authigenic clay mineral formation

Chemical weathering under arid and semi-arid conditions responds sensitively to changes in the hydrologic system of the basin, either through tectonically driven changes in the discharge system, or by climatically driven changes in the hydrological balance. A plot of the Bastau Formation sediment chemistry in the A-CN-K diagram (Fig. 11;

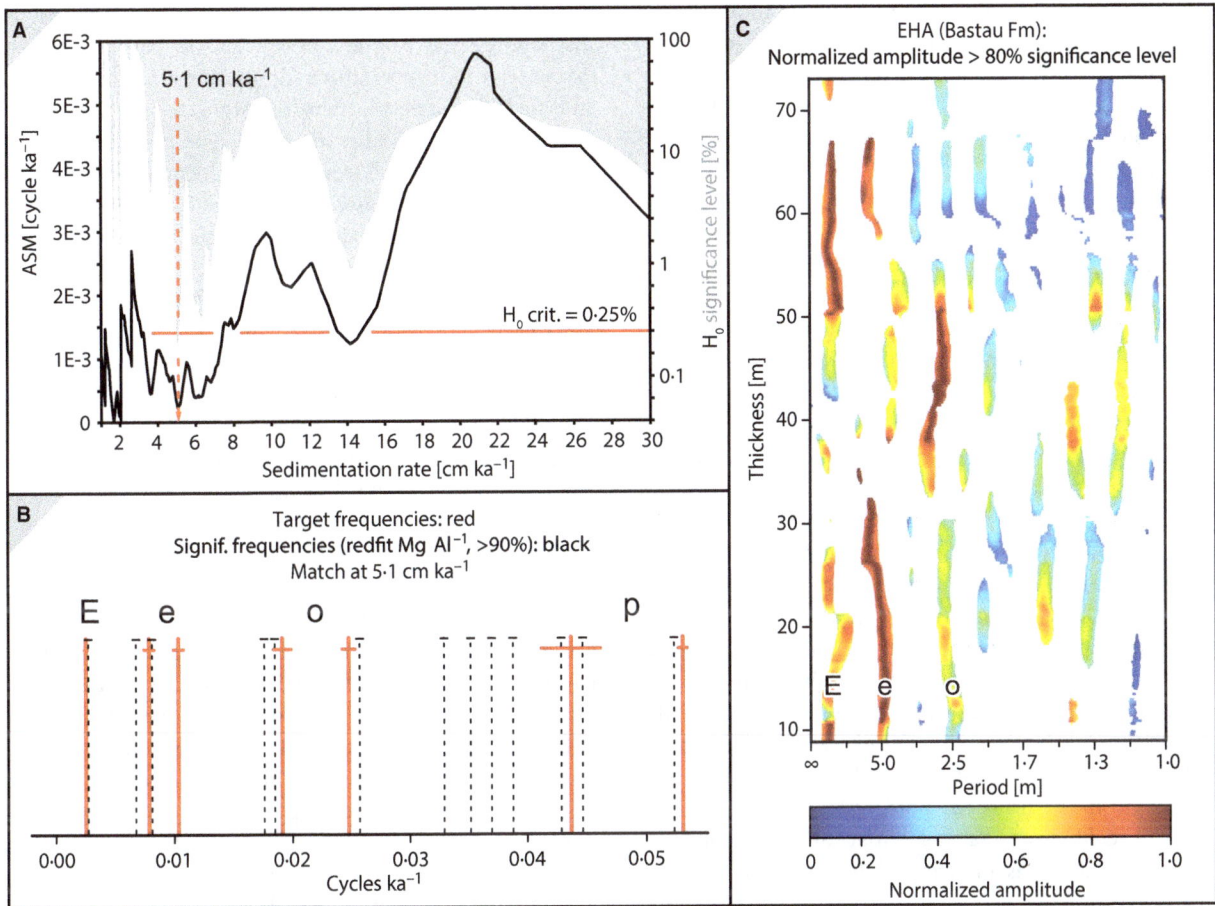

Fig. 10. (A) ASM of the Mg/Al time series calculated for sedimentation rates between 1 and 30 cm kyr^{-1} with 400 sedimentation rates investigated, resulting in a critical significance level H_0 for null hypothesis rejection of 0·25%. (B) Significant spectral peaks from Mg/Al redfit >90% CL (=candidate frequencies, black) and their match with seven orbital target frequencies (0·00247, 0·00786, 0·01035, 0·01913, 0·02475, 0·04366, 0·05305, red). The null hypothesis of no orbital forcing can be rejected for a mean sedimentation rate of 5·1 cm kyr^{-1}. (C) EHA amplitude spectrum of the Mg/Al time series at the 80% significance level.

Nesbitt & Young, 1984, 1989) shows the sediments to derive from source rocks with a composition typical for the upper continental crust in accordance with their origin from volcanic arc complexes (Jahn *et al.*, 2000). Mean CIA-Ca values of 71 ± 3 indicate that Bastau Formation sediments underwent sustained chemical weathering. Plagioclase weathering is the main mineral reaction causing the preferential removal of Ca and Na. Furthermore, the sediment chemistry data display in agreement with results of XRD analyses a trend towards elevated K concentrations, typical for the formation of illite. Illite can be directly formed from weathering of K-feldspar and muscovite or by post-depositional processes as diagenetic illitization. Illite can also be formed by low-temperature illitization of smectite (Baldermann *et al.*, 2015; Cuadros *et al.*, 2016) when K is fixed during repeated wetting and drying processes in the soil (Singer & Stoffers, 1980). In the Bastau Formation, the relative abundance of illite is negatively correlated with those of quartz and feldspar indicating a predominant detrital origin of illite, additionally supported by stratigraphic equivalent XRD data from a proximal site 40 km distant (Fig. 5C; Hellwig *et al.*, 2017). Illite formation by post-depositional alteration, as observed for Oligocene sediments in south-western Mongolia (Richoz *et al.*, 2017), is not considered a relevant process here. Illite percentages of mudflat sediments in the Bastau Formation are about 10% higher than those of the underlying Alakul Formation which underwent higher burial temperatures, but show a similar proximal-distal relationship in the abundance of unweathered quartz and feldspar to illite typical for detrital transport (Fig. 5C).

Sedimentary facies and geochemical indices in the Bastau Formation refer to five horizons of elevated weathering intensity (Figs 5 and 6). The horizons are characterized by hydromorphic features of soil formation (mottling), phreatic carbonate precipitation close to the groundwater table or by palustrine sedimentation, typical features in distal parts of endorheic basins. The higher abundance of clay minerals together with elevated CIA, CPA and Mg/Al, and low Ti/Al values suggest that these beds formed in times of low detrital sediment supply and landscape stability.

Intensified weathering can be explained either by (1) elevated water/moisture supply to the vadose zone by precipitation and/or surface and subsurface discharge; (2) longer periods of landscape stability with lower rates of sediment discharge from the hinterland; and (3) elevated rates of evaporation and capillary groundwater rise. An important feature of Bastau Formation sediments in addressing these processes is the covariance of CPA/CIA and Mg/Al which refers to more intense weathering under highly evaporative conditions (Fig. 7).

Magnesium fixation in silicates requires Mg enrichment in the aqueous environment by either by Ca removal by the early formation of carbonates and sulphates or by evaporative capillary groundwater rise (Calvo et al., 1999; Deocampo, 2015). Primary controls for the authigenic formation of Mg-rich clays (e.g. smectite, palygorskite) are high magnesia and silica activities and elevated pH (Deocampo, 2004; Deocampo et al., 2009). Studies in modern arid environments have shown that a strong enrichment of elements in pore waters takes place in areas sheltered from detrital input and subjected to strong evaporation as in marginal lacustrine settings, in

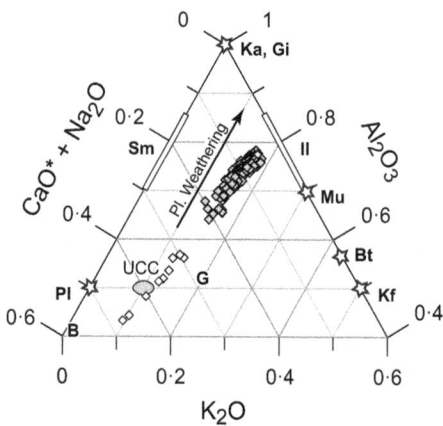

Fig. 11. A-CN-K ($Al_2O_3 - CaO^* + Na_2O - K_2O$) ternary diagram according to Nesbitt & Young (1984). The position of the upper continental crust (UCC), basalt (B), granite (G), plagioclase (Pl), kalifeldspar (Kf), muscovite (Ms), biotite (Bt), smectite (Sm), illite (Il) and kaolinite (Ka) is given for orientation. Data points of the Bastau Fm plot on a line typical for plagioclase weathering because of preferential removal of Ca and Na (grey diamonds) with a shift towards increased K-content (see text). The weathering line is distorted with respect to CaO* due to an overestimation of CaO* in the gypsum-rich upper Bastau Fm (white diamonds).

interdune clay pans and ponds where groundwater is discharging (Calvo et al., 1999). The K, Mg and Si are extracted from supersaturated solutions to form inter-stratified illite and trioctahedral Mg-rich smectite (Jones & Weir, 1983; Banfield et al., 1991). The presence of paly-gorskite in calcic soils results from incongruent dissolution precipitation (Jones & Galan, 1988), direct precipitation from oversaturated solutions or from transformation of inherited clay precursers (Cuadros et al., 2016). Badaut & Risacher (1983) observed authigenic Mg-smectite formation under conditions of silica saturation where the pH was above 8·2 in lakes in Bolivia. High salinity/alkalinity, dissolved silica enrichment and high pH values (>9·5 to 10) also favour Mg enrichment and trioctahedral clay formation in palaeolake Olduvai, Tanzania (Deocampo, 2004). Fibrous clays such palygorskite and chlorite degrade to smectite when the climate is more humid and annual rainfall exceeds 300 mm (Calvo et al., 1999).

The source of K, Mg and Si needed for clay mineral formation in Bastau Formation sediments is derived from feldspar weathering of the volcanic source rocks in the catchment area. The presence of fibrous palygorskite and abundant chlorite argue for their authigenic formation under more arid conditions. This process is most intense in the intervals of elevated Mg/Al ratios. Accordingly, it is possible to state that the highest degrees of weathering were achieved when the regional climate was at its driest and elevated rates of potential evapotranspiration supported capillary groundwater rise. On the other hand, hydromorphic soil features and phreatic carbonate precipitation argue for a higher groundwater table in the weathered horizons and thus for higher water supply to the vadose zone. The most probably explanation for this water supply is an elevated subsurface discharge from montane areas in the surroundings of the Ili Basin. In an analogue to modern observations, the highest weathering intensities probably occurred when mean annual precipitation values were at or below 300 mm, soil water pH reached values above 8·5 and the phreatic zone was at or close to the surface. Regarding the interpretation of weathering indices as proxies in palaeoclimatology, we can state, if covariance with CPA and/or CIA is achieved, the Mg/Al ratio serves as an aridity index.

Regional hydrology

The primary water source for arid to semi-arid basins is precipitation, and in a long-term steady-state system, precipitation is balanced by the water loss to evapotranspiration, underlying aquifer recharge, surface runoff and vadose infiltration of soil moisture (Maliva & Missimer, 2012). In the distal depositional setting of the Ili Basin, water supply is provided either by rainfall or by surface and subsurface discharge from the head of the alluvial systems of the Tian Shan mountain system (Hellwig *et al.*, 2017). Orogenic uplift reinforced in the late Oligocene to early Miocene at rather low and continuous rates (4 to 5 cm ka^{-1}) and created a hilly landscape (Burtman, 2012). Sedimentological and geochemical data of this study provide evidence for variable water availability with wetter and more arid phases within the alluvial plain of the Ili Basin.

Deposition of medium-grained to coarse-grained sandstones indicates episodes of rapid unchannelled surface drainage with transport of coarser detritus from local source areas. Episodes of distal sheet flood deposition are characterized by low CIA, CPA and Mg/Al ratios, larger grain sizes and poorly sorted fine-sized to medium-sized sandstones. These beds have the lowest Ti/Al ratios, characteristic of unweathered source rock. Sheet flood deposits are overlain by mudstones with hydromorphic features of soil formation (mottling) and phreatic carbonate precipitation close to the groundwater table. Mottling results from reduction of ferrous iron to ferric iron during periods of waterlogging and is indicative of pronounced seasonal wetting (Huggett & Cuadros, 2005; Gale *et al.*, 2006). In addition, high CIA, CPA and Mg/Al ratios and the occurrence of palygorskite indicate arid soil formation by capillary groundwater rise as well as starvation of detrital sediment supply. Elevated groundwater inflow was probably maintained through the underlying relatively permeable sandstones, an effective mechanism described for semi-arid alluvial systems in Miocene basins of Spain (Sanz *et al.*, 1995). Furthermore, incongruent dissolution of detrital Ti-oxides by reducing ground waters and alteration to anatase, as observed for the Jurassic Morrison Formation in Colorado (Adams *et al.*, 1974; Sanford, 1994), is a probably process which led to the observed lowering of Ti/Al ratios. The several metres thick structureless mudstones, instead, are well drained and represent higher rates of vertical mudflat accretion when ongoing detrital sediment production at moderate weathering rates occurred. The well-drained mudflat deposits show the highest Ti/Al ratios. It has been shown by several studies that Ti can be enriched in the fine fraction of detrital sediments (Taboada *et al.*, 2006; Yang *et al.*, 2006; Minyuk *et al.*, 2014). Detrital Ti enrichment occurred by fluvio-aeolian processes when heavy Ti-bearing minerals were concentrated by episodic flooding events and aeolian blowout of ancient surfaces (Figs 5 and 6). Furthermore, successive Ti enrichment in mudflats occurs by in situ weathering and oxidation and the formation of sedimentary anatase (Bestland, 1997).

The Ti/Al ratio is considered as a proxy for sedimentary discharge here. The average TiO$_2$ concentration

(0·7 ± 0·1%) in the Bastau Formation is typical for sediments sourced by weathering of igneous and metamorphic parent materials (Salminen *et al.*, 2005). The rather small variability in Ti/Al does not indicate significant provenance changes in the catchment. Climate exerts a major control on sediment supply and fan activity. Ahlborn *et al.* (2015) found a strong relationship between precipitation and sediment supply in a small catchment area on the southern-central Tibetan Plateau. Modelling results have shown that climate variability, in particular precipitation, produce extremely fast responses throughout the catchment–fan system and overprint lower frequency tectonic variations (Allen & Densmore, 2000). Higher rates of alluvial fan activity are directly linked to higher hinterland precipitation and water availability. A similar pattern emerges for sediment accumulation in the Ili Basin. Horizons with elevated Ti/Al correspond to periods of higher detrital sediment supply and thus to overall wetter conditions. Lowered Ti/Al ratios represent intervals of reduced detrital influx and intensified soil formation under more arid conditions (Horizons I to IV and GB).

Accordingly, sheet flood deposition with subsequent pedogenic mottling and deposition of unstructured mudflats represent two climate end-members in the water balance of the local hydrologic system. Flash floods, formed when the water supply exceeds the soil infiltration capacity, allow for the formation of short-term ponds and ground water recharge (Amiaz *et al.*, 2011). At a later stage, evaporative capillary ground water rise led to pore water enrichment and intensified authigenic clay mineral formation under arid conditions. Such a scenario is supported by strong seasonal gradients in both discharge and evaporation. Instead, periods of rather steady detrital supply during mudflat accretion represents a balance of continuous sediment production and in situ weathering without the formation of hydromorphic soil features during periods of less extreme climate. Such a scenario is supported by less pronounced seasonality of discharge and evaporation.

Vegetation

The palynological assemblages from samples KAZ-10 and KAZ-11 yield a consistent picture of the vegetation and environment characterizing the study site at the time of sediment deposition. Based on the ecological preferences of their nearest living parent plants, the identified pollen and spores can be attributed to different vegetation units.

The occurrences of *Sparganiaceaepollenites* (nearest living relative: *Sparganium* – common name: bur-reed) and *Cyperaceaepollis* (Cyperaceae – sedges) along with dinoflagellate cysts of presumably freshwater origin indicate the existence of perennial marshland and at least temporary open-water bodies. This aquatic/marshland setting was surrounded by riparian forests as documented by the occurrences of substantial amounts of *Alnuspollenites* (*Alnus* – alder) and *Ulmipollenites* (*Ulmus* – elm), and the co-occurrences of *Carpinuspollenites* (*Carpinus* – hornbeam), *Fraxinoipollenites* (*Fraxinus* – ash), *Pterocaryapollenites* (*Pterocarya* – wingnut) and notably *Taxodiaceapollenites* (*Taxodium* – swamp cypress). Ferns and Poaceae (grasses), represented by fern spores and *Graminidites*, respectively, thrived as part of the forest understorey.

Further away from the marshland and under drier conditions, a steppe vegetation prevailed that was characterized by Poaceae along with Asteraceae (represented by *Compositoipollenites*) and xerophytic herbs such as chenopods (goosefoot – represented by *Chenopodipollis*), and different taxa of *Ephedra* (joint-pine – represented by *Ephedripites* spp.).

Finally, the slopes of higher altitude settings in the surroundings supported – possibly patchy – montane, conifer-dominated forests represented by *Pinuspollenites* (*Pinus* – pine), *Piceapollenites* (*Picea* – spruce) and *Abiespollenites* (*Abies* – fir); they may have benefitted from enhanced soil moisture and air humidity in comparison to that available to the lower elevation steppe vegetation.

The palynological results derived from samples KAZ-10 and KAZ-11 are in excellent agreement with other palaeobotanical evidence from the upper Middle to lowermost Upper Miocene of Central Asia, such as from eastern Kazakhstan (Akhmetyev *et al.*, 2005) and the Qaidam Basin of the north-eastern Tibetan Plateau (Miao *et al.*, 2011).

Orbital control on mudflat deposition

The period of the Bastau Formation sediment deposition falls into the juvenile stage of Tian Shan's orogeny of weak crustal deformation and low uplift rates (Burtman, 2012). Corresponding sedimentation rates were in the range of 6 to 13 cm ka^{-1} in the inner and outer basins of the Tian Shan (Huang *et al.*, 2006; Heermance *et al.*, 2007; Charreau *et al.*, 2008). Estimates of average sedimentation rates in the Aktau succession (5·0 to 8·5 cm kyr^{-1}) based on biostratigraphic data presented here fall into this range. The spectral analysis results show two dominant cycle lengths (5·0 to 5·4 m and 20 to 22 m, 6·4 to 7·3 m and 27 to 30 m) at different levels in the Bastau Formation, which we interpret as the signals of short and long eccentricity based on its cycle-to-frequency ratio. The very significant 27 to 28 m cycle in the Ti/Al time series argues for the presence of roughly three 405 kyr cycles (Fig. 9) with an overall duration of

deposition of the Bastau Formation of 1·0 to 1·2 Myr. However, lower sedimentation rates in the lower Bastau together with the filter outputs of the 5·0 to 5·4 m and 20 to 22 m cycles show the presence of four 405 kyr cycles, which would extend the duration to 1·4 to 1·6 Myr.

Filter outputs of the two dominant frequencies appear noisy with the highest amplitudes of the 6·4 to 7·3 m cycle at horizons, where the 27 to 30 m cycle displays minima and a very weak signal in the intervals between (Fig. 9). Changes in the sedimentation rate, for example, by elevated/lowered clastic supply or by the longer presence of stable landscapes, can lead to signal distortion (Abels *et al.*, 2009, 2014; Hilgen *et al.*, 2014). Complex interactions between climate and depositional processes as described above involve non-linear feedbacks, which affect the significance of spectral peaks. Together with the low resolution of biostratigraphic age control, cyclostratigraphy is therefore not developed for the Bastau succession, instead the discussion is centred on cycle pattern and amplitudes, which might have been forced by orbitally controlled climate change.

Of interest here is the anti-phase correlation of MS and CPA. Maxima in MS correspond to higher supply rates of unweathered detrital components. Following the notion that higher rates of alluvial activity correspond to higher rates of precipitation and *vice versa*, maximum amplitudes of the 6·4 to 7·3 m cycle should represent the intervals of highest climate variability by recording the most pronounced extremes between evaporation and precipitation in the hydrological balance. The long 27 to 30 m cycle is best expressed in the Ti/Al ratio. The filter output shows minima in the horizons of most elevated climate extremes between pronounced water supply and aridity (Horizons I-IV, GB) and maxima in the interval where the 6·4 to 7·3 m cycle is hard to detect in CPA and MS (Fig. 9). Long-term variability of the Ti/Al paced by the 405 ka cycle describes variations between two climate stages. Minima in Ti/Al correspond to high amplitude climate shifts expressed by abundant discharge of sheet floods during times of elevated precipitation alternating with periods of low detrital supply and alkaline weathering in times of high evaporation. Maxima in Ti/Al reflect periods of more stable sediment production and moderate weathering intensity in times of less extreme climate change. Insolation-driven climate changes strongly affect seasonality, and the observed pattern argues for changes in the seasonal contrast as driving force.

The biostratigraphic age control for the Bastau succession does not allow for direct comparison of the observed orbitally driven climate pattern with the orbital solution (Laskar *et al.*, 2004). However, based on the observations presented here it is possible to speculate that maxima in the Ti/Al filter output could correspond to long eccentricity minima and the Ti/Al minima to long eccentricity maxima. Individual horizons of maximum aridity represent individual short eccentricity or obliquity cycles. Long eccentricity minima represent periods of higher precipitation and fan activity, corresponding to overall wetter conditions as a result of lower seasonality of precipitation in both the catchment area and the site of deposition. Long eccentricity maxima, instead, refer to lower rates of fine-grained sediment supply and drier periods. At the same time, there is a higher probability for the deposition of coarse-grained sheet floods. This is best expressed around the GB horizon. Stronger seasonal gradients promote seasonally intensified precipitation in the catchment area which discharged into the basin by surface and subsurface flow. Higher MS values and a lower CPA (56 to 58 m) refer to the accumulation of coarser and less weathered clastic material. More subsurface discharge resulted in a groundwater table rise until the formation of palustrine conditions. Parallel to increased water supply, seasonally elevated rates of evaporation led to drying, capillary groundwater rise and gypsum formation. Evidence from palynology describe the low-lying landscape as dry steppe with patchy conifer-dominated forests at higher elevations. Such an open landscape without closed vegetation cover does not provide favourable conditions for the formation of thick soil horizons in the catchment area which could stabilize the erosion of freshly weathered material.

The observation provides support for the hypothesis of Zachos et al. (2010) who suggested that the long eccentricity variations in the global carbon cycle are controlled by seasonality of precipitation on land. More year-round precipitation favours the areal spread of humid conditions and wetlands during times of long eccentricity minima, which in turn led to increased terrestrial carbon sequestration. More seasonal precipitation, instead, supports monsoonal and dry climates and increased steppe and dry grasslands. Abels et al. (2014) observed a well-developed cyclicity in fluvial sediments of the lower Eocene Willwood Formation of the Bighorn Basin in North America and showed precessional control for overbank avulsion. Furthermore, at the 100 kyr and 405 kyr scales, the bundling of well-developed simple pedofacies cycles can be linked to eccentricity maxima and, thus, to intervals of mature palaeosol development. The 405 kyr cyclicity may have originated from subsequent relatively wet conditions related to high amplitude precession cycles during eccentricity maxima. Such a pattern is also similar to that observed in the Bastau succession. Although it is not possible to discuss phase relationships in terms of the orbital solution, based on the observed pattern, it is possible to suggest that arid to semi-arid terrestrial mudflat sedimentation in the mid-Miocene Ili Basin in Central Asia was strongly controlled by seasonal changes in moisture availability paced by long eccentricity.

Mid-Miocene palaeoenvironment

Proxy records from various basins provide evidence that Central Asia's climate was warmer and wetter in the mid-Miocene in comparison to the long-term Cenozoic average that displayed pronounced aridity and desertification from the Oligocene to early Miocene (Guo et al., 2002; Sun & Windley, 2015; Zheng et al., 2015). Sedimentary facies in the northern Junggar Basin (Fig. 12) comprise fluvial and lacustrine deposits between 17·5 and 13·5 Ma with the onset of aeolian red clay deposition after 13·5 Ma (Sun et al., 2010). Evidence for the presence of perennial fluvial drainage and lakes comes also from the southern Junggar Basin with the development of modern-like desert vegetation after 13·5 Ma (Tang et al., 2011b; Charreau et al., 2012). South of the Tian Shan, reddish mudstones of the Jidike Formation indicate the prevalence of arid conditions since 13·5 Ma (Sun et al., 2015), and palynological evidence from basins in the north-eastern Tibetan Plateau, such as Qaidam and Tianshui, argue for a warmer, moister period between 14 and 15 Ma with substantial cooling and drying afterwards (Hui et al., 2011; Miao et al., 2011). These findings are consistent with the assumption that on longer timescales mid-Miocene warming and late Miocene cooling correspond to the global climate evolution (Zachos et al., 2008; De Vleeschouwer et al., 2017).

Results from climate modelling with Miocene boundary conditions and lower than present topography suggest significant warming in Inner Asia compared to today (Henrot et al., 2010; Tang et al., 2011a). Climate warming is most pronounced in the winter with more zonal climate and increased moisture supply by westerly winds. Strong low-level westerlies were generated as a result of a strong N-S pressure gradient between atmospheric high pressure above the Tibetan Plateau and a low-pressure cell above the northern lowlands (Tang et al., 2011a; Fig. 12). Westerly wind-driven moisture supply is documented for many sites in Inner Asia based on the oxygen isotopic composition of pedogenic carbonates (Caves et al., 2015). In addition, the lack of oxygen isotopic fractionation along the trajectories argues for a high degree of regional moisture recycling by evapotranspiration (Caves et al., 2015).

The palynological results from the Bastau Formation presented here describe an open landscape covered by steppe vegetation with small riparian forests around smaller ponds typical of semi-arid climates. The geochemical data show a strong sedimentary response to regional surface and subsurface water availability relative to orbital

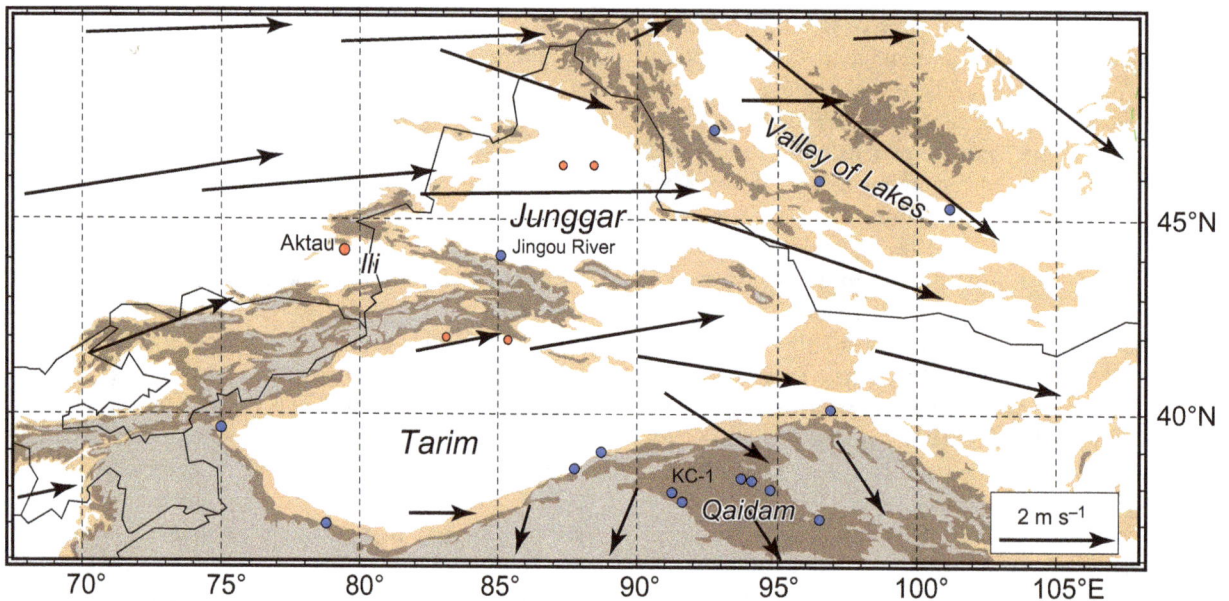

Fig. 12. Topographic map of Inner Asia showing present-day mountain ranges together with results of a regional Miocene climate simulation with lower topography than at present (Tang *et al.*, 2011a). Also shown are localities with Miocene palaeoclimate proxy data from literature and this study (Sun *et al.*, 2010, 2015; Miao *et al.*, 2011; Tang *et al.*, 2011b; Charreau *et al.*, 2012; Caves *et al.*, 2015). Black arrows are modelled vectors of low level winter winds at 850 hPa. Blue sites show localities with pedogenic oxygen isotopic evidence for westerly wind sourced moisture supply (Caves *et al.*, 2015). Topographic colour coding represents the current 1500 m (light brown), 2500 m (dark brown) and 4000 m (grey) isoheights.

extremes. Insolation changes affect the length of the seasons and, thus, the intensity of winter precipitation. Most of the precipitation is probably trapped by the low-relief mountain ranges of the Tian Shan, which served as a regional source for runoff, aquifer recharge and detrital sediment supply. The intensity of westerly winter winds are assumed to have responded sensitively to orbital forcing, hence controlling the amount of moisture available for regional recycling in both the catchment area and the aquifers. In times of pronounced regional evapotranspiration, a stronger summer–winter gradient of westerly wind intensity probably reduced the amount of precipitation within the basin and additionally increased the probability for heavy rain storms and flash flood deposition due to more intense precipitation events in the catchment area. In times of wetter conditions, a lower summer–winter gradient of westerly intensity increased the amounts of year-round surface runoff and activated mudflat aggradation.

The overall successions of both sedimentary facies and geochemical data of the Bastau Formation represent enhancement of aridity on longer time scales. The establishment of alkaline palustrine conditions marks a transition in the Miocene evolution of the Ili Basin when hypersaline conditions prevailed in the basin with limited runoff. Incoming water from floods evaporated and

as the brine concentrates gypsum precipitated either directly from solution or by forming pedogenic crystallites and crusts on top of the playa floor. The appearance of gypsum occurred rather abruptly and could be explained either by the formation of endorheic conditions or by climate change. Since endorheic conditions were already prevalent from the onset of mudflat deposition in the Bastau Formation, the first accumulation of gypsum is related to a severe change in the regional climate system from wetter semi-arid conditions to more pronounced aridity. A dry climate persisted from this time onwards.

The sudden appearance of gypsum in the Miocene Ili Basin is difficult to explain and the source of the sulphur is questionable since gypsum is completely absent in the succession below the palustrine GB horizon. There are no local sulphur sources in the catchment area since it consists entirely of rhyolitic and andesitic volcanic rocks (Jahn *et al.*, 2000, 2004). It is possible that the sulphur originated from a marine source. Deposition of Bastau Formation sediments lasted for about 1·0 to 1·2 Myr with the biostratigraphic data indicating a middle Miocene, possibly Langhian age (Fig. 3). At this time, extensive evaporites were deposited at the margins of the Eastern Paratethys (Rögl, 1999; Popov *et al.*, 2004; Bruch *et al.*, 2007). Although, this area is located more than 500 km west of the Ili Basin based on palaeogeographic reconstructions (Popov *et al.*, 2004), the precise position of the Eastern Paratethys shoreline is not well constrained from geological data because of the absence of Neogene sediments in central Kazakhstan. However, assuming the prevalence of westerly winds evident from Miocene climate modelling (Henrot *et al.*, 2010; Tang *et al.*, 2011a), the high degree of regional moisture recycling by evapotranspiration evident from the low fractionation of pedogenic oxygen isotopes (Caves *et al.*, 2015), sulphur could have been transported via several precipitation/infiltration/deflation steps from the Eastern Paratethys. Such a mechanism is also supported by oceanographic box modelling results, which show increased salinity and lowered temperature in the Mediterranean–Eastern Paratethys system as a result of gateway closure to the Indian Ocean in the late Langhian to early Serravallian (Karami *et al.*, 2009). Furthermore, the rather sudden occurrence of gypsum could be related to transgressive eastward extension of the Eastern Paratethys (Popov *et al.*, 2004). Although a precise dating of the facies shift is not yet possible, the mid-Miocene Tchokrakian transgression is a good candidate.

CONCLUSIONS

Middle Miocene mudflat and marginal playa lake sediments were deposited at the margin of low-gradient alluvial systems in the endorheic Ili Basin, south-east Kazakhstan within the Tian Shan mountain system. The 91 m thick Bastau Formation exposed in the Aktau Mountains was studied for its bulk-sediment geochemistry, MS and sediment colour to characterize its composition and to determine changes in weathering, pedogenesis and alluvial fan activity. A positive correlation between weathering indices (CIA, CPA) and the Mg/Al ratio documents evaporative enrichment and authigenic Mg fixation by clay mineral formation in the vadose zone in highly alkaline/saline settings. Four major periods of arid soil formation and one palustrine interval represent the highest degrees of weathering in periods when pedogenesis exceeded sedimentary discharge. Periods of higher moisture availability and higher rates of hinterland precipitation are indicated by successive mudflat accretion by higher alluvial fan activity recorded by elevated Ti/Al ratios and MS.

Time series analysis of chemical weathering indices, MS and sediment colour data show cycle-to-frequency ratios typical of Milankovitch cyclicity, with dominant periodicities interpreted as representing short and long eccentricity cycles. In particular, the Ti/Al ratio demonstrates a pacing of pedogenic mottling and detrital mudflat accretion by long eccentricity, thus suggesting an orbital control on regional moisture availability and mudflat deposition. It is assumed that more frequent bundling of sheet flood deposition and subsequent pedogenic alteration correspond to long eccentricity maxima, and longer lasting periods of elevated fan activity to long eccentricity minima. However, a better-constrained age model is necessary to relate these changes to the orbital solution.

The overall sedimentary succession shows a long-lasting increase in aridity in the studied area. Of particular interest is the abrupt onset of gypsum formation at the time the GB horizon was deposited. The sudden appearance of gypsum beds indicates a climate and possibly orbital trigger analogous to the initiation of the Messinian salinity crisis in the Mediterranean. Such a trigger could have been the transition to the mid-Miocene cooling. On longer timescales, the closure of the eastern gateway of the Mediterranean to the Indian Ocean enhanced restriction within Eastern Tethys, providing the necessary boundary conditions.

In light of the results above, the quasi-continuous, terrestrial Miocene succession of the Aktau Hills emerges as a sensitive recorder of changes in atmospheric moisture supply. This makes it a highly promising terrestrial archive for palaeoclimate research, ideally located in order to address the role of Central Asia in the global climate evolution during the Miocene.

ACKNOWLEDGEMENTS

We thank the Deutsche Forschungsgemeinschaft for granting several field campaigns (DFG grants VO 687/13, VO 687/16 and PR 651/13). In particular, we thank Konstantin Kossov, Julia Zhilkina and Marat Ainsonow for logistical support and their warm and friendly company in the field. The administration and rangers of the State National Park Altyn Emel are thanked for providing access to the Aktau Mountains for geological field work. We thank André Baldermann and an anonymous reviewer for their constructive and insightful reviews of this work.

References

Abdrakhmatov, K.Y., Aldazhanov, S.A., Hager, B.H., Hamburger, M.W., Herring, T.A., Kalabaev, K.B., Makarov, V.I., Molnar, P., Panasyuk, S.V., Prilepin, M.T., Reilinger, R.E., Sadybakasov, I.S., Souter, B.J., Trapeznikov, Y.A., Tsurkov, V.Y. and Zubovich, A.V. (1996) Relatively recent construction of the Tien Shan inferred from GPS measurements of present-day crustal deformation rates. *Nature*, **384**, 450–453.

Abels, H.A., Aziz, H.A., Ventra, D. and Hilgen, F.J. (2009) Orbital climate forcing in mudflat to marginal lacustrine deposits in the Miocene Teruel Basin (Northeast Spain). *J. Sed. Res.*, **79**, 831–847.

Abels, H.A., Kraus, M.J. and Gingerich, P. (2014) Precession-scale cyclicity in the fluvial lower Eocene Willwood Formation of the Bighorn Basin, Wyoming (USA). *Sedimentology*, **60**, 1467–1483.

Adams, S.S., Curtis, H.S. and Hafen, P.L. (1974) Alteration of detrital magnetite-ilmenite in continental sandstones of the Morrison Formation, New Mexico. In: *Symposium on the Formation of Uranium Ore Deposits*, International Atomic Energy Agency, Vienna, Proceedings series, 219–252.

Ahlborn, M., Haberzettl, T., Wang, J.B., Alivernini, M., Schlutz, F., Schwarz, A., Su, Y.L., Frenzel, P., Daut, G., Zhu, L.P. and Mäusbacher, R. (2015) Sediment dynamics and hydrologic events affecting small lacustrine systems on the southern-central Tibetan Plateau – the example of TT Lake. *Holocene*, **25**, 508–522.

Akhmetyev, M.A., Dodoniv, A.E., Sornikova, M.V., Spasskaya, I.I., Kremenetsky, K.V. and Klimanov, V.A. (2005). Kazakhstan and Central Asia (Plains and Foothills). In: *Cenozoic Climatic and Environmental Changes in Russia* (Eds A.A. Velichko and V.P. Nechaev), *Geol. Soc. Am. Spec. Paper*, **382**, 139–161.

Allen, P.A. and Densmore, A.L. (2000) Sediment flux from an uplifting fault block. *Basin Res.*, **12**, 367–380.

Amante, C. and Eakins, B.W. (2009) *ETOPO1 1 Arc-Minute Global Relief Model: Procedures, Data Sources and Analysis*. NOAA Technical Memorandum NESDIS NGDC-24, Nat. Geophys. Data Center, NOAA. https://doi.org/10.7289/v5c 8276m.

Amiaz, Y., Sorek, S., Enzel, Y. and Dahan, O. (2011) Solute transport in the vadose zone and groundwater during flash floods. *Water Resour. Res.*, **47**, W10513. https://doi.org/10.1029/2011wr010747.

Badaut, D. and Risacher, F. (1983) Authigenic smectite on diatom frustules in Bolivian saline lakes. *Geochim. Cosmochim. Acta*, **47**, 363–375.

Baldermann, A., Warr, L.N., Letofsky-Papst, I. and Mavromatis, V. (2015) Substantial iron sequestration during green-clay authigenesis in modern deep-sea sediments. *Nat. Geosci.*, **8**, 885–890.

Banfield, J.F., Jones, B.F. and Veblen, D.R. (1991) An AEM-TEM Study of Weathering and Diagenesis, Abert Lake, Oregon. 2. Diagenetic Modification of the Sedimentary Assemblage. *Geochim. Cosmochim. Acta*, **55**, 2795–2810.

Bazhanov, V.S. and Kostenko, N.N. (1961) Geologicheskiy razrez Dzhungarskogo Alatau i ego paleozoologicheskoye obosnovanie [Geological section of Dzhungarian Alatau and its paleontological basis]. In: *Materialy po istorii fauny i flory Kazakhstana* (Ed. I.G. Galuzo), *Akademia Nauk Kazakhskoy SSR, Alma Ata*, **3**, 47–52.

Bestland, E.A. (1997) Alluvial terraces and paleosols as indicators of early Oligocene climate change (John Day Formation, Oregon). *J. Sed. Res.*, **67**, 840–855.

Bodina, L.E. (1961) Ostrakody tretichnykh otlozhenii Zaisanskoi I Iliiskoi depressi [Ostracods of Tertiary deposits in the Zaisan and Ili depressions]. *Trudy VNIGRI*, **170**, 43–153.

Bruch, A.A., Uhl, D. and Mosbrugger, V. (2007) Miocene climate in Europe – patterns and evolution – a first synthesis of NECLIME. *Palaeogeogr. Palaeoclimatol. Palaeoecol.*, **253**, 1–7.

Buggle, B., Glaser, B., Hambach, U., Gerasimenko, N. and Markovic, S. (2011) An evaluation of geochemical weathering indices in loess-paleosol studies. *Quatern. Int.*, **240**, 12–21.

Burtman, V.S. (2012) Geodymanics of Tibet, Tarim, and the Tien Shan in the Late Cenozoic. *Geotectonics*, **46**, 185–211.

Calvo, J.P., Blanc-Valleron, M.M., Rodríguez-Arandía, J.P., Rouchy, J.M. and Sanz, M.E. (1999) Authigenic clay minerals in continental evaporitic environments. In: *Paleoweathering, Palaeosurfaces and Related Continental Deposits* (Eds M. Thiry and R. Simon-Coincon), *Spec. Publ. Int. Ass. Sedimentol.*, **27**, 129–151.

Caves, J.K., Winnick, M.J., Graham, S.A., Sjostrom, D.J., Mulch, A. and Chamberlain, C.P. (2015) Role of the westerlies in Central Asia climate over the Cenozoic. *Earth Planet. Sci. Lett.*, **428**, 33–43.

Charreau, J., Avouac, J.P., Chen, Y., Dominguez, S. and Gilder, S. (2008) Miocene to present kinematics of fault-bend folding across the Huerguosi anticline, northern Tianshan (China), derived from structural, seismic, and magnetostratigraphic data. *Geology*, **36**, 871–874.

Charreau, J., Kent-Corson, M.L., Barrier, L., Augier, R., Ritts, B.D., Chen, Y., France-Lannord, C. and Guilmette, C. (2012) A high-resolution stable isotopic record from the Junggar Basin (NW China): implications for the paleotopographic evolution of the Tianshan Mountains. *Earth Planet. Sci. Lett.*, **341**, 158–169.

Cuadros, J., Diaz-Hernandez, J.L., Sanchez-Navas, A., Garcia-Casco, A. and Yepes, J. (2016) Chemical and textural controls on the formation of sepiolite, palygorskite and dolomite in volcanic soils. *Geoderma*, **271**, 99–114.

De Vleeschouwer, D., Vahlenkamp, M., Crucifix, M. and Pälike, H. (2017) Alternating Southern and Northern Hemisphere climate response to astronomical forcing during the past 35 m.y. *Geology*, **45**, 375–378.

Deocampo, D.M. (2004) Authigenic clays in East Africa: regional trends and paleolimnology at the Plio-Pleistocene boundary, Olduvai Gorge, Tanzania. *J. Paleolimnol.*, **31**, 1–9.

Deocampo, D.M. (2015) Authigenic clay minerals in lacustrine mudstones. In: *Paying Attention to Mudrocks* (Eds D. Larsen, S.O. Egenhoff and N.S. Fishman), *Geol. Soc. Am. Spec. Paper*, **515**, 45–64.

Deocampo, D.M., Cuadros, J., Wing-Dudek, T., Olives, J. and Amouric, M. (2009) Saline Lake diagenesis as revealed by coupled mineralogy and geochemistry of multiple ultrafine clay phases: pliocene Olduvai Gorge, Tanzania. *Am. J. Sci.*, **309**, 834–868.

Dettman, D.L., Fang, X.M., Garzione, C.N. and Li, J.J. (2003) Uplift-driven climate change at 12 Ma: a long delta O-18 record from the NE margin of the Tibetan plateau. *Earth Planet. Sci. Lett.*, **214**, 267–277.

Dupont-Nivet, G., Krijgsman, W., Langereis, C.G., Abels, H.A., Dai, S. and Fang, X.M. (2007) Tibetan plateau aridification linked to global cooling at the Eocene-Oligocene transition. *Nature*, **445**, 635–638.

Dzhamangaraeva, A.K. (1997) Pliocene charophytes from Aktau Mountain, southeastern Kazakhstan. *Geobios*, **30**, 475–479.

Eberl, D.D., Srodon, J. and Northrop, H.R. (1986) Potassium fixation in smectite by wetting and drying. *ACS Sym. Ser.*, **323**, 296–326.

Gale, A.S., Huggett, J.M., Pälike, H., Laurie, E., Hailwood, E.A. and Hardenbol, J. (2006) Correlation of Eocene-Oligocene marine and continental records: orbital cyclicity, magnetostratigraphy and sequence stratigraphy of the Solent Group, Isle of Wight, UK. *J. Geol. Soc. London*, **163**, 401–415.

Goldberg, K. and Humayun, M. (2010) The applicability of the Chemical Index of Alteration as a paleoclimatic indicator: an example from the Permian of the Parana Basin, Brazil. *Palaeogeogr. Palaeoclimatol. Palaeoecol.*, **293**, 175–183.

Gromova, V. (1952) Primitívnye tapiroobrazye iz Paleogena Mongolij [Primitive tapirs from the Paleogene of Mongolia]. *Akad. Nauk SSSR Trud. Paleontolog. Inst.*, **41**, 99–119.

Guo, Z.T., Ruddiman, W.F., Hao, Q.Z., Wu, H.B., Qiao, Y.S., Zhu, R.X., Peng, S.Z., Wei, J.J., Yuan, B.Y. and Liu, T.S. (2002) Onset of Asian desertification by 22 Myr ago inferred from loess deposits in China. *Nature*, **416**, 159–163.

Hammer, Ø., Harper, D.A.T. and Ryan, P.D. (2001) PAST: paleontological Statistics software package for education and data analysis. *Palaeontol. Electr.*, **4**, 9.

Han, F., Rydin, C., Bolinder, K., Dupont-Nivet, G., Abels, H.A., Koutsodendris, A., Zhang, K.X. and Hoorn, C. (2016) Steppe development on the Northern Tibetan Plateau inferred from Paleogene ephedroid pollen. *Grana*, **55**, 71–100.

Harzhauser, M., Daxner-Höck, G., López-Guerrero, P., Maridet, O., Oliver, A., Piller, W.E., Richoz, S., Erbajeva, M.A., Neubauer, T.A. and Göhlich, U.B. (2016) Stepwise onset of the Icehouse world and its impact on Oligo-Miocene Central Asian mammals. *Sci. Rep.*, **6**, 36169.

Heermance, R.V., Richard, V., Chen, J., Burbank, D.W., Douglas, W. and Wang, C. (2007) Chronology and tectonic controls of late Tertiary deposition in the Southwestern Tian Shan Foreland, NW China. *Basin Res.*, **19**, 599–632.

Hellwig, A., Voigt, S., Mulch, A., Frisch, K., Bartenstein, A., Pross, J., Gerdes, A. and Voigt, T. (2017) Late Oligocene–early Miocene humidity change recorded in terrestrial

sequences in the Ili Basin (SE Kazakhstan, Central Asia). *Sedimentology*. https://doi.org/10.1111/sed.12390.

Hendrix, M.S., Dumitru, T.A. and Graham, S.A. (1994) Late Oligocene Early Miocene unroofing in the Chinese Tien-Shan – an early effect of the India-Asia collision. *Geology*, **22**, 487–490.

Henrot, A.J., Francois, L., Favre, E., Butzin, M., Ouberdous, M. and Munhoven, G. (2010) Effects of CO_2, continental distribution, topography and vegetation changes on the climate at the Middle Miocene: a model study. *Clim. Past*, **6**, 675–694.

Herb, C., Koutsodendris, A., Zhang, W.L., Appel, E., Fang, X.M., Voigt, S. and Pross, J. (2015) Late Plio-Pleistocene humidity fluctuations in the western Qaidam Basin (NE Tibetan Plateau) revealed by an integrated magnetic-palynological record from lacustrine sediments. *Quatern. Res.*, **84**, 457–466.

Hilgen, F.J., Hinnov, L.A., Aziz, H.A., Abels, H.A., Batenburg, S., Bosmans, J.H.C., De Boer, B., Husing, S.K., Kuiper, K.F., Lourens, L.J., Rivera, T., Tuenter, E., Van De Wal, R.S.W., Wotzlaw, J.F. and Zeeden, C. (2014) Stratigraphic continuity and fragmentary sedimentation: the success of cyclostratigraphy as part of integrated stratigraphy. *Geol. Soc. Spec. Publ.*, **404**, 157–197.

Holbourn, A., Kuhnt, W., Lyle, M., Schneider, L., Romero, O. and Andersen, N. (2014) Middle Miocene climate cooling linked to intensification of eastern equatorial Pacific upwelling. *Geology*, **42**, 19–22.

Holbourn, A., Kuhnt, W., Kochhann, K.G.D., Andersen, N. and Meier, K.J.S. (2015) Global perturbation of the carbon cycle at the onset of the Miocene Climatic Optimum. *Geology*, **43**, 123–126.

Hoorn, C., Straathof, J., Abels, H.A., Xu, Y.D., Utescher, T. and Dupont-Nivet, G. (2012) A late Eocene palynological record of climate change and Tibetan Plateau uplift (Xining Basin, China). *Palaeogeogr. Palaeoclimatol. Palaeoecol.*, **344**, 16–38.

Huang, B.C., Piper, J.D.A., Peng, S.T., Liu, T., Li, Z. and Zhu, R.X. (2006) Magneto stratigraphic study of the Kuche Depression, Tarim Basin, and Cenozoic uplift of the Tian Shan Range, Western China. *Earth Planet. Sci. Lett.*, **251**, 346–364.

Huggett, J.M. and Cuadros, J. (2005) Low-temperature illitization of smectite in the late Eocene and early Oligocene of the Isle of Wight (Hampshire basin), UK. *Am. Mineral.*, **90**, 1192–1202.

Hui, Z.C., Li, J.J., Xu, Q.H., Song, C.H., Zhang, J., Wu, F.L. and Zhao, Z.J. (2011) Miocene vegetation and climatic changes reconstructed from a sporopollen record of the Tianshui Basin, NE Tibetan Plateau. *Palaeogeogr. Palaeoclimatol. Palaeoecol.*, **308**, 373–382.

Jahn, B.M., Wu, F.Y. and Chen, B. (2000) Massive granitoid generation in Central Asia: Nd isotope evidence and implication for continental growth in the Phanerozoic. *Episodes*, **23**, 82–92.

Jahn, B.M., Windley, B., Natal'in, B. and **Dobretsov, N.** (2004) Phanerozoic continental growth in central Asia – Preface. *J. Asian Earth Sci.*, **23**, 599–603.

Jones, B.F. and **Galan, E.** (1988) Sepiolite and Palygorskite. *Rev. Mineral.*, **19**, 631–674.

Jones, B.F. and **Weir, A.H.** (1983) Clay-Minerals of Lake Abert, an alkaline, saline lake. *Clays Clay Miner.*, **31**, 161–172.

Karami, M.P., Meijer, P.T., Dijkstra, H.A. and **Wortel, M.J.R.** (2009) An oceanic box model of the Miocene Mediterranean Sea with emphasis on the effects of closure of the eastern gateway. *Paleoceanography*, **24**, PA4203. https://doi.org/10.1029/2008pa001679.

Kent-Corson, M.L., Ritts, B.D., Zhuang, G.S., Bovet, P.M., Graham, S.A. and **Chamberlain, C.P.** (2009) Stable isotopic constraints on the tectonic, topographic, and climatic evolution of the northern margin of the Tibetan Plateau. *Earth Planet. Sci. Lett.*, **282**, 158–166.

Kober, M., Seib, N., Kley, J. and **Voigt, T.** (2013) Thick-skinned thrusting in the northern Tien Shan foreland, Kazakhstan: structural inheritance and polyphase deformation. *Geol. Soc. Spec. Publ.*, **377**, 19–42.

Kordikova, E.G. (2000) Insectivora (Mammalia) from the Lower Miocene of the Aktau Mountains, South-Eastern Kazakhstan. *Senckenb. Lethaea*, **80**, 67–79.

Kordikova, E.G. and **de Bruijn, H.** (2001) Early Miocene Rodents from the Aktau Mountains (South-Eastern Kazakhstan). *Senckenb. Lethaea*, **81**, 391–405.

Kordikova, E.G. and **Mavrin, A.V.** (1996) Stratigraphy and Oligocene-Miocene mammalian biochronology of the Aktau Mountains, Dzhungarian Alatau range, Kazakhstan. *Palaeovertebrata*, **25**, 141–174.

Laskar, J., Robutel, P., Joutel, F., Gastineau, M., Correia, A.C.M. and **Levrard, B.** (2004) A long-term numerical solution for the insolation quantities of the Earth. *Astron. Astrophys.*, **428**, 261–285.

Li, D., He, D., Maa, D., Tang, Y., Kong, Y. and **Tang, J.** (2015) Carboniferous-Permian tectonic framework and its later modifications to the area from eastern Kazakhstan to southern Altai: insights from the Zaysan-Jimunai Basin evolution. *J. Asian Earth Sci.*, **113**, 16–35.

Licht, A., Van Cappelle, M., Abels, H.A., Ladant, J.B., Trabucho-Alexandre, J., France-Lanord, C., Donnadieu, Y., Vandenberghe, J., Rigaudier, T., Lecuyer, C., Terry, D., Adriaens, R., Boura, A., Guo, Z., Soe, A.N., Quade, J., Dupont-Nivet, G. and **Jaeger, J.J.** (2014) Asian monsoons in a late Eocene greenhouse world. *Nature*, **513**, 501–506.

Liu, W.G., Liu, Z.H., An, Z.S., Sun, J.M., Chang, H., Wang, N., Dong, J.B. and **Wang, H.Y.** (2014) Late Miocene episodic lakes in the arid Tarim Basin, western China. *Proc. Natl Acad. Sci. USA*, **111**, 16292–16296.

Lucas, S.G., Bayshashov, B.U., Tyutkova, L.A., Zhamangara, A.K. and **Aubekerov, B.Z.** (1997) Mammalian biochronology of the Paleogene-Neogene boundary at Aktau Mountain, eastern Kazakhstan. *Paläontol. Z.*, **71**, 305–314.

Macaulay, E.A., Sobel, E.R., Mikolaichuk, A., Kohn, B. and **Stuart, F.M.** (2014) Cenozoic deformation and exhumation history of the Central Kyrgyz Tien Shan. *Tectonics*, **33**, 135–165.

Maliva, R. and **Missimer, T.** (2012) *Arid Lands Water Evaluation and Management, Environmental Science and Engineering*. Springer-Verlag, Berlin, Heidelberg, 148 pp.

Maynard, J.B. (1992) Chemistry of modern soils as a guide to Interpreting Precambrian Paleosols. *J. Geol.*, **100**, 279–289.

Meyers, S.R. (2014) Astrochron: an R package for astrochronology. Available at: https://cran.r-project.org/package=astrochron.

Meyers, S.R. and **Sageman, B.B.** (2007) Quantification of deep-time orbital forcing by average spectral misfit. *Am. J. Sci.*, **307**, 773–792.

Meyers, S.R., Bradley, B.S. and **Michael, A.A.** (2012) Obliquity forcing of organic matter accumulation during Oceanic Anoxic Event 2. *Paleoceanography*, **27**, PA3212. https://doi.org/10.1029/2012PA002286.

Miao, Y.F., Fang, X.M., Herrmann, M., Wu, F.L., Zhang, Y.Z. and **Liu, D.L.** (2011) Miocene pollen record of KC-1 core in the Qaidam Basin, NE Tibetan Plateau and implications for evolution of the East Asian monsoon. *Palaeogeogr. Palaeoclimatol. Palaeoecol.*, **299**, 30–38.

Miao, Y.F., Fang, X.M., Liu, Y.S., Yan, X.L., Li, S.Y. and **Xia, W.M.** (2016) Late Cenozoic pollen concentration in the western Qaidam Basin, northern Tibetan Plateau, and its significance for paleoclimate and tectonics. *Rev. Palaeobot. Palynol.*, **231**, 14–22.

Minyuk, P.S., Borkhodoev, V.Y. and **Wennrich, V.** (2014) Inorganic geochemistry data from Lake El'gygytgyn sediments: marine isotope stages 6-11. *Clim. Past*, **10**, 467–485.

Molnar, P. and **Tapponnier, P.** (1975) Cenozoic tectonics of Asia – effects of a continental collision. *Science*, **189**, 419–426.

Nesbitt, H.W. and **Young, G.M.** (1982) Early proterozoic climates and plate motions inferred from major element chemistry of lutites. *Nature*, **299**, 715–717.

Nesbitt, H.W. and **Young, G.M.** (1984) Prediction of some weathering trends of plutonic and volcanic-rocks based on thermodynamic and kinetic considerations. *Geochim. Cosmochim. Acta*, **48**, 1523–1534.

Nesbitt, H.W. and **Young, G.M.** (1989) Formation and diagenesis of weathering profiles. *J. Geol.*, **97**, 129–147.

Popov, S.V., Rögl, F., Rozanov, A.Y., Steiniger, F.F., Shcherba, I.G. and **Kovac, M.** (eds.) (2004) Lithological-Paleogeographical maps of Paratethys. *Cour. Forsch.-Inst. Senckenberg*, **250**, 1–46.

Popov, S.V., Antipov, M.P., Zastrozhnov, A.S., Kurina, E.E. and **Pinchuk, T.N.** (2010) Sea-level fluctuations on the northern shelf of the Eastern Paratethys in the Oligocene-Neogene. *Stratigr. Geo. Correl.*, **18**, 200–224.

Ramstein, G., Fluteau, F., Besse, J. and **Joussaume, S.** (1997) Effect of orogeny, plate motion and land sea distribution on

Eurasian climate change over the past 30 million years. *Nature*, **386**, 788–795.

Richoz, S., Baldermann, A., Frauwallner, A., Harzhauser, M., Daxner-Höck, G., Klammer, D. and Piller, W.E. (2017) Geochemistry and mineralogy of the Oligo-Miocene sediments of the Valley of Lakes, Mongolia. *Palaeobio. Palaeoenv.*, **97**, 233–258.

Rögl, F. (1999) Mediterranean and Paratethys. Facts and hypotheses of an Oligocene to Miocene paleogeography (short overview). *Geol. Carpathica*, **50**, 339–349.

Salminen, R. (Chief-editor), Batista, M.J., Bidovec, M., Demetriades, A., De Vivo, B., De Vos, W., Duris, M., Gilucis, A., Gregorauskiene, V., Halamic, J., Heitzmann, P., Lima, A., Jordan, G., Klaver, G., Klein, P., Lis, J., Locutura, J., Marsina, K., Mazreku, A., O'Connor, P.J., Olsson, S., Ottesen, R.T., Petersell, V., Plant, J.A., Reeder, S., Salpeteur, I., Sandström, H., Siewers, U., Steenfeldt, A. and Tarvainen, T. (2005) *FOREGS Geochemical atlas of Europe, Part 1: Background Information, Methodology and Maps.* Geological Survey of Finland, Espoo, 525 pp, 36 figures, 362 maps.

Sanford, R.F. (1994) Hydrogeology of Jurassic and Triassic wetlands in the Colorado Plateau and the origin of tabular sandstone uranium deposits. *U.S. Geol. Surv. Prof. Paper*, **1548**, 1–40.

Sanz, M.E., Alonsozarza, A.M. and Calvo, J.P. (1995) Carbonate pond deposits related to semiarid alluvial systems – examples from the Tertiary Madrid Basin, Spain. *Sedimentology*, **42**, 437–452.

Schulz, M. and Mudelsee, M. (2002) REDFIT: estimating red-noise spectra directly from unevenly spaced paleoclimatic time series. *Comput. Geosci.*, **28**, 421–426.

Sheldon, N.D. and Tabor, N.J. (2009) Quantitative paleoenvironmental and paleoclimatic reconstruction using paleosols. *Earth-Sci. Rev.*, **95**, 1–52.

Singer, A. (1988) Illite in aridic soils, desert dusts and desert loess. *Sed. Geol.*, **59**, 251–259.

Singer, A. and Stoffers, P. (1980) Clay mineral diagenesis in two East-African Lake-Sediments. *Clay Miner.*, **15**, 291–307.

Sobel, E.R., Hilley, G.E. and Strecker, M.R. (2003) Formation of internally drained contractional basins by aridity-limited bedrock incision. *J. Geophys. Res.-Soil Earth*, **108**, 2344. https://doi.org/10.1029/2002jb001883.

Sobel, E.R., Chen, J. and Heermance, R.V. (2006) Late Oligocene-Early Miocene initiation of shortening in the Southwestern Chinese Tian Shan: implications for Neogene shortening rate variations. *Earth Planet. Sci. Lett.*, **247**, 70–81.

Sun, J.M. and Windley, B.F. (2015) Onset of aridification by 34 Ma across the Eocene-Oligocene transition in Central Asia. *Geology*, **43**, 1015–1018.

Sun, J.M., Ye, J., Wu, W.Y., Ni, X.J., Bi, S.D., Zhang, Z.Q., Liu, W.M. and Meng, J. (2010) Late Oligocene-Miocene mid-latitude aridification and wind patterns in the Asian interior. *Geology*, **38**, 515–518.

Sun, J.M., Gong, Z.J., Tian, Z.H., Jia, Y.Y. and Windley, B. (2015) Late Miocene stepwise aridification in the Asian interior and the interplay between tectonics and climate. *Palaeogeogr. Palaeoclimatol. Palaeoecol.*, **421**, 48–59.

Taboada, T., Cortizas, A.M., Garcia, C. and Garcia-Rodeja, E. (2006) Particle-size fractionation of titanium and zirconium during weathering and pedogenesis of granitic rocks in NW Spain. *Geoderma*, **131**, 218–236.

Tang, H., Micheels, A., Eronen, J. and Fortelius, M. (2011a) Regional climate model experiments to investigate the Asian monsoon in the Late Miocene. *Clim. Past*, **7**, 847–868.

Tang, Z.H., Ding, Z.L., White, P.D., Dong, X.X., Ji, J.L., Jiang, H.C., Luo, P. and Wang, X. (2011b) Late Cenozoic central Asian drying inferred from a palynological record from the northern Tian Shan. *Earth Planet. Sci. Lett.*, **302**, 439–447.

Torres, M.A. and Gaines, R.R. (2013) Paleoenvironmental and Paleoclimatic Interpretations of the Late Paleocene Goler Formation, Southern California, USA, Based on Paleosol Geochemistry. *J. Sed. Res.*, **83**, 591–605.

Wang, B. (2006) *The Asian Monsoon*. Springer Praxis Books, Berlin, 788 pp.

Weedon, G. (2003) *Time-series Analysis and Cyclostratigraphy*. Cambridge University Press, Cambridge, 259 pp.

Yang, S.L., Ding, F. and Ding, Z.L. (2006) Pleistocene chemical weathering history of Asian arid and semi-arid regions recorded in loess deposits of China and Tajikistan. *Geochim. Cosmochim. Acta*, **70**, 1695–1709.

Zachos, J.C., Dickens, G.R. and Zeebe, R.E. (2008) An early Cenozoic perspective on greenhouse warming and carbon-cycle dynamics. *Nature*, **451**, 279–283.

Zachos, J.C., Mccarren, H., Murphy, B., Röhl, U. and Westerhold, T. (2010) Tempo and scale of late Paleocene and early Eocene carbon isotope cycles: implications for the origin of hyperthermals. *Earth Planet. Sci. Lett.*, **299**, 242–249.

Zheng, H.B., Wei, X.C., Tada, R.J., Clift, P.D., Wang, B., Jourdan, F., Wang, P. and He, M.Y. (2015) Late Oligocene-early Miocene birth of the Taklimakan Desert. *Proc. Natl Acad. Sci. USA*, **112**, 7662–7667.

9

Interpreting ambiguous bedforms to distinguish subaerial base surge from subaqueous density current deposits

BENJAMIN L. MOORHOUSE and JAMES D. L. WHITE

Geology Department, University of Otago, Leith St, Dunedin 9054, New Zealand (E-mail: blm687@gmail.com)

Keywords

Ambiguous bedforms, Northeast Otago, subaerial dunes, subaqueous dunes, Surtseyan volcanism, volcanogenic sedimentary structures.

ABSTRACT

Bedform geometries of volcanogenic sedimentary structures such as dune, low- to high-angle cross-stratification and planar stratification produced in subaqueous and subaerial environments can be very similar. This has historically created difficulties in unambiguously distinguishing primary from reworked deposits, and gas-deposited versus water-deposited ones. The origins of dunes and associated structures within the pyroclastic deposits of basaltic Surtseyan-style eruptions exposed in seacliffs along the Cape Wanbrow coastline of the Northeast Otago region in the South Island of New Zealand are such an example. Careful analysis of contextual information of these deposits, including granulometry, dune geometry, grading, sorting and particle hydraulic equivalence, has been completed to distinguish between major flow types in different environments. To determine the depositional setting of the Cape Wanbrow dune-bearing deposits, a number of related sediments have been examined in more detail including well-described examples of (i) subaerial dry pyroclastic deposits, (ii) subaerial moist pyroclastic deposits, (iii) aeolian deposits, (iv) deposits of unidirectional fluid-gravity water flows (e.g. rivers) and cyclic tidal flows, and (v) aqueous sediment-gravity flow deposits encompassing both pyroclastic material and those with non-pyroclastic material. From this compendium, a framework has been created that allows users to compare the physical controls that shape each example with the factors controlling bedform deposition in that environment. Identifying key characteristics of the Cape Wanbrow dunes and comparing multiple flow types and environments using the designed framework indicates that they were deposited by subaerial dry pyroclastic density currents. This conclusion has wider implications for the entire Surtseyan stack at Cape Wanbrow, because it indicates that at least this volcano (Rua[2]) became emergent and fully subaerial during its lifespan.

INTRODUCTION

Distinguishing between subaqueous and subaerial ancient volcanogenic deposits can be difficult if rock features are poorly preserved, exposures are limited, and independent palaeoenvironmental indicators are absent or ambiguous (Fisher & Schmincke, 1984; Cole & Decelles, 1991; Trofimovs *et al.*, 2007). In particular, assigning a subaqueous or subaerial interpretation to sedimentary structures such as dune, low- to high-angle cross-stratification and planar stratification has proven challenging because they can form in many settings and from many types of flow. This

means bedforms and stratigraphy with similar, if not identical, bedform geometries have been described from both subaerial and subaqueous environments. It is important to study such structures in great detail because recognizing a subaqueous or subaerial setting of deposition and distinguishing among major flow types is crucial for our understanding of volcano-sedimentary processes and environments (White, 1991, 1996) and for assessment of volcanic hazards during and after eruptions (Lorenz, 2007).

The issue of distinguishing between subaerially versus subaqueously formed dunes has arisen before. Wohletz &

Sheridan (1983) identified dunes at Pahvant Butte, Utah, as being of subaerial, base-surge origin and built this into a general model for subaerial emplacement of 'tuff rings' beneath tuff cones. The same beds were reinterpreted by White (1996) as subaqueously formed dunes deposited under combined-flow conditions prior to tuff cone emergence. Before pyroclastic density currents (surge; base-surge) dunes became widely recognized in the 1970s, many pyroclastic beds were inferred to have been reworked by water or wind simply on the basis of the presence of cross-bedding (Denny, 1940). Wentworth (1937) disagreed with Stearns (1935) about an aeolian versus primary and pyroclastic origin for the black tuff from Diamond Head tuff cone. Smith & Katzman (1991) noted the ease of incorrectly identifying aeolian tuffs as surge deposits in proximal settings. Nocita (1988) inadvertently described soft-sediment deformation in the Puye fan, New Mexico, as affecting pyroclastic surge deposits, but later retracted his interpretation in reply to a comment by McPherson *et al.* (1989) that demonstrated a fluvial origin for those deposits. This long history of difficulties in unambiguously distinguishing primary from reworked deposits, and gas-deposited versus water-deposited ones, illustrates the significance of the problem.

Herein, a framework has been developed using examples of dune bedforms produced by different flow types in both subaqueous and subaerial settings that will allow users to compare key characteristics of such ambiguous sedimentary structures in order to interpret depositional setting and flow type. Dunes and associated structures will be described from subaqueous and subaerial environments with the aim of introducing a set of generally applicable criteria for distinguishing subaerially from subaqueously formed dunes in ancient volcanogenic successions.

To test the robustness of the framework, dunes and associated structures have been described in pyroclastic deposits of basaltic Surtseyan-style eruptions exposed in seacliffs along the Cape Wanbrow coastline of Northeast Otago, in the South Island of New Zealand (Fig. 1). These deposits lie within a monogenetic volcanic field known as the Waiareka-Deborah Volcanics (Gage, 1957; Coombs *et al.*, 1986; Hoernle *et al.*, 2006) and were formed by eruptions between 38 and 33 Ma that occurred on a continental shelf during a period when it was generally submerged, and at a site 10s of kilometres offshore from the nearest inferred contemporary shorelines. The stratigraphy of Cape Wanbrow indicates that eruptions produced multiple volcanoes whose edifices overlapped within a small area, but separated by millions of years. The small Cape Wanbrow highland includes the remains of six volcanoes that are distinguished by discordant to

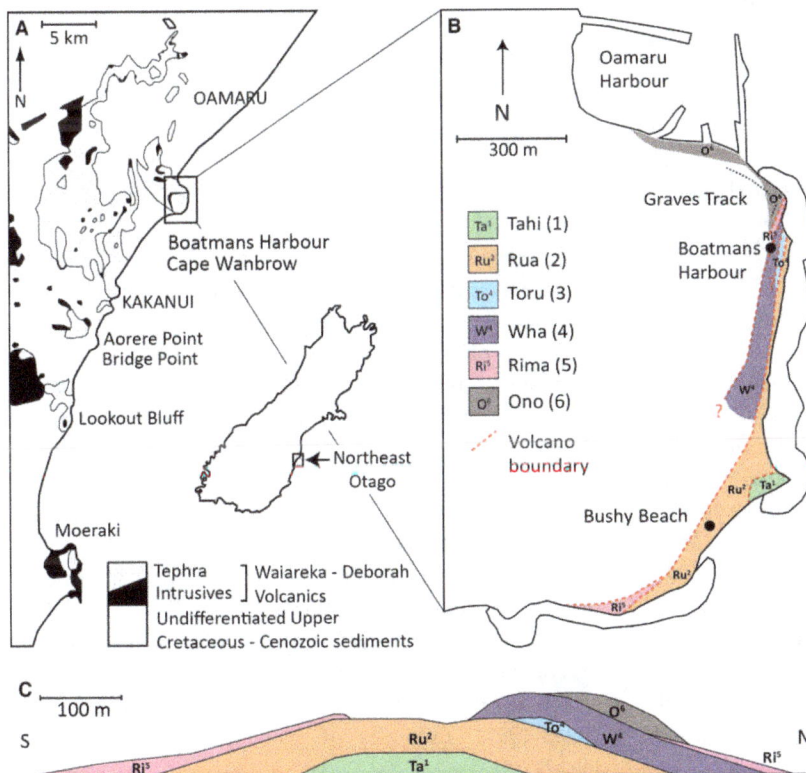

Fig. 1. (A) Location and geological setting of the Cape Wanbrow volcanic succession, Northeast Otago, South Island, New Zealand, (B) location points at Cape Wanbrow and projected intervolcano boundaries, (C) relationships of the remains of six submarine volcanoes that erupted in more or less the same location.

locally concordant intervolcano contacts marked by biogenic accumulations or other slow-formed features (Fig. 1) (Moorhouse *et al.*, 2015). Dunes and associated sedimentary structures, including low- to high-angle cross-stratification and planar stratification, only occur within the top 40 m of the second volcano of the Cape Wanbrow stack. The deposits of this volcano, named Rua[2], are located at the north end of the Cape (Fig. 1). By determining whether this key suite of bedforms were emplaced subaqueously or subaerially will aid in better understanding the Cape Wanbrow succession, and particularly an evolution of Rua[2] Volcano.

DUNE-FORMING CURRENTS IN SUBAQUEOUS VERSUS SUBAERIAL SETTINGS

Before interpreting the depositional setting and flow type responsible for the emplacement of the ambiguous dune-bearing Cape Wanbrow deposits, it was necessary to first evaluate similar structures described from both subaqueous and subaerial environments that have a primary (e.g. pyroclastic density currents) or re-worked (e.g. aeolian re-sedimentation) origin.

To do this, key features of well-described examples of dunes are considered from (i) subaerial dry pyroclastic deposits, (ii) subaerial moist pyroclastic deposits, (iii) deposits of aeolian flows, (iv) subaqueous unidirectional and cyclic flow deposits (e.g. rivers, tides) and (v) subaqueous sediment-gravity flow deposits encompassing those comprising pyroclastic material and those with non-pyroclastic material. This procedure includes examination of the physical controls that shape each example and the factors controlling bedform deposition in each environment with the goal of using key deposit characteristics to determine a specific flow type and environment for deposition.

Subaerially developed dunes

First we consider dunes in volcanic subaerial environments developed from pyroclastic sediment-gravity flows (i.e. pyroclastic density currents) or the re-sedimentation by aeolian currents. Dunes in volcanic subaerial environments deposited by the former are thought to be formed by low-density, high-velocity and turbulent pyroclastic density currents (surges; Wright *et al.*, 1980; Cole, 1991; Valentine & Fisher, 2000; Bridge & Demicco, 2008) with a first-order distinction proposed between moist currents, in which water droplets promote particle cohesion and aggregation, and dry currents in which no particle aggregation takes place (Allen, 1982; Bridge & Demicco, 2008).

Subaerial dry deposits

Dry dilute pyroclastic density currents (pyroclastic surges; 'dry' base surges) are low-density, high-velocity and dominantly turbulent (Walker, 1984; Cole, 1991; Bridge & Demicco, 2008; Douillet *et al.*, 2013). Deposits are characteristically fines-poor and therefore tend to be better-sorted than their wet equivalents (most commonly have a sorting value of 1·5 σ). A key feature is the absence or lack of accretionary lapilli and/or vesicular tuff, reflecting a lack of condensed water droplets during emplacement (Sheridan & Updike, 1975; Fisher & Schmincke, 1984; Walker, 1984; Cole, 1991; Valentine & Fisher, 2000; Bridge & Demicco, 2008).

Commonly described characteristics of dunes are alternating coarser-grained and finer-grained sand layers that grade upwards from a planar stratification into dunes with cross-beds of relatively low angles. These dunes are broadly similar in shape to those in aeolian deposits, but comparatively higher sedimentation rates are reflected in their poorly sorted nature (Valentine & Fisher, 2000). The laminae within dunes of dry pyroclastic density currents usually have progressively steepening dips on the lee sides of dunes and build up in a down-current direction from either planar beds or previously formed dunes. Layers are usually truncated and overlain by continuous stoss- to lee-side layers with a sigmoidal shape that tend to thicken on the leeward side. Dunes have down-current-migrating crests with very few displaying up-current-migrating ones. It is uncommon to see bed bases that behaved plastically beneath ballistic particle impacts, and deposits generally reflect a tractional depositional style. Ash grade deposits in particular can look similar to those of wind-blown sand and dust, but the pyroclastic dunes are not as well-sorted (Valentine & Fisher, 2000).

Particles in dry dilute pyroclastic density currents, commonly depositing at the base of the current, move in traction carpets (Valentine & Fisher, 2000). Thick sequences of alternating fine- to coarse-grained pyroclastic material represent deposition by a highly pulsatory current formed by its inherent turbulence or multiple closely timed explosions. Common erosional truncations associated with these layers indicate that currents were at times erosional (Schmincke *et al.*, 1973; Lorenz, 1986; Cole, 1991; White, 1991; Sohn, 1996; Dellino *et al.*, 2004). The progressive or regressive nature of dunes is thought to be due to changes in velocity and flow regime, which have been suggested as an important ways of controlling bedform migration direction (Schmincke *et al.*, 1973; Valentine, 1987; Chough & Sohn, 1990; Cole, 1991). Examples of geometries can be seen in Fig. 2.

Examples of dune architecture from various flow types

Cape Wanbrow (1)
Thickening lee-side layers
indicate migration away
from source = progressive

Overlying sigmoidal thin
beds (1–2 cm) with a fine-
to medium-grain-size

Cape Wanbrow (2)

Asymmetric lenticular sets with convex tops
that migrate in direction of current

Alternating layers of fine
lapilli to fine-grained ash form
steepening lee-sides which fine upwards

Some thickening
and thinning of layers around crests and swales

Subaerial		Subaqueous	
Dry surge	Moist surge	Sediment-gravity flow	Fluid-gravity water flow
Type a of Cole, 1991		**Pyroclastic:**	**Rock record: (e.g. fluvial)**
	Type VI of Schmincke *et al.*, 1973		
	Type d of Cole, 1991	White, 1996	Type 1 of Fielding, 2006
Type c of Cole, 1991			
	Sheridan and Updike, 1975	**Non-pyroclastic:**	Type 4 of Fielding 2006
		Mulder, 2009	**Flume experiments:**
Type e of Cole, 1991		Prave and Duke, 1990	Alexander *et al.*, 2001
	Fisher and Waters, 1970		
	Crowe and Fisher, 1973		Yokokawa, 2010
		Walker, 1967	
Fluid-gravity water flow (e.g. eolian; Smith and Katzman, 1991)			Cartigny *et al.*, 2014
		Skipper, 1971	Middleton., 1965
1 m	← Flow direction		

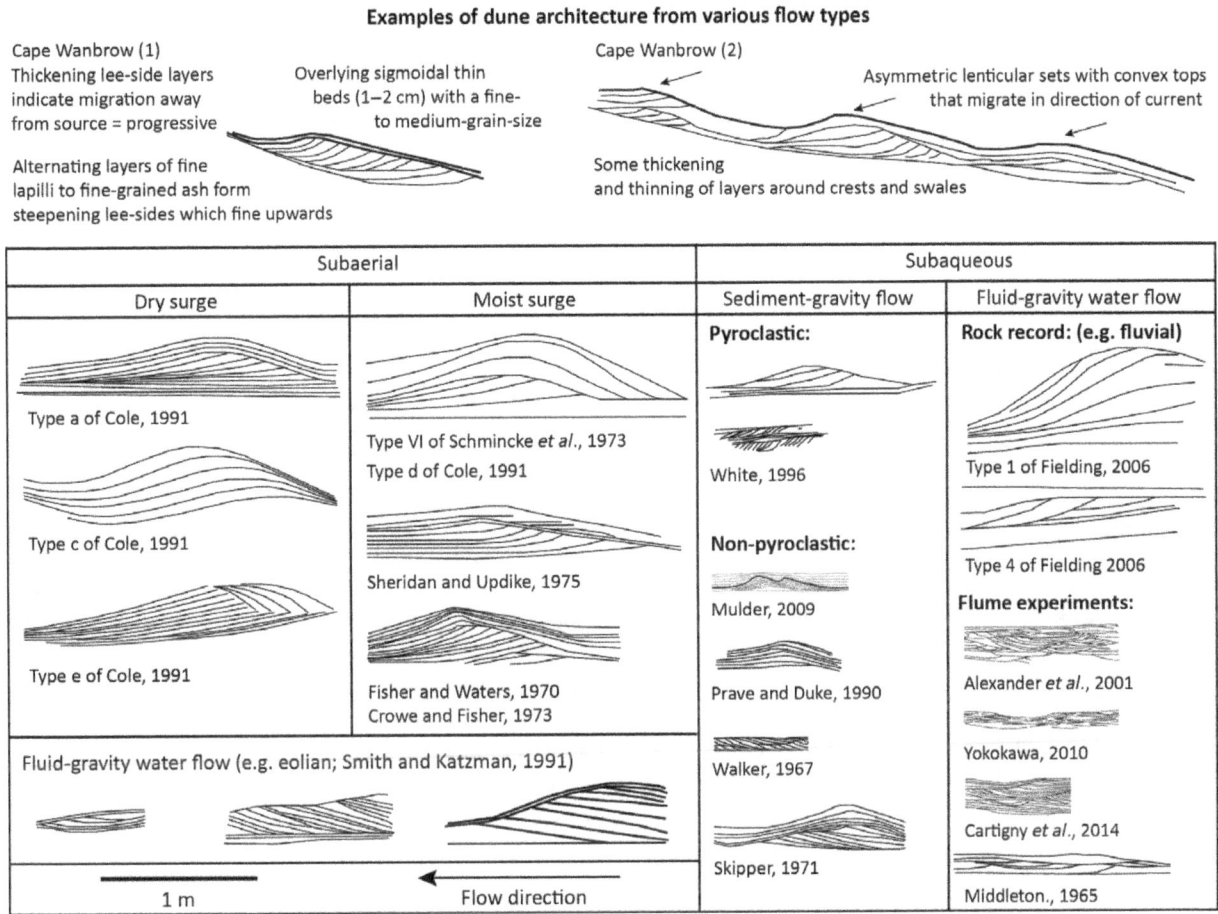

Fig. 2. Sedimentary structures in the literature that resemble elements of the architecture of the dunes found at Cape Wanbrow.

Subaerial moist deposits

Moist pyroclastic density currents are low-density, high-velocity currents (Walker, 1984; Cole, 1991; Bridge & Demicco, 2008) that have water droplets present which cause fine-ash grains to cohere or adhere to larger grains, leading to the formation of accretionary and armoured lapilli. Particle aggregation increases near-vent deposition of fine-grained pyroclastic material, which causes deposits to be more poorly sorted than their dry equivalents. Some authors have suggested that cohesion produces steep sided dunes and associated cross-stratification (Allen, 1982; Valentine & Fisher, 2000). Cohesion causes ash particles to enter the depositional system as clusters rather than as individual particles, leading to the aggradation of plastically behaved beds (Valentine & Fisher, 2000). Soft-state deformation is common within deposits of moist pyroclastic currents, with prominent bedding sags formed beneath emplaced ballistic fragments (Schmincke *et al.*, 1973; Cole, 1991; Valentine & Fisher, 2000). Similar to dry pyroclastic currents, deposits of moist currents display

alternating thin beds of fine- to medium-grained ash that are often truncated and overlain by a thin, fine-grained and continuous layer from stoss side to lee side.

Dunes of moist pyroclastic density currents have a large variation in wavelength across the literature due to the steepness of the density gradient. Steeper gradients will result in shorter wavelengths and may be due to the distance from the vent (Valentine & Fisher, 2000). Dunes in the deposits of moist pyroclastic density currents can display progressive, regressive or stationary migration directions due to changes in velocity and flow regime which have been suggested as important ways of controlling bedform migration direction (Schmincke *et al.*, 1973; Valentine, 1987; Chough & Sohn, 1990; Cole, 1991). Examples of geometries can be seen in Fig. 2.

Subaerial aeolian flow deposits

The entrainment of pyroclastic material in aeolian flows is an important mechanism for re-deposition of pyroclastic material. The dunes produced by aeolian re-sedimentation are significant as they have distinctive characteristics that aid discrimination between a primary versus reworked origin. Wind-blown volcaniclastic deposits are typically fine to medium ash and very well sorted, lacking dense lapilli, blocks and very fine-grained ash. An abundance of mixed of non-volcanic detritus can exist, and generally, fragments are better-rounded than those observed in pyroclastic density current deposits (Smith & Katzman, 1991; Valentine & Fisher, 2000; Hooper et al., 2012). Wind-blown dunes have migration and facies variations that do not radiate from a central point, for example a volcanic source (Valentine & Fisher, 2000). Typically, only lee-side structures are well-preserved and dune crests are truncated by planar erosional surfaces. The geometry of wavelengths and wave heights in aeolian flow deposits tends to be intermediate between dunes of dry subaerial pyroclastic density currents (low-angle cross-laminae) and moist subaerial pyroclastic density current deposits (high-angle cross-laminae). Examples of geometries can be seen in Fig. 2.

Subaqueously developed dunes

Dunes forming under water can be produced by flows ranging from turbidity currents in the deep sea (Prave & Duke, 1990; Mulder et al., 2009) to shallow streamflow in fluvial settings (Fielding, 2006). To understand dune formation in subaqueous settings, sediment-gravity flows (e.g. marine or lacustrine density currents), unidirectional flows (e.g. rivers, streams) and cyclic tidal flows have been examined.

Dunes in deposits of unidirectional water flows (e.g. streamflow)

The bedforms developed under unidirectional flows can be plotted as a function of flow strength. Velocity at a given depth of flow is plotted against (equivalent) grain size (Fielding, 2006). At low flow velocities, termed the lower flow regime, asymmetrical ripples form in finer sands and with greater flow power a lower plane bed forms in coarser-grained sediment. The lower plane bed phase condition then transforms into a dune phase with a further increase in flow strength. At higher flow strengths, termed the upper flow regime, dunes and ripples become washed out to form an upper plane bed, and at still higher flow velocity antidunes are formed in the antidune stability field. Dunes with larger grain sizes, however, may pass directly into the antidune stability field with increasing flow velocity (Fielding, 2006). When the flow strength reaches extreme values they produce chute-and-pool conditions in the uppermost flow regime, representing positive surges slowing down and

forming hydraulic jumps (Schmincke et al., 1973; Alexander et al., 2001; Fielding, 2006; Postma et al., 2009; Cartigny et al., 2014). The key features of bedforms produced by unidirectional currents in flume experiments, and their equivalents in the rock record are summarized below.

Examples from unidirectional-flow flume experiments

Flume experiments documenting the flow characteristics and resultant deposits from subcritical and supercritical flows have offered insight into the formation of dunes, antidunes, chutes-and-pools and cyclic steps (Middleton, 1965; Jopling & Richardson, 1966; Hand, 1974; Cheel, 1990; Best & Bridge, 1992; Alexander et al., 2001; Yokokawa et al., 2009; Cartigny et al., 2014). Few of the latter structures and facies have been recognized and analysed from the sedimentary record, perhaps because they form during high-energy events and have poor preservation potential. Alternatively, the scarcity may reflect common misidentification of such structures and facies (Fielding, 2006; Cartigny et al., 2014).

Subcritical and supercritical flows in flume experiments have produced repeatable deposits across numerous experiments (Middleton, 1965; Hand, 1974; Alexander et al., 2001; Duller et al., 2008; Yokokawa et al., 2009; Cartigny et al., 2014). In flume experiments, sedimentary structures such as sub-horizontal plane beds, lenticular sets with boundary-conformable laminae, lenticular sets with convex tops that increase with curvature at higher flow energies, and asymmetric dunes and antidunes with long wavelengths and low amplitudes are characteristic of deposition by increasing flow energies that pass from an upper flow regime stage through to unstable antidune stage and chute-and-pool conditions (Alexander et al., 2001; Cartigny et al., 2014). Increasing flow energies ultimately results in cyclic steps; a series of slowly upslope migrating bedforms (steps), where each downward step (the lee side of the bedform) is manifested by a steeply dropping flow passing through a hydraulic jump before re-accelerating on the flat stoss side (Middleton, 1965; Hand, 1974; Alexander et al., 2001; Duller et al., 2008; Cartigny et al., 2011, 2014). Examples of geometries can be seen in Fig. 2.

Asymmetric dunes with long wavelengths and low amplitudes that locally aggrade to form a stack of dune structures are similar to unstable antidune and chutes-and-pools described from deposits of subaqueous supercritical flows, and must therefore be added to the list of dune types to be considered during interpretation. Examples are known from deep marine deposits (Prave & Duke, 1990; Mulder et al., 2009), fluvial ones (Fielding, 2006) and density current flume experiments (Middleton,

1965; Hand, 1974; Alexander *et al.*, 2001; Duller *et al.*, 2008; Cartigny *et al.*, 2014).

Examples from the rock record

Unidirectional water flows are dominant in fluvial settings (Miall, 1985; Fielding, 2006). Characteristics of dunes produced in the dune, and in upper flow regime, antidune stability fields are most similar to the dune structures produced by pyroclastic density currents, so it is important to include these in the framework for determining flow type and depositional setting for dune formation. A wide spectrum of deposit architectures can be produced by streamflow, and Fielding (2006) proposed an Upper Flow Regime Sheets, Lenses and Scour Fills element (upper flow regime = UFR), established to encompass sand-dominated, large-scale, upper flow regime architectural elements. UFR elements in fluvial deposits have been rarely reported because until recently structures within the element have not been recognized as having formed under the upper flow regime. Fielding (2006) identified a multitude of such UFR elements, which include sigmoidal and low-angle cross-bedding, planar lamination, flat and low-angle lamination with minor convex-upward elements, convex-upward bedforms that dip both down-current and up-current with low-angle cross-bedding and symmetrical drapes, and backsets terminating up dip against an upstream-dipping erosion surface. These structures characteristically formed at the transition from dunes to upper plane bed, followed by a transition to the antidune stability field, and finally recording chute-and-pool conditions in the uppermost flow regime (Schmincke *et al.*, 1973; Alexander *et al.*, 2001; Fielding, 2006; Postma *et al.*, 2009; Cartigny *et al.*, 2014). These structures are therefore emplaced by flows that are broadly increasing in strength (Cheel, 1990; Fielding, 2006).

Particular structures described by Fielding (2006) are comparable with dunes at Cape Wanbrow. Type 1 bedforms (fig. 7; Fielding, 2006) were described as having a sigmoidal shape with low-angle cross-bedding that grades laterally and vertically into planar lamination and is interpreted as recording the transition between the dune and upper plane bed stability field. Type 4 bedforms (Fig. 7; Fielding, 2006) are minor, convex-upward and associated with low-angle cross-bedding and symmetrical drapes and are interpreted as having formed in the antidune stability field, therefore recording a flow broadly increasing in strength (Fielding, 2006). Examples of geometries can be seen in Fig. 2.

Preservation of structures formed in the upper flow regime is favoured by rapid changes in flow stage followed by a falling stage too short-lived to rework

sediment into lower regime bedforms (Jones, 1977; Fielding, 2006). This transition allows for rapid aggradation that records first, the dune phase conditions, followed by the abrupt transition into upper plane bed conditions and ultimately recording antidune conditions. Therefore, deposition of a UFR architectural unit records the progressive infilling of a single channel that is well-preserved due to the rapid rate of sediment aggradation (Rust & Gibling, 1990; Fielding, 2006).

Dunes in subaqueous deposits of sediment-gravity flows

Subaqueous pyroclastic density currents are a mixture of erupted particles and water driven across the sea floor by gravity, and result from either a subaqueous eruption or when a subaerial current travels into and mixes with water (Cole & Decelles, 1991; Carey, 2000; Freundt, 2003; Trofimovs *et al.*, 2007). There has been much more work on non-volcanic subaqueous density currents, or particulate sediment-gravity flows, in which sediment motion is in response to gravity and moves the interstitial fluid (Middleton & Hampton, 1973, 1976; Fisher, 1983). These offer insight into the behaviour and deposits of volcanic currents in subaqueous environments. Examples of flow types are discussed below.

Examples of non-pyroclastic material

Dunes are the most common type of bed wave in subaqueous sediment-gravity flows transporting sand and gravel. They are produced when sediment-transport rate increases on rippled beds or lower-stage plane beds when mean grain size exceeds 0·1 mm (Fisher, 1983; Bridge & Demicco, 2008). Dunes produced by subaqueous sediment-gravity flows comprising non-volcanic material can form in a large range of grain sizes (−2 to 3 ϕ) with most comprising grains between medium sand to silt (1 to 3 ϕ) deposited with moderate to poor sorting (1·0 to 2·0 σ), especially on lee-side slopes. Cross-stratification with low-angle, gently dipping lee sides and stoss sides that can be both asymmetrical and symmetrical usually build-up from conformable or erosional planar surfaces. Dunes have broad wavelengths (*ca* 20 to 60 cm) and low amplitudes (*ca* 2 to 6 cm) that can occur within bedsets that extend laterally for a few tens of metres (Walker, 1967; Skipper, 1971; Prave & Duke, 1990; Mulder *et al.*, 2009). Dunes discussed by Prave & Duke (1990) and Mulder *et al.* (2009) from the Upper Cretaceous calciclastic turbidites record a wave-like stratification within continuous thin layers (5 to 20 cm) that resembles hummocky or swaley cross-stratification.

Dunes within deposits of non-pyroclastic sediment-gravity flows have been interpreted as forming by subcritical flows (Cartigny *et al.*, 2011). The wave-like structures of Prave & Duke (1990) and Mulder *et al.* (2009) from the Upper Cretaceous calciclastic turbidites exposed in the western Basque Pyrenees have been interpreted as the product of Kelvin-Helmholtz instabilities formed at the upper flow interface of a turbidity current with surrounding ambient sea water. The turbidity current moves down-slope but, at certain times and places during deposition, there are bedforms that migrate up-current (Hand, 1974; Prave & Duke, 1990; Mulder *et al.*, 2009; Cartigny *et al.*, 2014). Large-scale dune-like structures on levees and canyon floors of submarine fan systems with symmetrical geometries combined with upslope migration directions described by Cartigny *et al.* (2011) are interpreted as antidunes formed by transitional to supercritical flows.

Pyroclastic material

Reports of dunes in pyroclastic subaqueous sediment-gravity flow deposits are remarkably rare with only a handful reported in the literature (Wright & Mutti, 1981; Heinrichs, 1984; White, 1996, 2001). Where documented they commonly display architectural characteristics such as, low-amplitude dunes with broad wavelengths and associated high-angle cross-stratification. These structures exist in poorly sorted, coarse to fine-ash beds that commonly have large clasts (>50 mm) present. Deposits of sediment-gravity flows made up of non-pyroclastic material display similar characteristics, including wavy or hummocky cross-stratification-like structures, but are commonly less poorly sorted than their pyroclastic equivalents and exist in mud to medium-grained sand. The absence or rarity of impact sags in pyroclastic material could be an indication of a subaqueous rather than subaerial setting. However, if volcanogenic sediments are known to be near glacial areas, glacial dropstones would need to be considered as they are common in deep water and impact sags do not always exist in subaerial deposits.

Deposits of both subaerial and subaqueous sediment-gravity flows exist within the pyroclastic material at Pahvant Butte, Utah, United States (White, 1996, 2001). First inferred by Gilbert (1890) and later White (1996, 2001), some pyroclastic deposits of Pahvant Butte have been shown to have a subaqueous origin based on palaeoenvironmental evidence. Subaqueously emplaced deposits are in places overlain by ripple-marked shoreline beds which are in turn capped by primary subaerial tuff cone deposits (White, 1996) and therefore record an evolution from emplacement by subaqueous currents through to emergence and subaerial deposition. Among deposits of four lithofacies that make up the subaqueously emplaced mound, lithofacies M3 of White (1996) comprises coarse sideromelane ash with dunes that display broad wavelengths (*ca* 1 m), low amplitudes (*ca* 15 cm) and an overall low-angle sigmoidal shape, as well as locally intercalated lenses of coarse ash showing high-angle cross-stratification. This lithofacies lacks any fine-ash layers, although locally ripples in finer-grained ash advance down the gently dipping dune foresets. Cross-sets dip outward from the cone and dunes grade laterally and vertically into sub-horizontally stratified lapilli ash and ash layers before commonly being truncated by low-relief erosion surfaces. No sag structures are observed (White, 1996). Examples of geometries can be seen in Fig. 2.

Subaqueous dune-bearing deposits dominated by pyroclastic material typically form from dilute turbulent eruption-fed currents formed entirely under water. In the case of Pahvant Butte, the structure of the dunes is interpreted as having developed by interaction between aqueous density currents (eruption-fed sediment-gravity flows) and oscillatory currents beneath surface waves that were probably long-wavelength waves formed by the eruption (White, 1996).

Description of dunes in Cape Wanbrow Deposits

Moorhouse *et al.* (2015) constructed a volcano-by-volcano overview of the Cape Wanbrow peninsula using distinct unconformable surfaces. Volcano names are in one of the national languages of New Zealand, te reo Māori, and the numbers in superscript aid identification of where in the sequence each volcano comes. Rua[2] Volcano was the second volcano built at Cape Wanbrow and its deposits buried the massive tuff breccias of Tahi[1] Volcano. The deposits of Rua[2] are the most extensively preserved and exposed of the volcanoes represented in Cape Wanbrow and are 300 m thick normal to bedding. The deposits display complex internal erosion features and sedimentary structures within interbedded fine- to coarse-grained ash and lapilli tuff. An abrupt, discordant but relatively planar erosion surface separates the deposits of Rua[2] from those of Toro[3] Volcano, of which only 5 m of massive tuff breccia containing rip-up clasts of underlying tuff and heavily deformed thin bedded fine- to medium-grained tuff is exposed. Three-dimensional asymmetric linguoid ripples are recorded in one bed towards the top of the Toro[3] deposits, only 1 m below the unconformity that separates Toro[3] deposits from the overlying ones of Wha[4] Volcano. The Toro[3]-Wha[4] boundary is a clearly discordant unconformable surface separating underlying bedded tuffs from overlying pillow lava. The surface is relatively planar and visible for 300 m northward from cliff top to beach level at Boatmans Harbour with an apparent dip of about 20° northwards. Where vegetation does not obstruct the contact, it is clear that underlying beds are truncated, and next to the boundary are slightly deformed (Moorhouse et al., 2015).

Thin interbeds of fine- to coarse-grained ash and fine lapilli tuff with isolated blocks (<10 cm in length) and units of massive tuff breccia make up much of the Cape Wanbrow succession that is described in detail as lithofacies E1 and E2 in Moorhouse et al. (2015) (Fig. 3). A dominant depositional feature that appears in deposits of Rua^2 Volcano, which lies near the top of the stack of Surtseyan volcanoes, comprises broad (ca 1 to 2 m), low-amplitude (ca 10 to 20 cm) dunes that grade laterally and vertically into sub-planar stratified tuff beds. Low-amplitude dunes often have a sigmoidal profile, or less commonly comprise lenticular bedsets with convex tops. These dunes occur within lithofacies E3 of Moorhouse et al. (2015) and are restricted to the top 40 m of the 340 m thick Rua^2 succession (Fig. 3). Within this 40 m interval, dunes with sigmoidal profiles occur most commonly towards the base and top, while dunes with obvious convex tops are restricted to the middle of the 40 m section (Fig. 3). No accretionary lapilli are observed in this lithofacies.

Dunes with a sigmoidal profile have progressively steepening lee-side layers that are built up from either planar beds or previously formed dunes. The steepening lee-side layers are truncated and overlain by beds that are continuous from stoss sides to lee sides and range from slightly asymmetrical to sigmoidal (Fig. 4). The steepening lee-side layers are 1 to 4 cm in thickness, but most do not exceed 2 cm. They comprise alternating layers of fine lapilli and coarse-grained to fine-grained ash which fine upwards across the structure. Therefore, a normal profile of a dune will have fine lapilli making up the shallow dipping lee-side laminae, progressing through to steeper lee-side layers of fine-grained ash (Fig. 5). The overlying sigmoidal shaped beds are always thin (1 to 2 cm), fine-grained to medium-grained and pass vertically and laterally into sub-parallel beds that are commonly thicker and of coarser grain size (Fig. 5). Continuous layers are commonly slightly thicker on the lee side, indicating they were migrating away from source and are therefore 'progressive' (Allen, 1982; Cole, 1991). Locally intercalated among dunes are lenses of planar bedding with low-angle

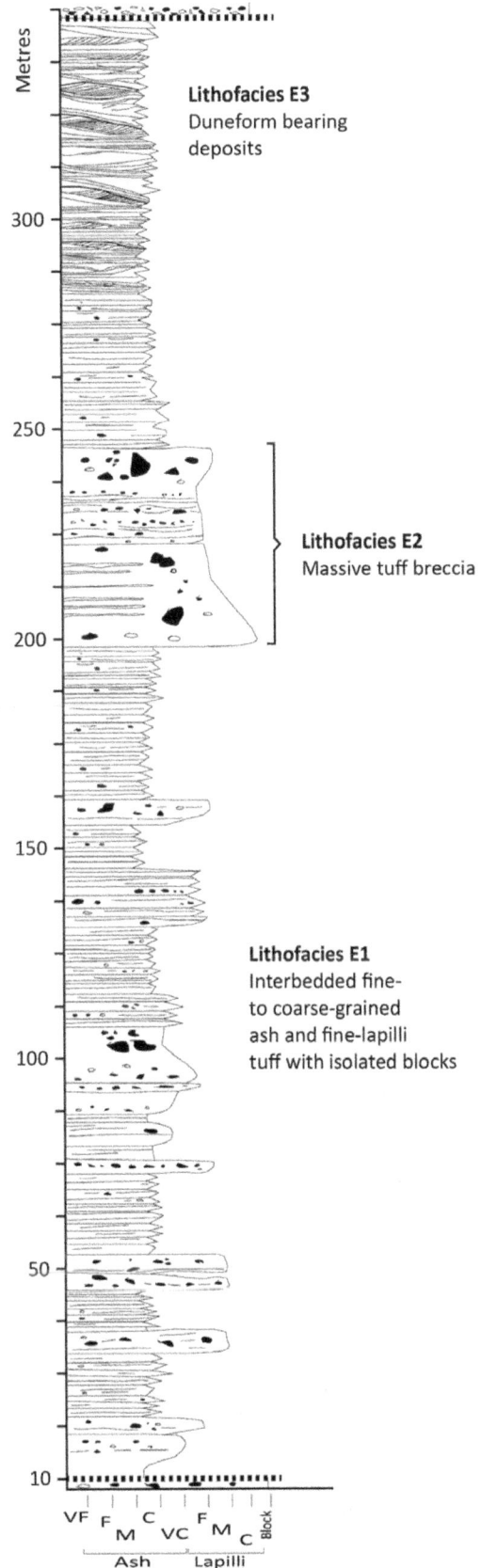

Fig. 3. Stratigraphic column of the Cape Wanbrow Rua^2 Volcano deposits. Thin interbeds of fine-grained to coarse-grained ash and fine lapilli tuff with isolated blocks is inferred to have been deposited subaqueously. These beds grade upwards into a lithofacies characterized by dune and low-angle cross-stratification.

Fig. 4. Dunes with a sigmoidal profile that comprise progressively steepening lee-side layers built up from either planar beds or previously formed dunes. Large hollow arrow indicates inferred palaeoflow direction.

cross-beds and rare examples of high-angle cross-stratification (Fig. 6). Dunes are present both in isolation, and in stacks or series that either aggraded or migrated down-slope (Fig. 7).

Rare lenticular sets with three-dimensionally convex tops are composed of thin beds (1 to 3 cm) of fine-grained to medium-grained ash and resemble hummocky cross-stratification with a wavy morphology where laminae thicken and thin to form convex-upward accretion hummocks and shallow depressions. These HCS-like structures are also non-symmetrical, with sets of laminae laterally prograding over short distances, displaying a uni-directional migration in the same direction as intercalated dunes with a sigmoidal profile (Fig. 8). They grade vertically into slightly wavy and planar lamination of a similar grain size that can also display a thickening and thinning around the dune crests. Dunes locally persist laterally for

several metres with bedform spacing averaging 40 to 50 cm in intervals that vary from 20 cm to 1 m, with most no longer than 30 cm (Fig. 8).

Constituents

The constituents of the volcaniclastic rocks of Cape Wanbrow are mainly juvenile particles (White & Houghton, 2006) with only traces of accidental material, such as quartz grains. The following constituents can be found in the rocks of Cape Wanbrow: vesicular pyroclasts originally of translucent basaltic glass (sideromelane) pyroclasts, now altered to a mixture of clay minerals and zeolites (palagonite); rare dense originally sideromelane particles; crystals and crystal fragments; lithic fragments; zeolite interstitial cement; and calcite interstitial cement. Variations in the abundance of pyroclasts, sideromelane

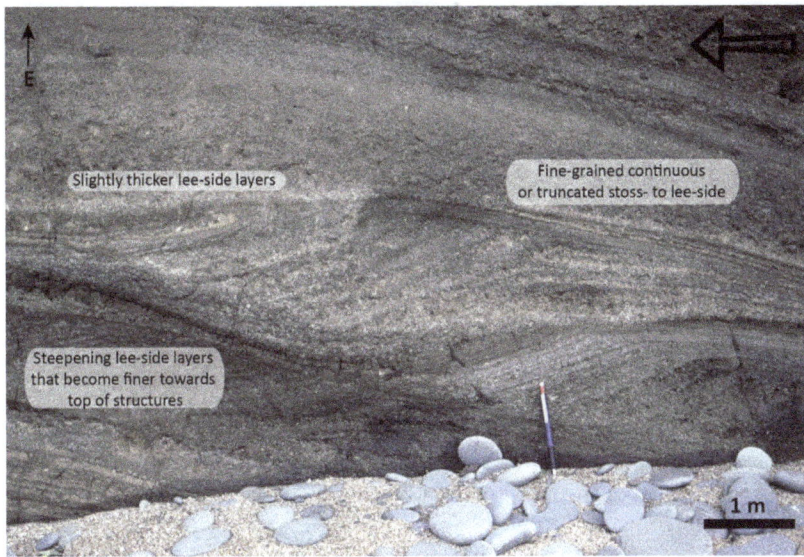

Fig. 5. Close-up of a dune with a sigmoidal profile. This structure represents a normal profile of a dune at Cape Wanbrow where fine lapilli make up the shallow dipping lee sides progressing through to steeper lee-side layers of fine-grained ash. Truncating and overlying these layers are sigmoidal shaped usually fine-grained thin beds. Large hollow arrow indicates inferred palaeoflow direction. Pen for scale is 15 cm.

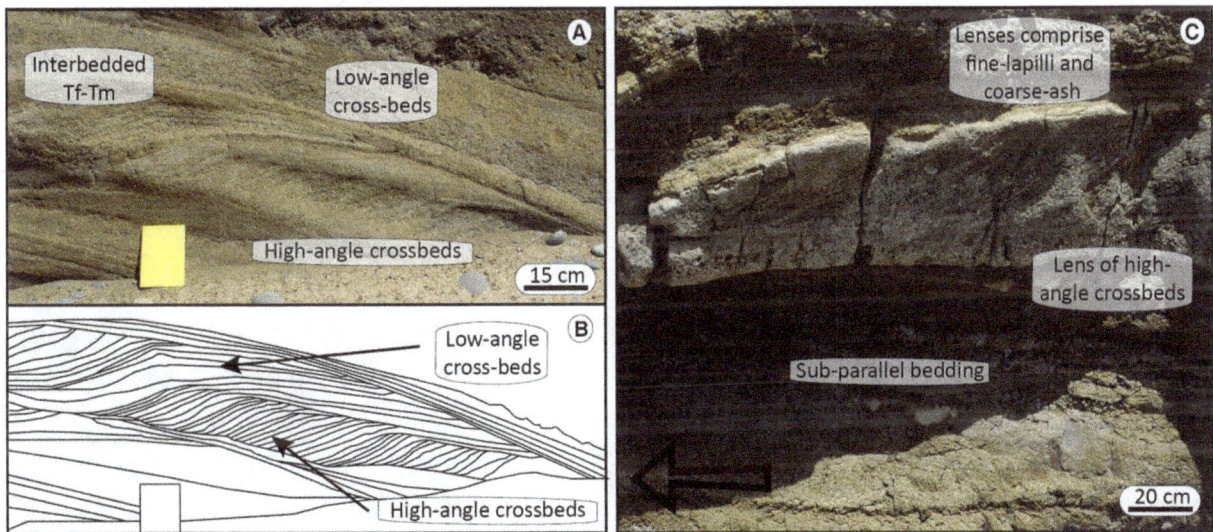

Fig. 6. (A) and (C Locally intercalated lenses of low-angle cross-bedded and planar-laminated layers along with rare examples of high-angle cross-stratification, (B) interpretation of panel (A) to highlight low-angle and high-angle cross-beds and planar stratification. Large hollow arrow indicates inferred palaeoflow direction for all panels.

pyroclasts, crystals and lithic fragments can be interpreted in terms of varying magmatic processes and eruption dynamics (Moorhouse *et al.*, 2015). Modification to original volcaniclastic populations (e.g. destruction of vesicular sideromelane pyroclasts and/or segregation of components) indicates reworking of volcanic source materials (Moorhouse *et al.*, 2015).

Granulometry

Samples were collected from three different dunes at Cape Wanbrow for standard thin section grain-size analyses in order to determine the size characteristics across dunes. Images of selected thin sections are shown in Fig. 9. The samples were collected from three points across the dune, the base, middle and top, in order to characterize visible particle size grading both vertically and laterally across the dune (Fig. 9). Unfortunately, the strong alteration of pyroclastic grains in the Cape Wanbrow deposit has destroyed details of original depositional textures, and in particular, there is local evidence for small matrix grains that seem elsewhere to have been obscured or destroyed. Cement has locally replaced even coarse ash, and therefore,

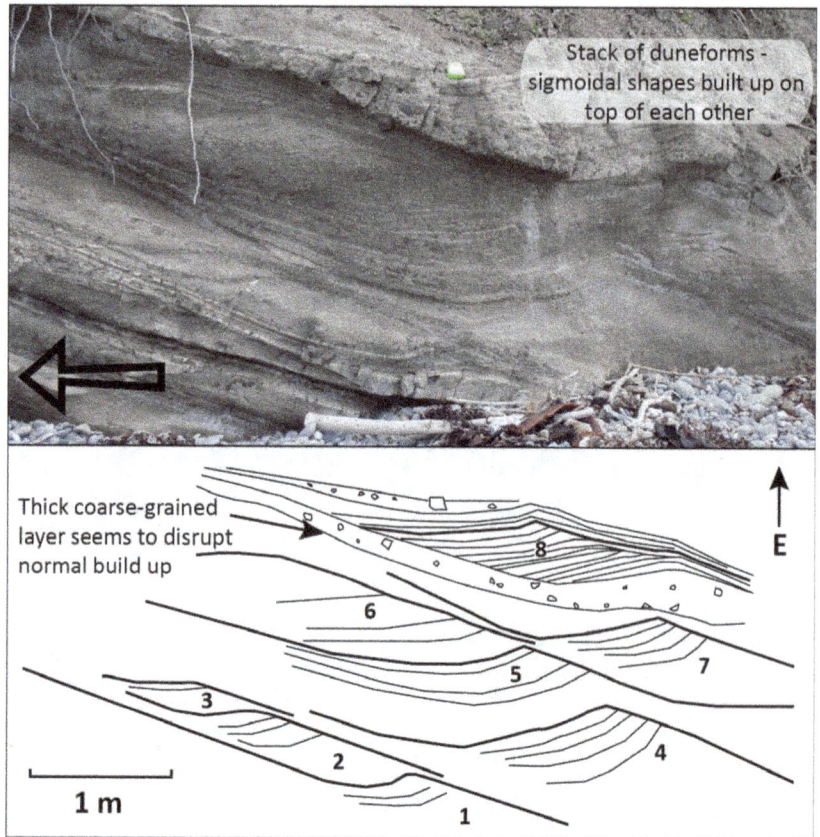

Fig. 7. A stack of dunes that are migrating down-current but also appear to be aggrading around every 3rd to 4th dune. Numbers represent the build-up of dune. Large hollow arrow indicates inferred palaeoflow direction.

Fig. 8. Continuous broad, low-amplitude and long wavelength dunes with convex tops. Small arrows point towards dune crests. Large hollow arrow indicates inferred consistent leftward transport.

granulometry may be missing a once-present fine-ash matrix (Moorhouse *et al.*, 2015). Image analysis was completed to determine quantitative grain-size distribution and sorting for nine samples in which sufficient textural detail was resolvable. Because of the alteration, the images had to be 'cleaned' by visual interpretation of grain boundaries. Grain-size bins were defined as the maximum horizontal intercept measured with a millimetre scale (Krumbein, 1935) and converted to phi (ϕ) units. Results from the samples, which include the base, middle and top of dunes, were plotted as cumulative curves (Fig. 9), and the median grain size (M_d)

and sorting coefficient (σ) were calculated (Folk, 1980) (Fig. 10):

$$M_d = \phi_{50}; \sigma = \frac{(\phi_{84} - \phi_{16})}{4} + \frac{(\phi_{95} - \phi_{5})}{6.6}$$

where ϕ_i is the grain size for which i % of the total material is smaller than the given grain size.

Measured vertical and lateral particle size grading and sorting are consistent with those apparent in the field and indicate that dunes display a fining upward trend across their internal layers, and are overall moderately to poorly sorted. Cumulative curves for each representative

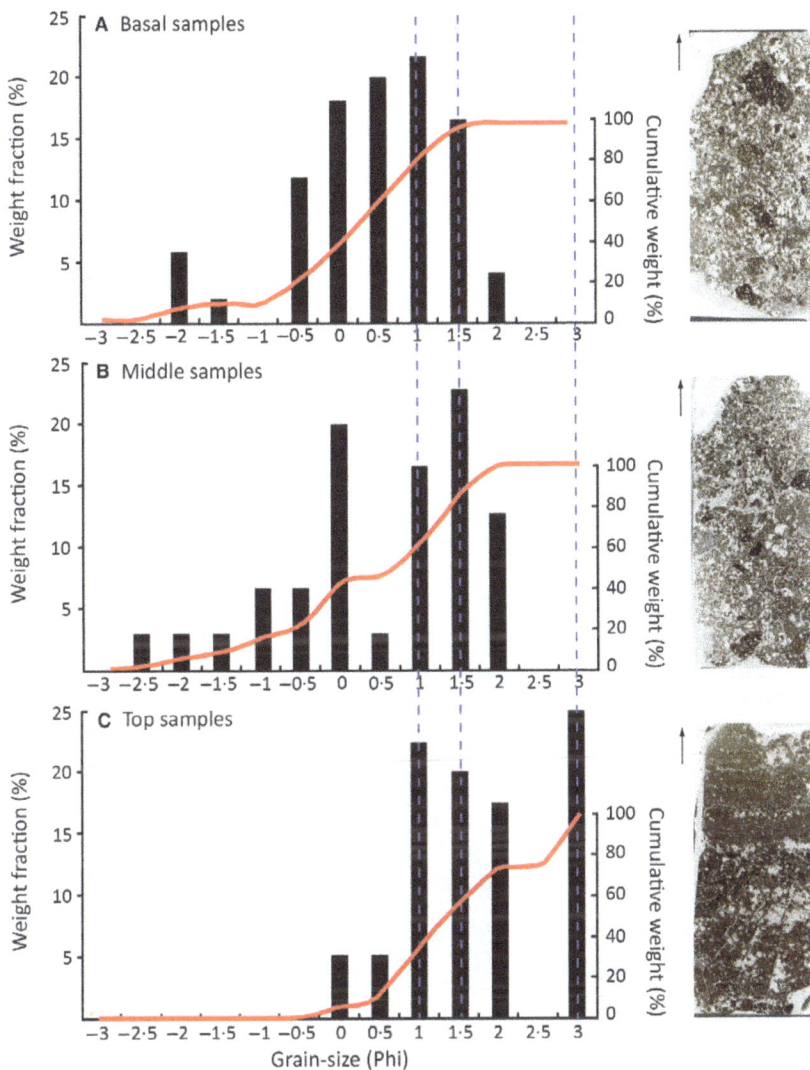

Fig. 9. Sample analysis of dunes. (A) Base, (B) middle and (C) top of dunes. The scale of the frequency curves on the left axis and of cumulative weight on the right axis. Thin section image of three selected samples from the (A) base, (B) middle and (C) top of dunes is displayed on the right. Arrow = way up.

sample across a dune show that the base, middle and top have distinctive size distributions and ranges (Fig. 10). Basal samples are better-sorted and coarser-grained (average σ, 0·96; M_d, 0·5) than those samples collected in the centres of dunes, which in turn are poorly sorted (average σ, 1·19; M_d, 1). Samples collected at the very tops of dunes are slightly better-sorted and finer-grained than the rest of the dune (average σ, 0·94; M_d, 1·5).

Superimposed onto a median diameter (M_d) versus sorting coefficient (σ) graph (Fig. 10) are the 'pyroclastic flow' and 'fall' fields of Walker (1971) and the 'base-surge' field defined by Crowe & Fisher (1973). The fields of subaqueous-gravity flows that produce dunes are also plotted from available non-pyroclastic data (Walker, 1967; Allen, 1970; Prave & Duke, 1990; Mulder et al., 2009) and pyroclastic data (Wright & Mutti, 1981;

Heinrichs, 1984; White, 1996) as well as a fluid-gravity flow field that encompasses both aeolian and fluvial data (Rust & Gibling, 1990; McLoughlin, 1993; Fielding et al., 1996; Fielding, 2006). The Cape Wanbrow samples overlap with the 'pyroclastic fall' field, the 'base-surge' field and both pyroclastic and non-pyroclastic sediment-gravity flow fields (Walker, 1967, 1971; Allen, 1970; Crowe & Fisher, 1973; Prave & Duke, 1990; Mulder et al., 2009).

Vesicular particles have a relatively low density when compared with standard quartz, feldspar or dense lithic grains, and as a result will behave differently. Settling-velocity data for the Cape Wanbrow samples cannot be obtained because the deposits are lithified. There are, however, settling-velocity data for similar vesicular particles from deposits of a shallow subaqueous basaltic eruption at Pahvant Butte determined from rapid

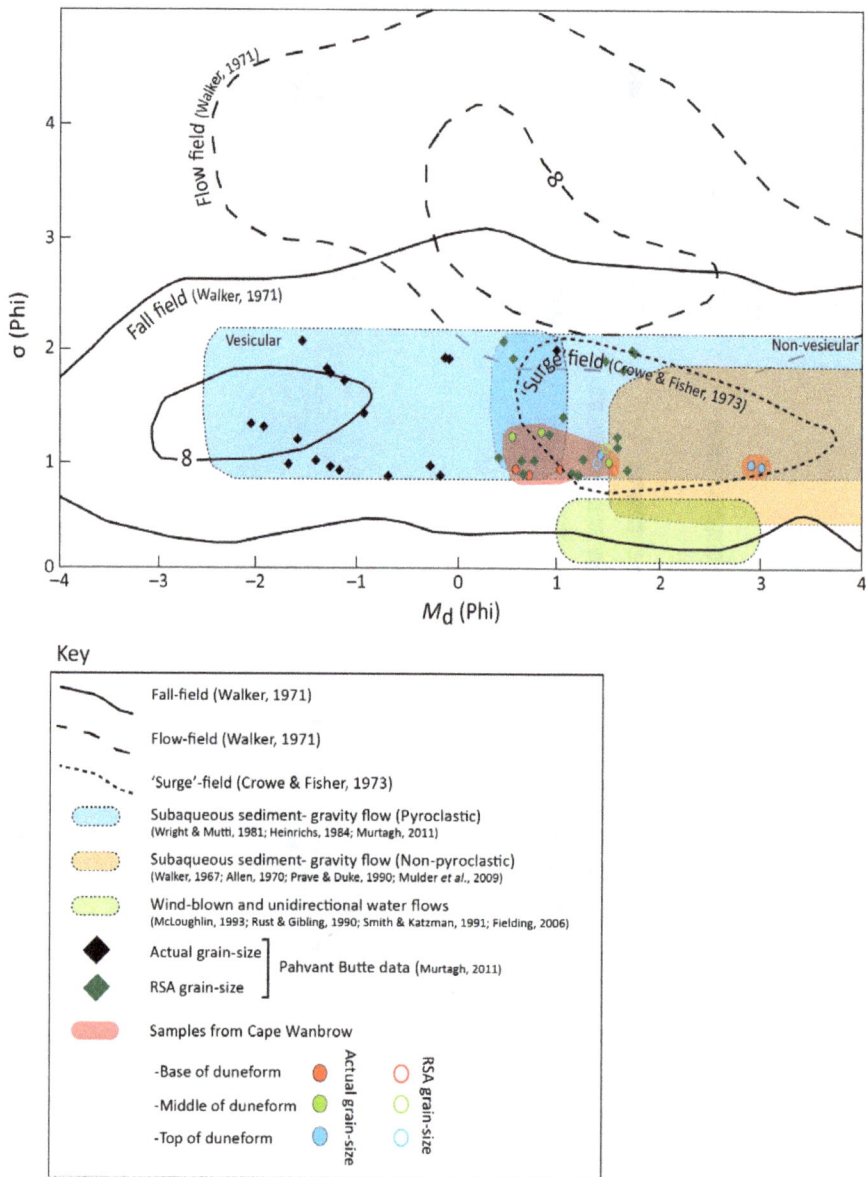

Fig. 10. Actual median diameter (M_d) and median diameter from settling-velocity equivalent grain sizes from rapid sediment analysis software (RSA) (M_d) versus sorting coefficient (σ) for all Cape Wanbrow samples. Superimposed onto graph are the flow fields and fall fields from Walker (1971), the surge field of Crowe & Fisher (1973) as well as fields created from the data on non-pyroclastic (Allen, 1970; Mulder *et al.*, 2009; Prave & Duke, 1990; Walker, 1967) and pyroclastic (Wright & Mutti, 1981; Heinrichs, 1984; White, 1996; Murtagh, 2011) subaqueous sediment-gravity flows and fluid-gravity flows (Rust & Gibling, 1990; McLoughlin, 1993; Fielding *et al.*, 1996; Fielding, 2006). The number '8' in 'flow' and 'fall' fields represents the greatest density of Walkers samples. Pyroclastic subaqueous-gravity flow fields are split into vesicular [data from Murtagh (2011)] and non-vesicular [data from Wright & Mutti (1981) and Heinrichs (1984)] groups. In order to compare pyroclastic and non-pyroclastic flow fields the granulometry data from Murtagh (2011) has been plotted twice, once with actual grain size (black diamonds) where they fall within the vesicular field and once using the settling-velocity equivalent grain sizes from rapid sediment analysis software (RSA) (green diamonds) where they plot within the non-vesicular field.

sediment analysis software (RSA) (Murtagh, 2011), and these data are used here to infer the 'quartz-equivalent' behaviour of the measured Cape Wanbrow deposits. Using data from Pahvant Butte and the scheme

presented by Oehmig & Wallrabe-Adams (1993) and Manville *et al.* (2002), settling-tube data were plotted against actual particle size to produce a settling-velocity curve. The median grain size of the Cape Wanbrow samples was then plotted against the curve to determine the size of quartz spheres that would settle at the same rate as the vesicular particles (Fig. 11). Coarse-sand-grade (*ca* −0·5 to 0 φ) vesicular pyroclasts have settling velocities equivalent to those of medium quartz sand (*ca* 1 to 1·5 φ) (Fig. 11). These values are consistent with the relationships between settling velocities of standard quartz grains versus vesicular particles recorded by Oehmig & Wallrabe-Adams (1993) and Manville *et al.* (2002).

Wavelength and wave height of dunes

Wavelengths versus wave heights of dunes are plotted in Fig. 12. The majority of dunes have wavelengths between 2 to 3 m and wave heights between 20 and 30 cm. Wave heights are determined by measuring the thickest part of a dune structure from flat base to the top of the crest due to undulations in an aggrading sequence commonly affecting the shapes of succeeding layers, resulting in the height of a single lamina changing upwards (Waters & Fisher, 1971; Crowe & Fisher, 1973; Schmincke *et al.*, 1973). For comparison, data on the dunes from Taal Volcano (Waters & Fisher, 1971; Crowe & Fisher, 1973), Ubehebe (Crowe & Fisher, 1973) and Laacher See (Schmincke *et al.*, 1973) have been plotted on Fig. 12. The Cape Wanbrow dunes appear to follow the same trend as those at Taal Volcano, Ubehebe and Laacher See (Waters & Fisher, 1971; Crowe & Fisher, 1973; Schmincke *et al.*, 1973).

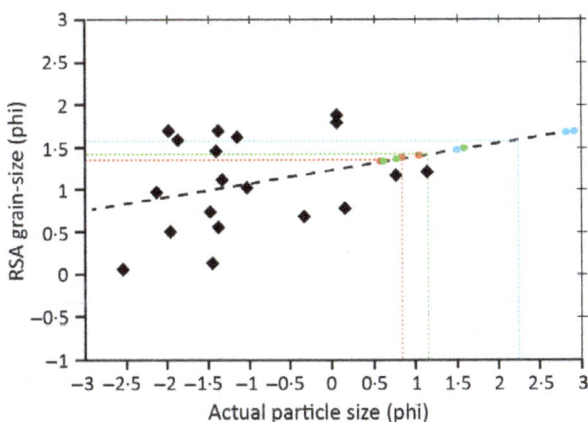

Fig. 11. Physical diameter (sieve-derived) versus hydraulic diameter (determined from rapid sediment analysis software – RSA) for Pahvant Butte deposits. Black dashed line – settling-velocity curve. Cape Wanbrow tuff samples from the base, middle and tops of dunes plotted along settling-velocity curve (colour coding: red base; green middle and blue top).

DISCUSSION

Having established generally applicable criteria for distinguishing subaerially from subaqueously formed dunes in ancient volcanogenic successions by examining well-described examples of deposits from five major flow types, it is now possible to evaluate characteristics of the Cape Wanbrow dunes and associated structures. Table 1 and Fig. 2 summarize the criteria created for deposit characteristics produced by each flow type.

Dunes at Cape Wanbrow have a sigmoidal profile, low-angle cross-stratification and progressively steepening lee-side layers in coarse-grained to fine-grained tuff and lapilli tuff. They built up from either planar beds or previously formed dunes and are truncated by asymmetric to sigmoidal layers that are continuous from stoss side to lee side (Moorhouse *et al.*, 2015).

Subaerially formed dunes

Characteristics that are consistent with deposition by subaerial dry pyroclastic density currents include; (i) a coarse to medium ash grain size (equivalent hydraulic grain size is medium-grained to fine-grained quartz) for most of the deposit (Fig. 9); (ii) poorly sorted nature (1 to 2 σ) (Fig. 10); (iii) consistent scale (wavelengths between 1 to 7 m) (Fig. 12) unlike those formed by non-pyroclastic sediment-gravity flows that can range from wavelengths of 50 cm (Prave & Duke, 1990; Mulder *et al.*, 2009) to 7 km (Cartigny *et al.*, 2011); (iv) no evidence of accretionary or armoured lapilli; (v) no soft-state deformation; (vi) no basal beds plastically deformed by ballistic blocks or bombs; and (vii) evidence for a traction-dominated sedimentation stage (Table 1). These characteristics provide robust support for the interpretation that the Cape Wanbrow dunes were deposited by dry pyroclastic density currents.

Although the Cape Wanbrow dunes have an overall coarse to medium ash grain size, they do fine upwards with tops of dunes being dominated by a fine-grain size. This feature supports deposition by moist pyroclastic density currents which usually contain more fine-grained ash material. However, the Cape Wanbrow deposits lack any evidence for accretionary or armoured lapilli, soft-state deformation or plastically behaving beds which is a strong argument against deposition by moist pyroclastic density currents and has therefore been ruled out as the flow type for these dunes (Table 1).

Dunes and associated structures deposited by wind share some features with those of Cape Wanbrow. Both have low-angle to intermediate-angle dips (25 to 35°) and an absence of soft-state deformation and plastically

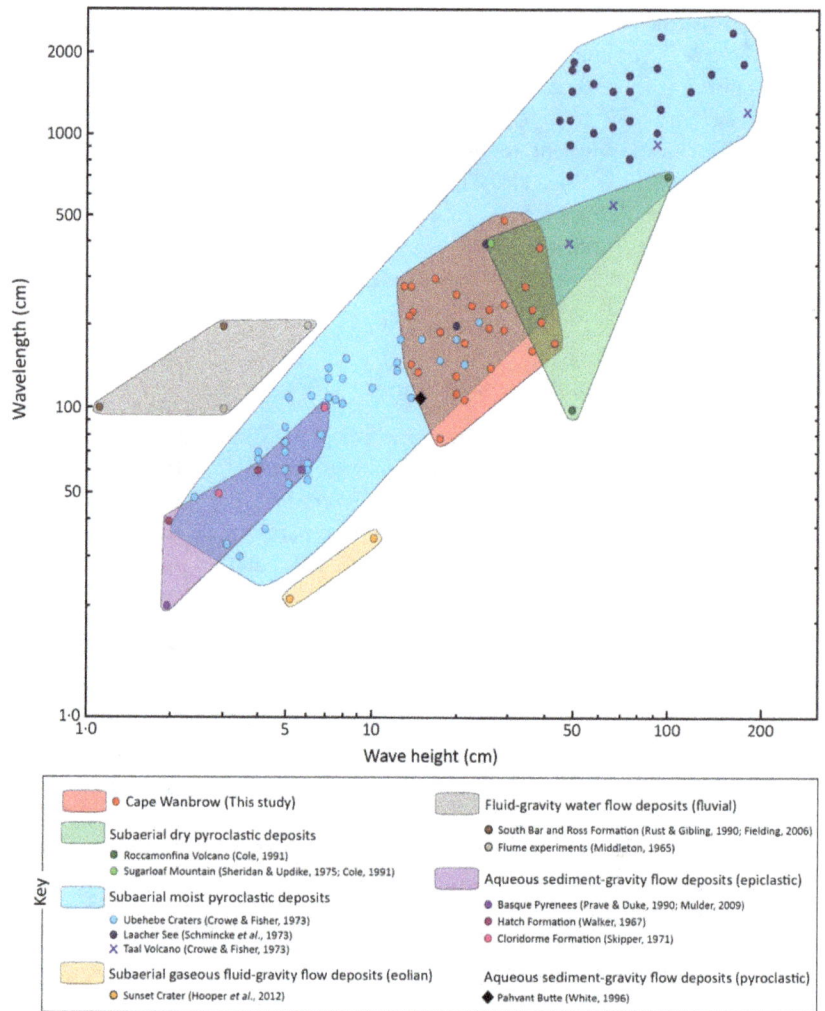

Fig. 12. Plot of wavelength of dunes versus wave height for the Cape Wanbrow dunes (red dots). There is a general increase of wavelength with increasing amplitude thickness, where available wavelengths and wave heights from the major flow types have also been plotted to create the most probable wavelength – wave height fields for that flow type. The most extensive data published come from Schmincke et al. (1973), from Laacher See and Crowe & Fisher (1973) and from both Taal Volcano and Ubehebe Craters. Dunes in pyroclastic aqueous sediment-gravity flow deposits are very rare and only one example from White (1996) has been plotted.

deformed bed bases from ballistics. However, key characteristics that rule out deposition by aeolian re-sedimentation are that aeolian deposits are characteristically fine-grained to medium-grained (1 to 3 ϕ), very well sorted (0·35 to 0·50 σ) and are often ungraded to inversely graded. Cape Wanbrow dunes, however, are coarse-grained to fine-grained, moderately to poorly sorted and tend to fine upwards.

Subaqueously formed dunes

Key characteristics that the Cape Wanbrow dunes share with deposits of subaqueous sediment-gravity flows like those that formed the Pahvant Butte deposits described by White (1996) include a sigmoidal shape, progressive migration, no accretionary lapilli, low-angle internal uni-directional cross-stratification and associated lenses of high-angle cross-stratification and dipping layers that grade laterally and vertically into sub-horizontally stratified layers (Table 1).

Smoothly undulating truncation of the tops of cross-sets, amalgamation of cross-sets and broad scours in sub-horizontal layers in the Pahvant Butte sequence are interpreted by White (1996) as being an example of form discordance produced by oscillatory motion superimposed on the density currents during deposition. Similar wave influence could be consistent with the presence of convex topped dunes at Cape Wanbrow, except the succession is erosionally truncated and thus leaves no evidence of very shallow-water deposition just prior to shoaling, such as provided by deposit context at Pahvant Butte.

Although the dunes of pyroclastic sediment-gravity flow deposits have many similarities with the Cape Wanbrow ones there are some differences that need to be considered. Deposits of pyroclastic material in sediment-gravity flows are dominantly, poorly sorted (1 to 3 σ), coarse-grained (1 to 2 ϕ) and contain very little fine-grained material (Table 1). For example, the lithofacies at Pahvant Butte lacks any fine-ash layers, although locally

Table 1. Summary of the common features of dunes in subaqueous and subaerial flow types

Common Features of Dunes	Scale				Erosion features
	WL	WH	Th	Sp	
Subaerial Dunes					
Dry pyroclastic density currents	1 to 7 m	0.2 to 1 m	0.1 to 1 m	*ca* 5 m	Little if any erosion occurs on the lee side
Grain size: Coarse-grained (0 to 1 φ) **Equivalent grain size in quartz:** Medium-grained (1 to 1·5 φ) **Sorting:** Poorly sorted but better sorted in comparison to their moist equivalents (1·0 to 2·0 but averaging *ca* 1·5σ) **Cross-bed dips:** Generally low-angle dips that resemble aeolian deposits. Progressively steepening lee sides in a down-current direction **Nature of continuous layers:** Overlying and truncating continuous stoss- to lee-side layers with a sigmoidal shape that tends to thicken on the leeward side **Nature of internal strata:** Alternating low-angle, unidirectional, sigmoidal shaped cross-stratified coarse-grained and fine-grained layers. Grade upwards from a planar stratification **Accretionary Lapilli:** Not common/None **Soft-state deformation:** Not common/None **Plastically deformed bed bases:** Not Common but localized sag structures may be present					Stoss side not always present due to erosion Overlying sigmoidal layers erode and truncate underlying low-angle cross-beds
Moist pyroclastic density currents	Large variation −1 to 20 m	Large variation −0·05 to 1 m	<3 m	3 to 20 m	Little if any erosion occurs on the lee side
Grain size: Fine-grained (2 to 3 φ) **Equivalent grain size in quartz:** Not much difference **Sorting:** Poorly sorted as well as more poorly sorted in comparison to their dry equivalents (2·0 to 4·0 σ) **Cross-bed dips:** Steep sided dunes and associated structures (cross-stratification) that become progressively steeper in a down-current direction **Nature of continuous layers:** Overlying fine-grained beds that are continuous from stoss- to lee side **Nature of internal strata:** Alternating thin beds of fine-grained to medium-grained ash displaying a constant thickness or has a slightly thicker stoss-side making dunes slightly asymmetric.					Stoss side not always present due to erosion Overlying sigmoidal layers erode and truncate underlying low-angle cross-beds

(Continued)

Table 1. Continued.

Common Features of Dunes	Scale				Erosion features
	WL	WH	Th	Sp	
Grade upwards from a planar stratification **Accretionary Lapilli:** Common **Soft-state deformation:** Common **Plastically deformed bed bases:** Common					
Aeolian flows **Grain size:** Fine- to medium-grained (1 to 3 φ) **Equivalent grain size in pyroclastics:** Coarse-grained (0 to 1 φ) **Sorting:** Very well sorted (0.35 to 0.50 σ) **Cross-bed dips:** Low- to intermediate-angle dips (25 to 35°) that are steeper than dry density current dips but shallower than moist density current dips **Nature of continuous layers:** Not common **Nature of internal strata:** Ungraded to inversely graded, thin beds (*ca* 2 to 10 mm) and laminae with intermediate dip angles **Accretionary Lapilli:** Not Common/None **Soft-state deformation:** Not Common/None **Plastically deformed bed bases:** Not Common/None	20 to 40 cm	0·5 to 1 m	0·5 to 1·0 m	20 to 40 cm	Typically only lee-side structures are well-preserved and dune crests are truncated by planar erosional surfaces Can be nearly monomict – epiclastic grains, reworked pyroclastic grains, or a mixture
Subaqueous Dunes *Aqueous sediment-gravity-flow – pyroclastic deposits* **Grain size:** Coarse-grained (0 to 1 φ) **Equivalent grain size in quartz:** Medium-grained (1 to 1·5 φ) **Sorting:** Poorly sorted to very poorly sorted (1·0 to 3·0 σ) **Cross-bed dips:** Low-angle dips and commonly locally intercalated with lenses of high-angle cross-stratification **Nature of continuous layers:** Low-angled sigmoidal shapes where fully preserved or smoothly undulating truncation of the tops of cross-sets **Nature of internal strata:** Low-angle sigmoidal shaped layers with internal unidirectional cross-stratification that grade laterally and vertically into sub-horizontally stratified ash layers **Accretionary Lapilli:** Not common/None **Soft-state deformation:** Not common	1 m	0·15 m	0·1 to 0·2 m	*ca* 1 m	These dunes are commonly truncated by low-relief erosion surfaces Very broadly undulating erosion surfaces Dune discordance – smooth undulating truncation of the tops of cross-sets producing similar dips of stoss-side and lee-side layers

(Continued)

Table 1. Continued.

	Common Features	Scale				Erosion features
		WL	WH	Th	Sp	
	Plastically deformed bed bases: Not common/None					
Aqueous sediment-gravity flow – epiclastic material	**Grain size:** Medium sand to silt (1 to 3 φ) **Equivalent grain size in pyroclastics:** Coarse-grained (0 to 1 φ) **Sorting:** Moderately to poorly sorted (0.5 to 2.0 σ) **Cross-bed dips:** Gently dipping with associated low-angle cross-stratification **Nature of continuous layers:** Wavy-like stratification within continuous thin layers (5 to 20 cm) that resemble hummocky cross-stratification. Overlain by swaley-fill layers that thicken above swales and thin above hummocks **Nature of internal strata:** Low-angle and symmetrical lee- to stoss dips that increase in dip angle in a down-current direction becoming non-symmetrical. Built up on a planar, conformable or erosional surface **Accretionary Lapilli:** Not common/None **Soft-state deformation:** Common **Plastically deformed bed bases:** Not common	0.2 to 0.6 m	0.02 to 0.06 m	0.1 to 1 m	0.2 to 1 m	Common to have erosional bases Structures often truncated by erosional surfaces Truncation between different generations of structures meaning only remnants of the hummocks retained No erosion at the base of the swale
Unidirectional and cyclic water flows	**Grain size:** Fine to medium-grained (1 to 2 φ) **Equivalent grain size in pyroclastics:** Coarse-grained (0 to 1 φ) **Sorting:** Well-sorted (0.35 to 0.50 σ) **Cross-bed dips:** Low-angle cross-bedding **Nature of continuous layers:** Sigmoidal to symmetrica drapes **Nature of internal strata:** Either a sigmoidal shape with low-angle cross-bedding or minor convex-upward bedforms **Grading upward from:** Grades laterally and vertically frcm and into planar lamination **Accretionary Lapilli:** Not common/None **Soft-state deformation:** Not common (rapid deposition) **Plastically deformed bed bases:** Not common	1 to 2 m	0.01 to 0.03 m	1 to 2 m	Variable up to 25 m	Different elements separated by erosion surfaces Overlapping erosion surfaces Increasing flow strengths and formation of chute-and-pools creates an upstream-dipping erosion surface

ripples in finer-grained ash advance down the gently dip-ping dune foresets (White, 1996). The dunes of Cape Wanbrow differ with a moderately to poorly sorted nature (0·5 to 1·2 σ), fine-grained ash tops and in particular a fine-ash sigmoidal drape. These differing features argue against aqueous sediment-gravity flows as the depositional mechanism for the Cape Wanbrow dunes.

The Cape Wanbrow dunes with their broad wave-length, low amplitude and in particular convex tops that resemble hummocky cross-stratification can be likened to deposits of non-pyroclastic subaqueous sediment-gravity flow deposits like those described by Prave & Duke (1990) and Mulder et al. (2009) (Table 1). Five major differences between the Cape Wanbrow structures and these HCS-like forms within the deposits of sediment-gravity flow deposits are notable (Table 1): (i) Cape Wanbrow deposits are overall coarser-grained, although this coarser-grained nature could be explained by the different settling velocities of low-density vesicular particles at Cape Wanbrow (Oehmig & Wallrabe-Adams, 1993; Manville et al., 2002) (Fig. 11); (ii) Cape Wanbrow dunes have a larger grain-size variation than those described from the HCS-like deposits; (iii) frequent small-scale syn-sedimentary deformation features, common within HCS-like deposits, are not common within the dunes of Cape Wanbrow; (iv) HCS-like structures frequently show migration trends with an up-slope motion but mostly do not display a dominant migration direction, whereas the dunes at Cape Wanbrow show a strong down-current migration; and (v) the average wave height and wave-length is much smaller than those within the Cape Wanbrow succession (Fig. 12) Based on these major differences, deposition by subaqueous sediment-gravity flow or combined-flow HCS seems unlikely (Table 1).

Unidirectional current deposits (e.g. fluvial currents) of the upper flow regime field seem to offer a good potential analogue for the Cape Wanbrow dunes with similar deposit architecture and characteristics (Table 1). In particular, types 1 and 4 of the sedimentary structures described by Fielding (2006) formed by unidirectional currents in upper flow regime are recognized in the Cape Wanbrow deposits (Fig. 2). These dune geometries make up part of eight main types of UFR elements which together form a hierarchy that is representative of an increasing flow strength and where preservation potential is high. Taking the features of UFR deposits and the high preservation potential into consideration, if the Cape Wanbrow dunes were deposited by unidirectional currents in the upper flow regime, as were those described by Fielding (2006) it might be expected that most, if not all eight types of UFR deposits would be present. Instead, only types 1 and 4 are recognized. There are also differences in grain size (silt to fine sand in upper flow regime

unidirectional currents as opposed to fine to coarse ash) and sorting (well sorted as opposed to moderately to poorly sorted) (Table 1), although part of this grain-size difference may be due to the settling velocity of vesicular particles being less than that of quartz, feldspar and litho-clasts (Oehmig & Wallrabe-Adams, 1993; Manville et al., 2002). These characteristics along with the lack of any hierarchical UFR deposit elements make deposition from unidirectional upper flow regime flows unlikely. There is also the question of where, in the context of a small island volcano, such UFR deposits could form other than in the sharply incised channels known from such volca-noes (Fisher, 1977; Verwoerd & Chevallier, 1987). Analysis of the architecture of dunes described from some flume experiments that produced broadly similar dunes to those in the Cape Wanbrow succession led to identification of significant differences between them. Key differences include (i) the average wavelength of dunes produced by Middleton's (1965) flume experiments are ca 50 to 80 cm with the largest 1 m long in an experimental tank 40 m long (Alexander et al., 2001; Spinewine et al., 2009; Cartigny et al., 2014); these wavelength values for experimental dunes are much shorter on average than those at Cape Wanbrow (ca 2 to 3 m); and (ii) these flume dunes lacked evidence for structures with steepening lee-side layers and a sigmoidal-shaped fine-grained drape, which is a key characteristic of the Cape Wanbrow dunes. Based on the significant differences discussed, deposition by subaqueous subcritical and/or supercritical flows would have to be ruled out for at least the majority of the Cape Wanbrow deposit.

Having used the created framework (Table 1) to compare the dunes and associated sedimentary structures found at the top of Rua[2] Volcano in the Cape Wanbrow succession to five major flow types in both subaerial and subaqueous environments, the geometrical and granulo-metric characteristics defined are most similar to the deposits of subaerial dry dilute pyroclastic density currents and pyroclastic subaqueous sediment-gravity flows.

Dunes deposited by dry dilute pyroclastic density currents (Table 1) are characteristically poorly sorted, coarse-grained and fines-poor with alternating coarse-grained and fine-grained layers that grade upwards from a planar stratification (Valentine & Fisher, 2000) (Table 1). Dunes in volcaniclastic subaqueous sediment-gravity flow deposits are dominantly coarse-grained, poorly sorted and have intercalated lenses of high-angle cross-stratification. Cape Wanbrow dunes comprise alternating layers of coarse-grained and fine-grained tuff that grade upwards from planar stratification and become finer-grained upwards and become fines-dominant at dune tops.

Due to dunes in both volcaniclastic subaqueous sediment-gravity flows and dry dilute pyroclastic density currents being characteristically coarse-grained the only remaining feature that matches the Cape Wanbrow dunes is the progressively steepening layers of alternating fine-grained and coarse-grained material found only in subaerial dry dilute pyroclastic density current deposits as described by Valentine & Fisher (2000). Based on this, the dunes of the Cape Wanbrow deposits are interpreted to have been emplaced subaerially by dry dilute pyroclastic density currents. This interpretation has major implications for the Cape Wanbrow stack as it demonstrates that at least one of the volcanoes (Rua[2]) became emergent during its lifespan and that subaerially formed deposits were preserved despite the offshore-island setting.

Rua[2] Volcano: an emergent volcano

The study by White (1996, 2001) of Pahvant Butte, a small volcano that erupted into a now absent pluvial lake in the Pleistocene is an excellent example of a subaqueous to emergent volcano and offers an opportunity to compare the Rua[2] Volcano deposits to the overall deposit characteristics of an emergent volcano.

Based on accounts of deposits at Pahvant Butte, which record the evolution from subaqueous to moist subaerial to dry subaerial, a well-preserved succession of subaqueous to emergent to fully subaerial should include examples of (i) clearly subaqueously deposited tephra followed by (ii) ripple-marked shoreline beds; (iii) subaerial cone moist density currents deposits with accretionary and armoured lapilli and potentially (iv) a transition to dry phreatomagmatic activity with subaerial grainflow and fallout processes.

The lower deposits of Rua[2] were deposited subaqueously by eruption-fed currents and offer no indications of becoming emergent, for example no ripple-marked shoreline beds or moist density currents deposits containing accretionary lapilli. The top 40 m of the Rua[2] Volcano succession is interpreted as being deposited by subaerial dry pyroclastic density currents based on the geometrical and granulometric characteristics of the deposits. Therefore, Rua[2] Volcano started its life subaqueously and became an emergent tuff cone to form an island and shoreline.

The expected deposits of an emergent volcano, including ripple-marked shoreline beds and moist density current deposits, are missing from the Rua[2] Volcano succession. There is also a general absence of shoreline and tuff cone deposits in emergent volcanoes recorded in the literature, other than Pahvant Butte. This situation is probably due to removal of most emergent deposits, because emergent tuff cone deposits of an island are

subject to slumping in response to collapse of the tephra pile and/or erosion and transportation by normal wave action; such slumping can shift originally subaerial deposits into a subaqueous position. Evidence for these processes acting on the emergent deposits of the Rua[2] volcano includes localized small-scale faults and truncations, onto which sedimentary sequences onlap (Moorhouse et al., 2015). These are the result of remobilized tephra by slumping as the pile over-steepens, which likely occurs during or shortly after deposition (Sohn et al., 2012; Moorhouse et al., 2015).

The conditions that allowed preservation of the subaerial pyroclastic succession of Rua[2] were probably sector collapse resulting in relocation of originally subaerial deposits to sites below wave base, where preservation potential is much greater. There were global sea-level changes of sufficient magnitude to make the site subaerial at times, but no published evidence for any non-volcanic subaerial deposits in the region in the Oligocene. Other volcanoes in the field may have emerged to form islands, but there is no direct evidence of emergence (Andrews, 2003; Maicher, 2003; Corcoran & Moore, 2008), and perhaps the best known of these, Bridge Point, has been specifically inferred to have stopped erupting before doing so (Cas et al., 1989).

CONCLUSION

Distinguishing between subaqueous and subaerial deposits in ancient volcanogenic successions remains a difficult task and becomes progressively harder when rock features are poorly preserved, exposures are limited, and independent palaeoenvironmental indicators are absent or ambiguous. However, careful attention to contextual information of sedimentary structures like dunes, low- to high-angle cross-stratification and planar stratification that are commonly reported in both subaerial and subaqueous environments has meant the uncertainty over the origin of Oligocene-Eocene aged dunes at Cape Wanbrow, Oamaru, has been partially removed. It is inevitable that questions remain regarding each interpretation, for example the relationship between subaqueously deposited material and dunes interpreted as deposited by dry dilute pyroclastic currents at Cape Wanbrow as well as the reason for a lack of emergent-style deposits similar to those reported at Pahvant Butte (White, 2001), but the comparisons made and distinctions drawn between deposits of the Cape Wanbrow lithofacies and broadly similar ones of other origins provide a framework for interpreting dunes and related strata of other explosive subaqueous to shallow marine, emergent and subaerial volcanoes.

References

Alexander, J., Bridge, J., Cheel, R. and Leclair, S. (2001) Bedforms and associated sedimentary structures formed under supercritical water flows over aggrading sand beds. *Sedimentology*, **48**, 133–152.

Allen, J.R.L. (1970) The sequence of sedimentary structures in turbidites, with special reference to dunes. *Scott. J. Geol.*, **6**, 146–161. doi:10.1144/sjg06020146.

Allen, J.R.L. (1982) *Sedimentary Structures. Their Character and Physical Basis*, Vol. 1. Elsevier, Amsterdam, 593 pp.

Andrews, B. (2003) Eruptive and depositional mechanisms of an Eocene shallow submarine volcano Moeraki Peninsula, New Zealand. In: *Explosive Subaqueous Volcanism* (Eds J.D.L. White, J.L. Smellie and D.A. Clague), *Geophysical Mongraph.*, **140**, 179–188.

Best, J.L. and Bridge, J.S. (1992) The morphology and dynamics of low amplitude bedwaves upon upper stage plane beds and the preservation of planar laminae. *Sedimentology*, **39**, 737–752. doi:10.1111/j.1365-3091.1992.tb02150.x.

Bridge, J.S. and Demicco, R.V. (2008) *Earth Surface Processes, Landforms and Sediment Deposits*. Cambridge University Press, Cambridge, 278 pp.

Carey, S. (2000) Volcaniclastic sedimentation around island arcs. In: *Encyclopedia of Volcanoes* (Eds H. Sigurdsson, B.F. Houghton, S. Mcnutt, H. Rymer and J. Stix), pp. 477–494. Elsevier, New York.

Cartigny, M.J.B., Postma, G., van den Berg, J.H. and Mastbergen, D.R. (2011) A comparative study of sediment waves and cyclic steps based on geometries, internal structures and numerical modeling. *Mar. Geol.*, **280**, 40–56. doi:10.1016/j.margeo.2010.11.006.

Cartigny, M.J.B., Ventra, D., Postma, G. and van Den Berg, J.H. (2014) Morphodynamics and sedimentary structures of bedforms under supercritical-flow conditions: new insights from flume experiments. *Sedimentology*, **61**, 712–748. doi:10.1111/sed.12076.

Cas, R., Landis, C. and Fordyce, R. (1989) A monogenetic, surtla-type, Surtseyan volcano from the Eocene-Oligocene Waiareka-Deborah volcanics, Otago, New Zealand: a model. *Bull. Volcanol.*, **51**, 281–298.

Cheel, R.J. (1990) Horizontal lamination and the sequence of bed phases and stratification under upper-flow-regime conditions. *Sedimentology*, **37**, 517–529. doi:10.1111/j.1365-3091.1990.tb00151.x.

Chough, S.K. and Sohn, Y.K. (1990) Depositional mechanics and sequences of base surges, Songaksan tuff ring, Cheju Island (Korea). *Sedimentology*, **37**, 1115–1135.

Cole, P. (1991) Migration direction of sand-wave structures in pyroclastic-surge deposits: implications for depositional processes. *Geology*, **19**, 1108–1111.

Cole, R. and Decelles, P. (1991) Subaerial to submarine transitions in early Miocene pyroclastic flow deposits, southern San Joaquin basin, California. *Geol. Soc. Am. Bull.*, **103**, 221–235.

Coombs, D., Cas, R., Kawachi, Y., Landis, C., McDonough, W. and Reay, A. (1986) Cenozoic volcanism in north, east and central Otago. *R. Soc. New Zealand Bull.*, **23**, 278–312.

Corcoran, P.L. and Moore, L.N. (2008) Subaqueous eruption and shallow-water reworking of a small-volume Surtseyan edifice at Kakanui, New Zealand. *Can. J. Earth Sci.*, **45**, 1469–1485. doi:10.1139/E08-068.

Crowe, B. and Fisher, R. (1973) Sedimentary structures in base-surge deposits with special reference to cross-bedding, Ubehebe Craters, Death Valley, California. *Geol. Soc. Am.*, **84**, 663–682.

Dellino, P., Isaia, R. and Veneruso, M. (2004) Turbulent boundary layer shear flows as an approximation of base surges at Campi Flegrei (Southern Italy). *J. Volcanol. Geotherm. Res.*, **133**, 211–228. doi:10.1016/S0377-0273(03)00399-8.

Denny, C.S. (1940) Santa Fe formation in the Española Valley, New Mexico. *Bull. Geol. Soc. Am.*, **51**, 677–694.

Douillet, G.A., Pacheco, D.A., Kueppers, U., Letort, J., Tsang-Hin-Sun, È., Bustillos, J., Hall, M., Ramon, P. and Dingwell, D.B. (2013) Dune bedforms produced by dilute pyroclastic density currents from the August 2006 eruption of Tungurahua volcano, Ecuador. *Bull. Volcanol.*, **75**, 762. doi:10.1007/s00445-013-0762-x.

Duller, R.A., Mountney, N.P., Russell, A.J. and Cassidy, N.C. (2008) Architectural analysis of a volcaniclastic jokulhlaup deposit, southern Iceland: Sedimentary evidence for supercritical flow. *Sedimentology*, **55**, 939–964. doi:10.1111/j.1365-3091.2007.00931.x.

Fielding, C.R. (2006) Upper flow regime sheets, lenses and scour fills: extending the range of architectural elements for fluvial sediment bodies. *Sed. Geol.*, **190**, 227–240. doi:10.1016/j.sedgeo.2006.05.009.

Fielding, C.R., Kassan, J. and Draper, J.J. (1996) Geology of the Bowen and Surat Basins, eastern Queensland. Australasian Sedimentologists Group Field Guide Series. *Geol. Soc. Aust. Sydney*, **8**, 126.

Fisher, R.V. (1977) Erosion by volcanic base-surge density currents: U-shaped channels. *Geol. Soc. Am. Bull.*, **88**, 1287–1297.

Fisher, R.V. (1983) Flow transformations in sediment gravity flows. *Geology*, **11**, 273–274. doi:10.1130/0091-7613(1983)11<273:FTISGF>2.0.CO.

Fisher, R.V. and Schmincke, H.-U. (1984) *Pyroclastic Rocks*. Springer, Berlin, Heidelberg, New York, Tokyo, 472 pp.

Folk, R.L. (1980) *Petrology of Sedimentary Rocks*, pp. 38–48. Texas Hemphill's Book Store, Austin, TX.

Freundt, A. (2003) Entrance of hot pyroclastic flows into the sea: experimental observations. *Bull. Volcanol.*, **65**, 144–164. doi:10.1007/s00445-002-0250-1.

Gage, M. (1957) *The Geology of Waitaki Subdivision*. No. 55. New Zealand Department of Scientific and Industrial Research, Wellington, New Zealand.

Gilbert, G.K. (1890) Lake Bonneville. *US Geol. Surv. Monogr.*, **1**, 1–438.

Hand, B.M. (1974) Supercritical flow in density currents. *SEPM J. Sed. Res.*, **44**, 637–648. doi:10.1306/74D72AB3-2B21-11D7-8648000102C1865D.

Heinrichs, T. (1984) The Umsoli chert, turbidite testament for a major phreatoplinian event at the onverwacht/fig tree transition (Swaziland supergroup, Archaean, South Africa). *Precambr. Res.*, **24**, 237–283. doi:10.1016/0301-9268(84)90061-5.

Hoernle, K., White, J.D.L., van den Bogaard, P., Hauff, F., Coombs, D.S., Werner, R., Timm, C., Garbe-Schönberg, D., Reay, A. and **Cooper, A.F.** (2006) Cenozoic intraplate volcanism on New Zealand: upwelling induced by lithospheric removal. *Earth Planet. Sci. Lett.*, **248**, 350–367. doi:10.1016/j.epsl.2006.06.001.

Hooper, D.M., Mcginnis, R.N. and **Necsoiu, M.** (2012) Volcaniclastic aeolian deposits at Sunset Crater, Arizona: Terrestrial analogs for Martian dune forms. *Earth Surf. Proc. Land.*, **37**, 1090–1105. doi:10.1002/esp.3238.

Jones, C.M. (1977) Effects of varying discharge regimes on bed-form sedimentary structures in modern rivers. *Geology*, **5**, 567–570. doi:10.1130/0091-7613(1977) 5<567:EOVDRO>2.0.CO;2.

Jopling, A.V. and **Richardson, E.V.** (1966) Backset bedding developed in shooting flow in laboratory experiments. *J. Sed. Res.*, **36**, 821–825.

Krumbein, W.C. (1935) Thin-section mechanical analysis of indurated sediments. *J. Geol.*, **43**, 482–496.

Lorenz, V. (1986) On the growth of maars and diatremes and its relevance to the formation of tuff rings. *Bull. Volcanol.*, **48**, 265–274.

Lorenz, V. (2007) Syn- and posteruptive hazards of maar–diatreme volcanoes. *J. Volcanol. Geoth. Res.*, **159**, 285–312. doi:10.1016/j.jvolgeores.2006.02.015.

Maicher, D. (2003) A cluster of Surtseyan volcanoes at Lookout Bluff, north Otago, New Zealand; aspects of edifice spacing and time. In: *Explosive Subaqueous Volcanism* (Eds J.D.L. White, J.L. Smellie and D.A. Clague), *Geophysical Mongraph.*, **140**, 167–178.

Manville, V., Segschneider, B. and **White, J.D.L.** (2002) Hydrodynamic behaviour of Taupo 1800a pumice: implications for the sedimentology of remobilized pyroclasts. *Sedimentology*, **49**, 955–976. doi:10.1046/j.1365-3091.2002.00485.x.

McLoughlin, S. (1993) Plant fossil distributions in some Australian Permian non-marine sediments. *Sed. Geol.*, **85**, 601–619. doi:10.1016/0037-0738(93)90104-D.

McPherson, J.G., Flannery, J.R. and **Self, S.** (1989) Soft-sediment deformation (fluid escape) features in a coarse-grained pyroclastic- surge deposit, north-central New Mexico. *Sedimentology*, **36**, 943–949.

Miall, A.D. (1985) Architectural-element analysis: a new method of facies analysis applied to fluvial deposits. *Earth Sci. Rev.*, **22**, 261–308.

Middleton, G.V. (1965) Antidune cross-bedding in a large flume. *J. Sed. Res.*, **35**, 922–927. doi:10.1306/74D713AC-2B21-11D7-8648000102C1865D.

Middleton, G.V. and **Hampton, M.A.** (1973) Sediment gravity flows: mechanics of flow and deposition. In: *Turbidites and Deep-Water Sedimentation* (Eds G.V. Middleton and A.H. Bouma), pp. 1–38. Short Course Lecture Notes. Society of Economic Paleontologists and Mineralogists, Los Angeles.

Middleton, G.V. and **Hampton, M.A.** (1976) Subaqueous sediment transport and deposition by sediment gravity flows. In: *Marine Sediment Transport and Environmental Management* (Eds D.J. Stanley and D.J.P. Swift), pp. 197–218. John Wiley & Sons, New York.

Moorhouse, B.L., White, J.D.L. and **Scott, J.M.** (2015) Cape Wanbrow: a stack of Surtseyan-style volcanoes built over millions of years in the Waiareka-Deborah volcanic field, New Zealand. *J. Volcanol. Geotherm. Res.*, **298**, 27–46.

Mulder, T., Razin, P. and **Faugeres, J.-C.** (2009) Hummocky cross-stratification-like structures in deep-sea turbidites: Upper Cretaceous Basque basins (Western Pyrenees, France). *Sedimentology*, **56**, 997–1015. doi:10.1111/j.1365-3091.2008.01014.x.

Murtagh, R.M. (2011) *An Investigation Into the Explosivity of Shallow Subaqueous Basaltic Eruptions.* PhD Thesis. University of Otago, Dunedin, 161 pp, 323 pp.

Nocita, B.W. (1988) Soft-sediment deformation (fluid escape) features in a coarse-grained pyroclastic-surge deposit, north-central New Mexico. *Sedimentology*, **35**, 275–285. doi:10.1111/j.1365-3091.1988.tb00949.x.

Oehmig, R. and **Wallrabe-Adams, H.** (1993) Hydrodynamic properties and grain-size characteristics of volcaniclastic deposits on the Mid-Atlantic Ridge north of Iceland (Kolbeinsey Ridge). *J. Sed. Petrol.*, **63**, 140–151.

Postma, G., Cartigny, M. and **Kleverlaan, K.** (2009) Structureless, coarse-tail graded Bouma Ta formed by internal hydraulic jump of the turbidity current? *Sed. Geol.*, **219**, 1–6. doi:10.1016/j.sedgeo.2009.05.018.

Prave, A. and **Duke, W.** (1990) Small-scale hummocky cross-stratification in turbidites: a form of antidune stratification? *Sedimentology*, **37**, 531–539.

Rust, B.R. and **Gibling, M.R.** (1990) Three-dimensional antidunes as HCS mimics in a fluvial sandstone: the Pennsylvanian South Bar Formation near Sydney, Nova Scotia. *J. Sed. Petrol*, **60**, 540–548. doi:10.2110/jsr.60.640.

Schmincke, H.-U., Fisher, R.V. and **Waters, A.C.** (1973) Antidune and chute and pool structures in the base surge deposits of the Laacher See area, Germany. *Sedimentology*, **20**, 553–574. doi:10.1111/j.1365-3091.1973.tb01632.x.

Sheridan, M.F. and **Updike, R.G.** (1975) Sugarloaf Mountain Tephra – a Pleistocene Rhyolitic Deposit of Base-Surge Origin in Northern Arizona. *Geol. Soc. Am. Bull.*, **86**, 571–581. doi:10.1130/0016-7606(1975) 86<571:SMTAPR>2.0.CO;2.

Skipper, K. (1971) Antidune cross-stratification in a turbidite sequence, cloridorme formation, Gaspé, Quebec. *Sedimentology*, **17**, 51–68. doi:10.1111/j.1365-3091.1971.tb01130.x.

Smith, G.A. and Katzman, D. (1991) Discrimination of eolian and pyroclastic-surge processes in the generation of cross-bedded tuffs, Jemez Mountains volcanic field, New Mexico. *Geology*, **19**, 465–468. doi:10.1130/0091-7613(1991) 019<0465:DOEAPS>2.3.CO.

Sohn, Y.K. (1996) Hydrovolcanic processes forming basaltic tuff rings and cones on Cheju Island, Korea. *Geol. Soc. Am. Bull.*, **108**, 1199–1211. doi:10.1130/0016-7606(1996) 108<1199:HPFBTR>2.3.CO;2.

Sohn, Y.K., Cronin, S.J., Brenna, M., Smith, I.E.M., Nemeth, K., White, J.D.L., Murtagh, R.M., Jeon, Y.M. and Kwon, C.W. (2012) Ilchulbong tuff cone, Jeju Island, Korea, revisited: a compound monogenetic volcano involving multiple magma pulses, shifting vents, and discrete eruptive phases. *Geol. Soc. Am. Bull.*, **124**, 259–274. doi:10.1130/B30447.1.

Spinewine, B., Sequeiros, O.E., Garcia, M.H., Beaubouef, R.T., Sun, T., Savoye, B. and Parker, G. (2009) Experiments on Wedge-Shaped Deep Sea Sedimentary Deposits in Minibasins and/or on channel levees emplaced by turbidity currents. Part II. Morphodynamic evolution of the wedge and of the associated bedforms. *J. Sed. Res.*, **79**, 608–628. doi:10.2110/jsr.2009.065.

Stearns, H.T. (1935) Geology and ground- water resources of the Island of Oahu, Hawaii, Territory of Hawaii. *Div. Hydrogr. Bull.*, **1**, 142.

Trofimovs, J., Sparks, R.S.J. and Talling, P.J. (2007) Anatomy of a submarine pyroclastic flow and associated turbidity current: July 2003 dome collapse, Soufrière Hills volcano, Montserrat, West Indies. *Sedimentology*, **55**, 617–634. doi:10.1111/j.1365-3091.2007.00914.x.

Valentine, G.A. (1987) Stratified flow in pyroclastic surges. *Bull. Volcanol.*, **49**, 616–630. doi:10.1007/BF01079967.

Valentine, G.A. and Fisher, R.V. (2000) Pyroclastic Surges and Blasts. In: *Encyclopedia of Volcanoes* (Ed. H. Sigurdsson), pp. 581–589. Academic Press, London.

Verwoerd, W.J. and Chevallier, L. (1987) Contrasting types of Surtseyan tuff cones on Marion and Prince Edward Islands, southwest Indian Ocean. *Bull. Volcanol.*, **49**, 399–414.

Walker, R.G. (1967) Upper flow regime bed forms in turbidites of the Hatch Formation, Devonian of New York State. *J. Sed. Res.*, **37**, 1052–1058. doi:10.1306/74D7182F-2B21-11D7-8648000102C1865D.

Walker, G.P.L. (1971) Grain-size characteristics of pyroclastic deposits. *J. Geol.*, **79**, 696–714.

Walker, G.P.L. (1984) Characteristics of dune-bedded pyroclastic surge bedsets. *J. Volcanol. Geotherm. Res.*, **20**, 281–296.

Waters, A.C. and Fisher, R.V. (1971) Base surges and their deposits – Capelinhos and Taal volcanoes. *J. Geophys. Res.*, **76**, 5596–5614. doi:10.1029/JB076i023p05596.

Wentworth, C.K. (1937) The Diamond Head black ash. *J. Sed. Res.*, **7**, 91–103. doi:10.1306/D4268FB8-2B26-11D7-8648000102C1865D.

White, J.D.L. (1991) Maar-diatreme phreatomagmatism at Hopi Buttes, Navajo Nation (Arizona), USA Non-fragment. *Bull. Volcanol.*, **53**, 239–258.

White, J.D.L. (1996) Pre-emergent construction of a lacustrine basaltic volcano, Pahvant Butte, Utah (USA). *Bull. Volcanol.*, **58**, 249–262.

White, J.D.L. (2001) Eruption and reshaping of Pahvant Butte volcano in Pleistocene Lake Bonneville. In: *Volcaniclastic Sedimentation in Lacustrine Settings* (Eds J.D.L. White and N.R. Riggs), *Spec. Publ. Int. Assoc. Sediment.*, **30**, 61–81.

White, J.D.L. and Houghton, B.F. (2006) Primary volcaniclastic rocks. *Geology*, **34**, 677. doi:10.1130/G22346.1.

Wohletz, K.H. and Sheridan, M.F. (1983) Hydrovolcanic explosions. II. Evolution of basaltic tuff rings and tuff cones. *Am. J. Sci.*, **283**, 385–413. doi:10.2475/ajs.283.5.385.

Wright, J.V. and Mutti, E. (1981) The Dali Ash, Island of Rhodes, Greece: a problem in interpreting submarine volcanigenic sediments. *Bull. Volcanol.*, **44**, 153–167. doi:10.1007/BF02597701.

Wright, J.V., Smith, A.L. and Self, S. (1980) A working terminology of pyroclastic deposits. *J. Volcanol. Geotherm. Res.*, **8**, 315–336.

Yokokawa, M., Okuno, K., Nakamura, A., Muto, T., Miyata, Y. and Naruse, H. (2009) *Aggradational Cyclic Steps: Sedimentary Structures Found in Flume Experiments*. Proceedings 33rd IAHR Congress, Vancouver, pp. 5547–5554.

PERMISSIONS

LIST OF CONTRIBUTORS

Christian Betzler, Sebastian Lindhorst and Thomas Lüdmann
Institut für Geologie, CEN, Universität Hamburg, Bundesstrasse 55, 20146 Hamburg, Germany

Christian Hübscher
Institut für Geophysik, CEN, Universität Hamburg, Bundesstrasse 55, 20146 Hamburg, Germany

John J. G. Reijmer
King Fadh University of Petroleum and Minerals, Kfupm Dhahran 31261, Saudi Arabia

Juan-Carlos Braga
Departamento de Estratigrafía y Paleontología, Universidad de Granada,Campus de Fuentenueva s.n., 18002 Granada, Spain

Richard Newport, Cathy Hollis and Jonathan Redfern
School of Earth and Environmental Sciences, Manchester University, Manchester M13 9PL, UK

Stéphane Bodin
Department of Geoscience, Aarhus University, Høegh-Guldbergs Gade 2, 8000 Aarhus C, Denmark

Timothy Mcintyre
Department of Earth and Atmospheric Sciences, University of Alberta, Edmonton, Alberta, Canada T6G 2R3

Philip Fralick
Department of Geology, Lakehead University, Thunder Bay, Ontario, Canada P7B 5E1

Christopher Bowie
Devon Energy Corporation, 333 West Sheridan Avenue, Oklahoma City, OK 73102, USA.

Mara Brady
Department of Earth and Environmental Sciences, California State University, Fresno, 2576 East San Ramon Avenue, M/S ST-24, Fresno, CA 93740, USA

Daniel Smrzka and Jennifer Zwicker
Department für Geodynamik und Sedimentologie, Universität Wien, 1090 Wien, Austria

Sadat Kolonic
Shell Petroleum Development Company of Nigeria, Port Harcourt, Nigeria

Daniel Birgel
Institut für Geologie, Universität Hamburg, Bundesstraße 55, 20146 Hamburg, Germany

Crispin T.S. Little
School of Earth and Environment, University of Leeds, Leeds, LS2 9JT, UK

Akmal M. Marzouk
Geology Department, Faculty of Science, Tanta University, 31527 Tanta, Egypt

El Hassane Chellai
Department of Geology, Faculty of Sciences Semlalia, Cadi Ayyad University, Marrakech, Morocco

Thomas Wagner
Lyell Centre, Heriot-Watt University, Edinburgh, EH14 4AS, UK

Jörn Peckmann
Department für Geodynamik und Sedimentologie, Universität Wien, 1090 Wien, Austria
Institut für Geologie, Universität Hamburg, Bundesstraße 55, 20146 Hamburg, Germany

Wesley C. Ingram and Christopher S. Martens
Department of Marine Sciences, University of North Carolina-Chapel Hill, Chapel Hill, NC 27599, USA

Stephen R. Meyers, Zhizhang Shen and Huifang Xu
Department of Geoscience, University of Wisconsin-Madison, Madison, WI 53706, USA

Wietse I. Van De Lageweg and Maarten G. Kleinhans
Faculty of Geosciences, Universiteit Utrecht, Utrecht, The Netherlands

Wout M. Van Dijk
Department of Geography, Durham University, South Road, Durham, DH1 3LE, UK

Darren Box
Exxon Mobil Russia, 31 Novinsky Boulevard, 123242, Moscow, Russia

Silke Voigt, Konstantin Frisch, Alexander Bartenstein, Alexandra Hellwig and Rainer Petschick
Institute of Geosciences, Goethe University Frankfurt, Altenhöferallee 1, 60438 Frankfurt, Germany

Yuki Weber
Institute of Geosciences, Goethe University Frankfurt, Altenhöferallee 1, 60438 Frankfurt, Germany
Department of Earth and Planetary Sciences, Harvard University, 20 Oxford Street, Cambridge, MA 02138, USA

André Bahr and Jörg Pross
Institute of Geosciences, Goethe University Frankfurt, Altenhöferallee 1, 60438 Frankfurt, Germany

Institute of Earth Sciences, Heidelberg University, Im Neuenheimer Feld 234-236, 69120 Heidelberg, Germany

Andreas Koutsodendris
Institute of Earth Sciences, Heidelberg University, Im Neuenheimer Feld 234-236, 69120 Heidelberg, Germany

Thomas Voigt
Institute of Geosciences, Friedrich Schiller University Jena, Burgweg 11, 07749 Jena, Germany

Verena Verestek and Erwin Appel
Institute of Geosciences, University Tübingen, Sigwartstrasse 10, Hölderlinstrasse 12, 72074 Tübingen, Germany

Benjamin L. Moorhouse and James D. L. White
Geology Department, University of Otago, Leith St, Dunedin 9054, New Zealand

Index